TARSIERS

THE RUTGERS SERIES IN
HUMAN EVOLUTION

Robert Trivers, *Founding Editor*

Lee Cronk, *Associate Editor*
Helen Fisher, *Advisory Editor*
Lionel Tiger, *Advisory Editor*

Kingsley R. Browne, *Biology at Work: Rethinking Sexual Equality*
David Haig, *Genomic Imprinting and Kinship*
John F. Hoffecker, *Desolate Landscapes: Ice-Age Settlement in Eastern Europe*
John T. Manning, *Digit Ratio: A Pointer to Fertility, Behavior, and Health*
Paul Rubin, *Darwinian Politics*
Patricia C. Wright, Elwyn L. Simons, and Sharon Gursky, eds., *Tarsiers: Past, Present, and Future*

TARSIERS

Past, Present, and Future

Edited by Patricia C. Wright,
Elwyn L. Simons, and
Sharon Gursky

RUTGERS UNIVERSITY PRESS
New Brunswick, New Jersey, and London

Library of Congress Cataloging-in-Publication Data

Tarsiers : past, present, and future / edited by Patricia C. Wright, Elwyn L. Simons, and Sharon Gursky.
    p.   cm. — (The Rutgers series in human evolution)
    Includes bibliographical references.
    ISBN 0-8135-3236-1 (cloth : alk. paper)
    1. Tarsiers. I. Wright, Patricia C., 1944– . II. Simons, Elwyn L. III. Gursky, Sharon, 1967– . IV. Series.
    QL737.P965 T37 2003
    599.8′3—dc21

                           2002012496

British Cataloging-in-Publication information is available from the British Library.

Manufactured in the United States of America

This book is dedicated to Mandarin.

Tarsius bancanus (Western tarsier). Photo by David H. Haring.

Contents

Figures

Tables

Acknowledgments

Our profound thanks to all the authors who contributed to this volume, and the reviewers who took the time to make each chapter better. We especially appreciate the editorial assistance of Sharon Pochron. Our gratitude to Lauren Block and the Institute for the Conservation of Tropical Environments, SUNY–Stony Brook, including Summer Arrigo-Nelson, Pablo Stevenson, and Misa Andriamihaja, for assistance in assembling the chapters into a coherent volume. We thank our spouses, Jukka Jernvall, Friderun Simons, and Michael Alvard, for guidance and patience. We are grateful to David Haring for sharing his many tarsier photos.

Patricia Wright and Elwyn Simons gratefully acknowledge the U.S. National Science Foundation for funding their tarsier research. Sharon Gursky would like to acknowledge Wenner Gren Foundation for Anthropological Research, Primate Conservation Inc. Douroucouli Foundation, Fulbright Fellowship, and the National Science Foundation for funding her fieldwork in Sulawesi.

We are grateful to Rutgers University Press, especially to our editors Helen Hsu and Audra Wolfe; to Adi Hovav; to Alice Calaprice for copyediting and to Carol Inskip for indexing.

TARSIERS

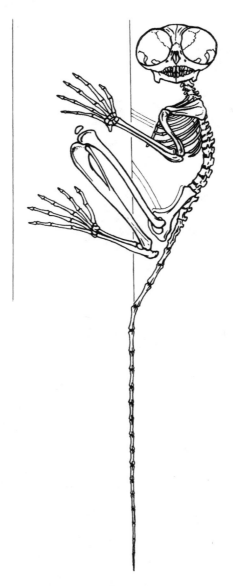

Figure I.1. Skeleton of *Tarsius*. Notice the postorbital plate, unfused mandibular symphysis, long tarsal bones, and large hands and feet. (Stephen D. Nash)

Introduction

Patricia C. Wright, Elwyn L. Simons, and Sharon Gursky

Why Tarsiers Interest Us

Today tarsiers are relict primates, with only 4–7 species living on a few islands in Southeast Asia: although they are small-bodied and nocturnal, tarsiers have provided a century of controversy. The combination of derived and ancient characters seen in tarsiers makes them pivotal to understanding the roots of primate evolution. The tarsiers' unusual diet and reproduction provide a unique perspective to understanding primate behavioral ecology, while their specialized anatomy shows us how primates can solve a wide range of functional needs.

Tarsiers provide one astounding fact after another. The world's most carnivorous primate, tarsiers exclusively eat live animals, predominantly insects, reptiles, and amphibians (Niemitz, 1984a; Gursky, 2000a). Among the smallest primates (80–150 g), tarsier eyes are bigger than their brains (Sprankel, 1965). Small and smooth, tarsier brains resemble carnivore rather than primate brains (Rosa et al., 1996). Estrous females have swollen red vulvas like Old World monkeys (Wright et al., 1986a), and tarsiers produce huge newborn infants, up to 25–30 percent of the mother's weight (Haring and Wright, 1989). Tarsiers gestate for six months—several weeks longer than macaques or capuchins, which are orders of magnitude larger than tarsiers (Izard et al., 1985; Gursky, 1997)—yet tarsier mothers wean infants within two months of birth—instead of 12–18 months like macaques and capuchins (Haring and Wright, 1989). Tarsier mothers give birth to one offspring at a time yet have 4–6 nipples (Wright et al., 1986b). Unlike other primates that produce large infants, tarsier fathers provide relatively little paternal care (Gursky, 2000b). Since infants are so heavy, mothers and other group members rarely transport them. Instead, a mother will park her infant on branches while she forages nearby (Gursky, 2000b). Tarsiers can turn their head 180 degrees in either direction (Ankel-Simons, 2000) enabling them to see both prey and predators. Owls can do this, but no other mammal can. Tarsiers have extremely long legs, and large hands and feet for their body size—adaptations for leaps of Olympic dimensions (Fleagle, 1998). Their anklebones are elongated—up to four times longer than other same-sized primates—and their leg bones (fibula and tibia) are fused, the only primate to have this special anatomy for extraordinary leaping (Wool-lard, 1925; Gebo, 1987).

Unresolved Questions

The relationship of tarsiers to other primates, both living and fossil, has been a source of debate for over a century. No other primate generates so much controversy. What is the tarsier's true phylogeny? Morphological evidence, including soft-tissue characteristics, suggests that tarsiers are closely related to small nocturnal prosimian primates (lemurs, lorises, and bush babies), called *strepsirhine primates*. However, another suite of anatomical and reproductive characters suggests that tarsiers may be more closely related to monkeys, apes, and humans, called *haplorhine primates*.

The controversy extends to the fossil record. Are tarsiers derived from omomyids, a group of Eocene prosimians? Are they from their own special branch? Did they share a common ancestry with the stock that produced monkeys, apes, and humans, the adapids? Recently the debate has taken a geographic twist: did tarsiers arise in Africa (Simons and Bown, 1985) or Asia (Beard, 1998)?

And as for modern tarsiers, how many extant species are there? The answers range from three to seven. How do these rarely studied species behave in the wild? What is their social organization? And should tarsiers be a conservation priority?

Aims of This Volume

Nearly two decades have passed since the publication of the last tarsier volume, Carsten Niemitz's (1984b) *Biology of Tarsiers*. Major advances have occurred since then in taxonomy, genetics research, and field biology. This collection of articles by tarsier experts brings together a large and representative sample of recent advances in tarsier studies, some using new theoretical approaches, others using new fossil finds or modern technological advances. We have organized this book into three sections, each broadening our awareness of this small primate that hunts in obscurity and haunts the dark island forests of southern Asia.

Section I, Past: Origins, Phylogeny, Anatomy, and Genetics

The past fifteen years have witnessed a series of discoveries concerning tarsier or tarsier-related fossils. A new fossil find from the Fayum of Egypt suggests an early tarsier distribution in Africa (Simons and Bown, 1985). More recently discovered fossil fragments suggest that Asia may have provided the earliest origin of tarsiers (Beard, 1998). These new glimpses into the ancestry of tarsiers, instead of clarifying the tarsier family tree, accelerated the controversy (Simons, chapter 1; Schwartz, chapter 3). While many scientists endorse the anthropoid affinity of *Tarsius* (Martin, 1993; Ross et al., 1998; Kay et al., 1997), several paleontologists disagree (see Simons, chapter 1). Accepting *Tarsius* as sharing a common stem with anthropoids suggests that living tarsioids are divorced from tarsier-like omomyids with their extensive and ancient fossil record dating mainly to the Eocene epoch of about 55– 34 million years ago (Simons, chapter 1; Schwartz, chapter 3).

For the first time, Jablonski (chapter 2) presents a broad view of tarsier habitats, combining a hypothetical reconstruction of Asian Eocene rain forests and current tarsier ecologies. We gain a new appreciation for this primate once we understand the unusual and long history of tarsier ecology.

Anatomical studies also highlight tarsier uniqueness. In chapter 4, Anemone and Nachman present a synthesis of the specialized postcranial anatomy that allows tarsiers to leap with the speedy propulsion of frogs. In chapter 5, Ankel-Simons and Simons present a comparative view of the vertebral columns of primates, elucidating tarsier ability to swivel its head 180 degrees. These studies give us a structural understanding of the athletic prowess of tarsiers.

Nearly explosive advances in the field of genetics have developed better techniques for examining phylogenetic questions, although the questions remain unanswered. In chapter 6, Meireles and coauthors suggest, using globin DNA sequences, that tarsiers are more aligned with monkeys, apes, and humans. Conversely, a recent comprehensive genetic analysis of pla- cental mammals places tarsiers with prosimians (Murphy et al., 2001). Yo- der (chapter 7), examining data from nuclear and mitochondrial DNA, finds equivocal evidence for the phyletic position of *Tarsius*. Genetic studies agree only on one thing—that the tarsier line reaches deep into the past.

Section II, Present: Taxonomy, Behavioral Ecology, and Vocalizations

When the last tarsier book was published (Niemitz, 1984b), scientists rec- ognized three species: *Tarsius bancanus, T. spectrum,* and *T. syrichta.* But in 1987 a new pygmy tarsier (*T. pumilus*) was described from a mountainous re- gion of Sulawesi (Musser and Dagosto, 1987), and in 1991, Niemitz and col- leagues described a new species (*T. dianae*) from central Sulawesi (Niemitz et al., 1991). Today, we recognize six or seven tarsier species from eleven is- lands in four countries: Indonesia, the Philippines, Brunei, and East Malay- sia. In chapter 8, Groves presents new data on external morphology that helps resolve some of the taxonomic issues.

Since Niemitz's book, technological advances have allowed scientists to observe these small, night-active primates. Radio telemetry, infrared binoculars, and GIS methods allow fieldworkers to obtain increasingly precise data on wild tarsiers. For the first time, scientists can apply theoretical perspectives based on standardized data-collection techniques to nocturnal primate studies (Gursky, 1997). Light-weight, portable, high-quality recording equipment now allows researchers to study communication and to use vocalizations to elucidate taxonomy. In chapter 9, Nietsch compares the calls of Sulawesi tarsiers and presents the first technical description of *Tarsius spectrum* vocalizations. Gursky in chapter 10 describes territorial behavior after monitoring seven groups of radio-collared *T. spectrum*. She puts these data into a theoretical framework that compares tarsiers to other primates. Taking advantage of improved logistics and changed politics in the Philippines, Dagosto and Gebo (chapter 11) present new fieldwork on unstudied *T. syrichta,* the first study of the Philippine tarsier in the wild. In chapter 12, Wright and colleagues try to resolve the conflicting reports concerning social organization in wild tarsiers by comparing testes from captive *T. bancanus* and *T. syrichta.*

Section III, Future: Conservation

In chapter 13, Fitch-Snyder examines the history of captive conservation efforts in Europe and the United States. New stocks of wild tarsiers brought into captivity in the 1980s provided an opportunity to better understand husbandry in captivity, but the prospects for future captive populations now look dim. Captive-born tarsiers survive very poorly.

In the final chapter, Wright looks to the future of tarsier survival. This rain forest species lives on only a few islands that are heavily deforested and increasingly susceptible to cutting and wildfires. Wright states that the International Union for the Conservation of Nature (IUCN), the world's experts on endangered and threatened species, has classified most tarsier species as "Data Deficient." This means that IUCN lacks sufficient information on wild populations to classify them. To provide for the future of tarsier species, additional surveys of population densities and geographic ranges (Gursky, 1998) must be conducted and coordinated into a management plan. The plan then needs to be implemented successfully. The extinction within our own time of this controversial and unique primate taxon, whose ancestors reach back into our most ancient primate origins, would bring shame upon us all.

With this volume we aim to update the scientific community on tarsier research. However, this book does not answer all questions about tarsiers. Indeed, many answers are not currently available. The ongoing debate on tarsier phylogeny remains unresolved, and new fossils and more genetics-based research may again shift our understanding of the origins of the genus *Tarsius.* More behavioral fieldwork will enlighten our understanding of tarsier behavior and the tarsiers' niche, and new surveys and censuses must aim to clarify the path to preserving these curious primates. It is our hope that the present collection will inspire new directions for future research.

References

Ankel-Simons F. 2000. Primate anatomy: an introduction. San Diego: Academic Press.

Beard KC. 1998. A new genus of Tarsiidae (Mammalia: Primates) from the Middle Eocene of Shanxi Province, China, with notes on the historical biogeography of tarsiers. Bull Carnegie Mus Nat Hist 34: 260–277.

Fleagle JG. 1998. Primate adaptation and evolution. New York: Academic Press.

Gebo DL. 1987. The functional anatomy of the tarsier foot. Am J of Phys Anthro 73: 9–31.

Gursky S. 1997. Modeling maternal time budgets: the impact of lactation and gestation on the behavior of the Spectral Tarsier, *Tarsius spectrum*. Ph.D. dissertation, SUNY-Stony Brook.

Gursky S. 1998. The conservation status of the Spectral Tarsier, *Tarsius spectrum*, in Sulawesi Indonesia. Folia Primatol 69: 191–203.

Gursky S. 2000a. The effects of seasonality on the behavior of an insectivorous primate. Int J Primatol 21: 477–495.

Gursky S. 2000b. Allocare in a nocturnal primate: data on the spectral tariser, *Tarsius spectrum*. Folia Primatol 71: 39–54.

Haring DM, Wright PC. 1989. Hand-raising a Philippine tarsier, *Tarsius syrichta*. Zoo Biol 8: 265–274.

Izard MK, Wright PC, Haring DM. 1985. Gestation length of *Tarsius*. Am J Primat 9: 327–331.

Kay RF, Ross C, Williams CA. 1997. Anthropoid origins. Science 275: 797–804.

Martin RD. 1993. Primate origins: plugging the gaps. Nature 363: 223–234.

Murphy WJ, Eizirik E, Johnson WE, Zhang YP, Ryder OA, O'Brien SJ. 2001. Molecular phylogenetics and the origins of placental mammals. Nature 409: 614–618.

Musser GG, Dagosto M. 1987. The identity of *Tarsius pumilus:* a pygmy species endemic to the montane mossy forests of central Sulawesi. Am Mus Nov 2867: 1–53.

Niemitz C. 1984a. Taxonomy and distribution of the genus *Tarsius* Storr, 1780. In Niemitz C, editor, Biology of tarsiers, 1–16. New York: Gustav-Fischer-Verlag, Stuttgart.

Niemitz C. 1984b. Vocal communication of two tarsier species (*Tarsius bancanus* and *Tarsius spectrum*). In Niemitz C, editor, Biology of tarsiers, 129–141. New York: Gustav-Fischer-Verlag.

Niemitz C, Nietsch A, Warter S, Rumpler Y. 1991. *Tarsius dianae:* a new primate species from Central Sulawesi (Indonesia). Folia Primatol 56: 105–116.

Rosa MP, Pettigrew JD, Cooper HM. 1996. Unusual pattern of retinogeniculate projections in the controversial primate, *Tarsius*. Brain Behav Evol 48: 121–129.

Ross C, Williams B, Kay RF. 1998. Phylogenetic analysis of anthropoid relationships. J Hum Evol 35: 221–306.

Simons EL, Bown TM. 1985. *Afrotarsius chatrathi,* first tarsiiform primate (Tarsiidae) from Africa. Nature 315: 477.

Sprankel H. 1965. Untersuchungen an *Tarsius.* I. Morphologie des Schwanzes nebst ethologischen Bemerkungen. Folia Primatol 3: 135–188.

Woollard JJ. 1925. The anatomy of *Tarsius spectrum.* Proc Zool Soc Lond: 1071–1184.

Wright PC, Izard MK, Simons EL. 1986a. Reproductive cycles in *Tarsius bancanus.* Am J Primatol 11: 207–215.

Wright PC, Toyama L, Simons EL. 1986b. Courtship and copulation in *Tarsius bancanus.* Folia Primatol 46: 142–148.

PART 1

PAST:
Origins, Phylogeny, Anatomy, and Genetics

CHAPTER 1

The Fossil Record
of Tarsier Evolution

Elwyn L. Simons

Of all major living primate groups, tarsiers provide the least diversity and represent the most distinct clade. Extant tarsiers share a suite of derived features not seen in other primates. These include (1) comparatively large eyeballs surrounded by bony flanges extending well away from the skull around the eye sockets; (2) cervical vertebrae structured so that the head can rotate through more than 180 degrees (Ankel, 1967); (3) hyper-elongated calcaneus and navicular bones; (4) a tibia and fibula that are fused in their distal two-thirds; (5) grooming claws on the second and third digit of the hind foot; (6) an uninflated mastoid region; (7) comparatively forward placement of the foramen magnum; (8) conical incisors; and (9) only one pair of lower incisors (fig. 1.1). Tarsiers exhibit locomotor behavior that almost universally involves vertical postures instead of the more typical pronograde quadrupedal locomotion of most other primates. Also, tarsiers feed exclusively on living animals, a behavioral trait shared by no other primate.

Since the end of the nineteenth century, scientists have considered tarsiers unusually interesting for reconstructing primate phylogeny, and their relationship to other primates has been a subject of controversy. Tarsiers show a strange combination of both archaic and uniquely derived characters and, until recently, their fossil record was not available. Tarsiers combine—among many other features—a primitive insectivore-like cheek-tooth structure, a simple brain, and a primitive pattern of retinogeniculate projections (Rosa et al., 1996) with derived nasal, orbital (Starck, 1984; Hofer, 1977), auditory (MacPhee and Cartmill, 1986), and postcranial adaptations. The tarsier astragalus, however, as in early anthropoideans, remains primitive. The very shallow mandibular horizontal ramus together with the low, slightly developed mandibular coronoid process are likely derived conditions as they are not seen to the same extent in any early primate (fig. 1.1).

Pocock (1918) reviewed tarsier anatomy in detail and compared it with that of lemurs and lorises. He concluded (p. 51) from the structure of the upper lip and nose, the placenta, the presence of the postorbital partition, and other well-known features "that Hubrecht was quite right in removing *Tarsius* from the Lemurs and placing it in the higher grade of Primates." He

9

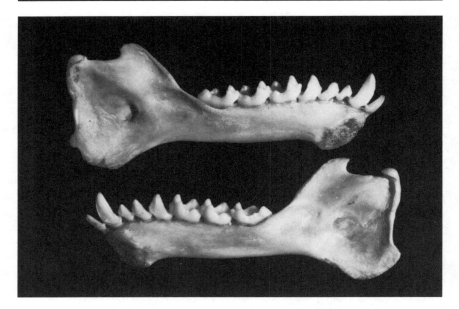

Figure 1.1. Lateral and internal views of the tarsier mandible. Note the shallowness of the mandible and the somewhat forward directed symphyseal region and incisor. The shape of the symphyseal cross section and somewhat squared-off posterior angle resemble those of *Eosimias* (fig. 1.4).

also suggested (based on the morphology of the soft tissue of upper lip and nostrils) that haplorhines could then contain two subdivisions: Tarsioidea and "Pithecoidea" (now Anthropoidea).

Pocock (1918) preferred the term "Pithecoidea" over "Anthropoidea" because "anthropoid" had become associated specifically with manlike apes. For the same reason, I proposed the term "anthropoidean" (Simons, 1987) because "anthropoid" often refers to apes alone, excluding New and Old World monkeys.

One would suppose that the considerable number of traits linking Tarsioidea with Anthropoidea would settle the characterization, but this has not been the case. Many supposed shared, derived anatomical features cited in earlier work (e.g., the forward position of the foramen magnum under the brain case, shared with *Homo*) are clearly a convergence due to the orthal element in the locomotion of the two groups. A similar erroneously chosen and alleged synapomorphy concerns the tubular ectotympanic of tarsiers and catarrhines. Basal catarrhines do not share this trait. Another character of Pocock's (the dry rhinarium) shared by anthropoideans and *Tarsius* could be coincidental—an independent loss due to the tarsier's strong emphasis on visual rather than olfactory predation.

Amusingly, Hofer (1980) discovered that an individual tarsier in Dr. Sprankel's collection had an exact strepsirhine nasal notch on one side, while the other of the same individual's nares had a haplorhine opening. Hence, Hofer considered the shape of the nostrils irrelevant for primate taxonomy.

The similarities of the placenta of *Tarsius* and the anthropoideans stressed by Luckett (1971, 1974, 1976, 1982, 1993) may be more apparent than real. Luckett never addressed Starck's earlier conclusion (1955) that only in the final stages of placentation do tarsier placentas resemble superficially those of anthropoideans (see also Schwartz and Tattersall, 1987). Two remarks from Luckett (1982) seem particularly relevant here: (1) "It is unscientific to deliberately ignore or dismiss findings that are contrary to one's own hypothesis of evolutionary relationships," and (2) "There is no infallible method for using developmental evidence in evaluating character state polarity." Both Luckett and Starck relied heavily on Hubrecht's preserved tarsier embryological collections housed at Utrecht. Starck also had his own tarsier colony and examined that material.

According to Starck (1955, p. 283; trans. F. A. Ankel-Simons):

> Historically *Tarsius* was interpreted as an intermediate between prosimian primates and anthropoids because there are similarities to Anthropoidea [among others, in] . . . the final (mature) stages of placentation. . . . *Tarsius* has central implantation (similar to lemurs). The early embryo is quickly incorporated into the trophoblast. . . . Even though the trophoblast continues to grow and develops a fingerlike extension from its basal aspect, it never shows any invasive tendencies and thus documents biological characteristics that are totally different from the trophoblast of higher primates, which intrudes into the decidua in a diffuse manner. . . . The ontogenetic development of the *Tarsius* placenta is uniquely, highly derived and in spite of the ostensible similarities only of the mature stage with placentae of anthropoids it cannot be interpreted as being an evolutionary transition between prosimian and anthropoid placentae.

The question of whether the manner of placentation links tarsiers and anthropoids is not currently resolved. It may remain difficult to settle in the future since new material will be extremely difficult to secure.

One of the strongest shared characters for Pocock (1918) was the presence of the postorbital partition in *Tarsius,* and this has continued to indicate an important derived similarity with anthropoids (Ross, 1994). Nevertheless, I believe that this is not a shared derived feature. Because scientists argue about this feature, I show it in detail in Figure 1.2.

As figure 1.2 demonstrates, *Tarsius* does not actually have a partition. A large orbital fissure exists—larger than in any anthropoidean—connecting the orbital and temporal fossae (outlined with dashed lines in fig. 1.2). Second, slight flanges from the maxilla and alisphenoid cause such closure as exists in *Tarsius.* The jugal is not expanded inward at all. Contrary to this, in

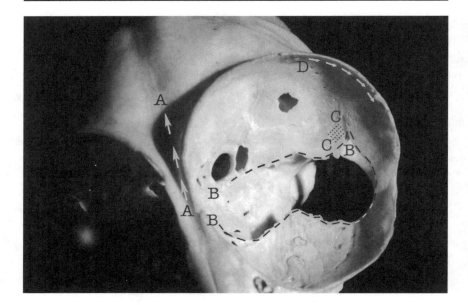

Figure 1.2. Photograph of the left orbit of *Tarsius.* Midline at A-A (arrows). B-B-B indicates the large postorbital opening, window, or fenestra between the orbital and temporal areas. Stippling at C-C shows the place of contact of a thin splint from the alisphenoid, which crosses laterally under the frontal to touch the zygomatic. In the earliest-known anthropoidean skulls of *Catopithecus* and *Proteopithecus,* the zygomatic forms a broad, spoonlike expansion that reaches medially across the postorbital region to touch the alisphenoid at the edge of the brain case. At D, white arrows indicate part of the extracranial flange that composes the rim of the eyesocket of *Tarsius.*

the earliest anthropoidean skulls from the Fayum, in Egypt, belonging to *Catopithecus* and *Proteopithecus,* the major part of the closure behind the eyeball is accomplished by a huge, spoonlike expansion of the jugal, which reaches inward alongside the alisphenoid and closes off most of the space behind the eyeball, leaving only a very small orbital fissure. The orbit of *Tarsius* would not be labeled significantly "closed" if it were not that in order to hold the enormous eyeball, a flange of bone encircles the eyeball, jutting well out from the skull (fig. 1.2). This helps anchor the eyeball—which is the largest, relatively, of any mammal. Each eyeball has about the same volume as the brain in some individuals (Sprankel, 1965), and, in turn, this means that the brain case is little larger than each eyeball. Thus, the partitioning in *Tarsius* is likely due to support of the eyeballs next to a comparatively small brain case rather than a derived feature shared with higher primates.

Pocock (1918) argued that only in tarsiers and anthropoideans does the alisphenoid touch the jugal. Nevertheless, the contact is accomplished quite differently in tarsiers than in other primates. The contact lies in a small area rather than along a broad front. (See the stippling at C-C in fig. 1.2.) In some tarsier individuals, no contact with the jugal occurs.

Ross and Covert (2000) recently dismissed the existence of a structural difference between the postorbital septum of *Tarsius* and anthropoids, citing Cartmill (1980) and Ross (1994). However, in these references, Ross (1994) merely echoes Cartmill (1980) by stating, "I am like minded." In turn, Cartmill (1980) disagrees with Simons and Russell (1960), who state that the septum contributes relatively little to the malar or jugal. Simons and Russell (1960) also suggest that the existing closure or septum in *Tarsius* can be attributed functionally to the bony expansion of the eye socket around the enormous eyeball rather than a shared-derived character with anthropoids.

Just as the periorbital cone extends outward from the cranium as far as the equator of the eyeball, the alisphenoid and maxillary bones extend slightly inward at the margins of the fenestra to form a socket for an unusually large eyeball. In tarsiers, unlike Anthropoidea, no expansion exists of the jugal inward along a long front with the alisphenoid, and hence no real postorbital plate exists behind the eyeball. No one has convincingly demonstrated that the partial closure of the postorbitum in *Tarsius* is a derived trait shared with anthropoids (Rosenberger and Szalay, 1980).

Simons and Rasmussen (1989) discuss a series of hypothetical shared-derived features linking *Tarsius* and anthropoideans. Apart from postorbial closure these similarities are (1) interorbital septum, (2) perbullar pathway of the internal carotid, and (3) division of the intrabullar cavity into separate tympanic and anterior accessory cavities. Through comparisons with the archaic anthropoid *Aegyptopithecus*, Simons and Rasmussen conclude that these features are not haplorhine synapomorphies.

Since Pocock (1918), two primatological camps formed, one using traditional terminology "Prosimii" and "Anthropoidea," and the other using "Strepsirrhini" for Prosimii excluding tarsiers and "Haplorhini" for anthropoideans including tarsiers. The latter division might be acceptable if almost all anatomical structures in *Tarsius* he cited had not subsequently been questioned as truly homologous.

Simons (1974) demonstrates another problem with characterizing fossils primates. Many early primate fossils are difficult to categorize as haplorhine or strepsirhine—or even as anthropoidean—because the categorization depends on soft-tissue features not observable in fossils. Beard (1988) presents a cogent argument, however, that strepsirhinism is reflected in fossil and living primates by a gap between the upper central incisors. In conjunction with this continuity between the rhinarium and vomeronasal organ, the cleft upper lip is tethered to the gums between the upper central incisors.

Such a condition causes a distinct space between the upper incisors or their roots where the rhinarium is tethered.

Tarsiers lack such spacing but it, and presumably a cleft upper lip, may have been present among omomyids (such as *Necrolemur*), which many paleontologists consider more closely related to *Tarsius* than *Tarsius* to Anthropoidea (Simons, 1961; Szalay, 1976; Rosenberger, 1985; Beard and MacPhee, 1994). Thus, according to Beard (1988), the gap in *Necrolemur,* other microchoerines, and omomyids may indicate that, "the haplorhine oronasal configuration of extant tarsiids and anthropoids was acquired independently, or that tarsiids and anthropoids form a clade to the exclusion of omomyids." However, strong evidence supports the idea that Tarsiidae arose from omomyids (Teilhard de Chardin, 1916–21; Simons, 1961, 1972; Szalay and Delson, 1979; Beard et al., 1991).

An extensive literature concerns derived dental features of tarsiers (Simons, 1961; Beard and MacPhee, 1994; Rasmussen et al., 1995). Some of these are (1) enlarged canines, (2) the loss of one pair of lower incisors, (3) apparent loss of the P_4 metaconid seen in the Eocene tarsiid *Xanthorhysis,* and (4) the loss of the double rooted P_3 seen in *Xanthorhysis* and *Afrotarsius.*

In comparison to the early Tertiary omomyids, tarsier premolars and molars do not seem derived. If any significant derived features exist, they appear to lie in the anterior teeth. However, if an Eocene tarsier radiation occurred, as new data from China suggest, tarsiers could not easily have only one set of features; they must have once been diverse. However, none of the fossils discussed below support this. Interestingly, Cartmill (1980) proposes that the common ancestor of monkeys and tarsiers was more monkeylike than modern tarsiers, and, hence, present-day tarsiers have moved away from anthropoideans as their very early ancestors were.

Recently a taxonomic placement at variance with the concept of haplorhini surfaced again. While analyzing genes and gene segments, Murphy and colleagues (2001) report: "Within primates, lemur (Strepsirrhini) and tarsier (Tarsiiformes) were found to be sister taxa (bootstrap $\geq$ 80%) that were separated from anthropoids by a deep divergence." More recently, Murphy (pers. com.) reports that, having added additional taxa, their work continues to sustain support for an ancient divergence between Prosimii (containing tarsiers) and the Anthropoidea. This strongly contrasts the groupings made by Kay et al. (1997) and Ross et al. (1998).

The Fossil Record

Unfortunately, known fossil tarsiers are limited in their completeness and in the number of specimens. Thus, scholars find agreement difficult. No evidence exists for the antiquity of any distinctive or possibly shared-derived features. In this section, I discuss the known record of fossil tarsiers from the earliest to the most recent forms. In addition, I review two other early primates—*Pseudoloris* and *Eosimias*—which have been ranked in other families but have possible tarsier affinities. Because fossil tarsiers are few and imperfectly known, and since living members appear closely related, geographically restricted and derived, we can be sure that we know very little about the diversity and morphological complexity that once existed in this family.

In addition, various authors have considered dental elements in present-day tarsiers as secondarily derivations (Maier, 1980). For example, the reduction of lower incisors to a single pair cannot be primitive. Moreover, since few students have examined large numbers of modern tarsier dentitions, the degree of intraspecific and intrageneric variation in tooth structure is seldom considered. Instead, many authors emphasize distinctions involving only very slight differences of the sort that might be included within specific or individual variation if almost any other mammal were concerned.

These features are often encoded for parsimony analysis as "present" or "absent" when actually many form a morphological continuum. One major problem with encoding character states such as these is that many features used to describe morphology of tarsioids and anthropoids differ so slightly that they are impossible to objectively enter into a parsimony analysis. For example, how far left or right of a protoconid does a cristid obliqua have to be on a lower molar to constitute a different character state? Recent evidence from developmental biology and the study of variational properties of teeth suggest that developmental processes themselves may play a large role in biasing tooth variation (Polly, 1998; Jernvall, 2000; Jernvall and Jung, 2000; Sauther et al., 2001). Quantifying typological divisions of character states may be of little use if some states are developmentally linked or are substantially more likely to occur than others.

The dentition of living tarsiers is clearly adapted to faunivorous feeding. Tarsiers have long, pointed incisors and canines for seizing prey, and they have cheek teeth with sharp cusps interconnected with sectoral crests for slicing and macerating chitin of prey. Many investigators believe that all primates began with an insectivorous feeding adaptation, and since the tarsier feeding pattern has not changed, this may explain why primate teeth have so little altered from a general insectivore pattern.

Fossil Tarsiers

Tarsius eocaenus

First described by Beard and colleagues in 1994, this forty-million-year-old fossil is based on material from Fissures A and C at Shanghuang, from the Middle Eocene of Jiangsu Province, China. This unprecedented temporal extension of the genus *Tarsius,* from the Middle Eocene to present, was not explicitly justified by the authors. No one has ever considered a primate genus as having a temporal extension even a fifth as long. Considering typical rates of evolutionary change among mammals, that this species actually belongs in the modern genus seems unlikely. Within all placental mammals, one only finds two extant genera of bat, *Hipposideros* and *Rhinolophus,* recorded as extending back as far as the Middle Eocene (M. McKenna, pers. com.). Not being a bat taxonomist, I have no idea whether the temporal range extensions of these two genera are valid or whether they are equally unsupportable. Possibly their dental morphology has changed little through time. Simpson (1953) calculated that the average longevity of carnivore genera was nine million years. Survivorship of genera in the lineage from *Hyracotherium* to *Equus* is somewhat less, and proboscidean genera have an even shorter temporal longevity.

The several isolated teeth that comprise the hypodigm of this species come from Fissures A and C. The fissure-fill sites are located near the village of Shanghuang in southern Jiangsu Province, China. In addition to tarsiers, these fissures yield primate fossils representing adapids, omomyids, and the distinctive genus *Eosimias.* The teeth of *T. eocaenus* (illustrated by Beard et al., 1994; Beard and MacPhee, 1994) resemble teeth of modern *Tarsius,* but not any more than those of *Afrotarsius,* long considered a probable tarsiid from north Africa. Perhaps *T. eocaenus* belongs to a different genus. In *T. eocaenus,* the M_2 hypoconid and M_3 protoconid are more laterally divergent than in modern *Tarsius* (Beard et al., 1994; Beard and MacPhee, 1994), but this ignores variability in the modern form. The M_3 entoconid of *T. eocaenus* is more mesial than is the hypoconid. This particular feature is of slight significance as it is also shared with *Eosimias* and some early anthropoideans, but it is not clearly present in *Tarsius, Afrotarsius,* or *Xanthorhysis.* Beard et al. (1994) and Beard and MacPhee (1994) conclude that the Middle Eocene occurrence of this species "brings the fossil record for Tarsiidae into much greater concordance with predictions based on cladistic reconstructions of the phylogenetic position of this group."

Assigning *Tarsius eocaenus*—represented by only five isolated teeth—to the modern genus seems unjustified (Simons, 1997b). Unknown parts of dentition may justify placement outside this genus or even outside the family. In fact, researchers have classified *Afrotarsius* using similarly incomplete data. The fossil was originally classified as a probable tarsiid (Simons and Bown, 1985) based on a lower jaw holding three molars and the bases of P_{3-4}. In the case of *T. eocaenus,* a prudent approach would have been to describe it as cf. [*conferro*] *T. eocaenus.* This would be particularly advisable because the presented extension of the genus across most of the Cenozoic is based on the retention of primitive dental characters. However, during the same period, mammalian tooth morphology was very homoplastic (Hunter and Jernvall, 1995; Sánchez-Villagra and Kay, 1996). The necessary unique lines of evidence required to justify the extension of a mammalian genus across a space of forty million years remains questionable.

Xanthorysis tabrumi

Beard (1998a) first described this species, which comes from the Late Middle Eocene Heti formation, Yuanqu Basin, southern Shanxi Province, China. The exact age of this deposit is not established. The published fossil record consists of a single left mandible containing P_3 to M_3 and the alveoli of the lower canine and P_2. The generic name comes from Greek for Yellow River, and the species name is coined for A. R. Tabrum, who discovered the only known specimen. Beard (1998a) considers this genus a sister group of all other tarsiers and hence, because of its age, concludes that it pinpoints the radiation of tarsiids and their relatives from no later than Early Paleogene. This fossil, taken in conjunction with *T. eocaenus* and Oligocene *Afrotarsius,* may establish Tarsiidae as having the longest temporal range of any of the modern primate families.

In the diagnosis of *Xanthorhysis,* several features differ from living *Tarsius.* These include (1) relatively longer P_{3-4} with divergent, rather than mesiodistally compressed roots, (2) a distinct and lower situated P_4 metaconid, and (3) lower, comparatively longer, narrower, and lower crowned molars with entoconids more distal in position (Beard, 1998a). Actually, such an entoconid location resembles *Afrotarsius,* contra Beard (1998a), who states that *Xanthorhysis* has more cuspidate paraconids and less reduced M_3 length than *Afrotarsius.* The position and relative size of the paraconid is quite similar in the two genera, nor does the size of M_3 and M_3 hypoconulid differ significantly between the two. The M_3 entoconid in *Xanthorhysis* is more distinct than in *Afrotarsius.* Distinct M_3 entoconids also characterize *Eosimias, Plesiopithecus,* and other primates. *Afrotarsius* resembles many modern *Tarsius* individuals in that the M_3 entoconid is poorly, or not at all, expressed.

Afrotarsius chatrathi

Simons and Bown (1985) describe this species, which is thought to be about 32 million years old. This first-named fossil tarsier was assigned tentatively to Tarsiidae. The sole specimen, CGM 42830, was collected by Prithijit Chatrath in Oligocene sediments at Quarry M, in the upper sequence, Jebel Qatrani Formation, Fayum Province, Egypt. Unfortunately, the original illustration (Simons and Bown, 1985) that accompanied the description of this species is inaccurately drawn, and the stereophoto pairs were printed at far too small a scale to be properly viewed. These poor illustrations may have contributed to Ginsburg and Mein's (1987) conclusion that the species had "simiiform" affinities. Certainly, drawings of *Afrotarsius* teeth fail to show the salient nonanthropoidean features of this species, such as trenchant crests, elevated trigonid, and pyramidal, uninflated cusps. These authors were not aware that features like these poorly illustrated characteristics would later be proposed as belonging to "stem" anthropoids (Beard et al., 1994, 1996; Beard and MacPhee, 1994). Their comments document the inadequacy of opinions based on illustrations rather than specimens.

Ginsburg and Mein (1987) base their view of anthropoidean affinity on two main points. The first concerns the position of the paraconid. They considered the paraconid cusp situated rather more labially in the trigonid, as in many omomyids and parapithecids. Actually, the paraconid cusp in *Afrotarsius* lies near the tooth midline—just where it is in extant *Tarsius*. In *Xanthorhysis*, it lies even farther lingually than in *Afrotarsius*.

All such slight positional distinctions have little meaning if developmental basis and variational properties of characters are not understood (Jernvall and Jung, 2000; Sauther et al., 2001). Only among living tarsiers, where numerous individuals can be studied, can uniformity in cusp and crest position be documented. In addition, most parapithecids can hardly be described as having a distinct paraconid; they rather have a crest with no clear-cut apex. Ginsburg and Mein (1987) also consider the smaller size of M_3 in *Afrotarsius* significant. They propose *Afrotarsius* as a new family, with M_{1-2} of the tarsier type and with M_3 identical in shape but smaller than the anterior molars.

We now know that in the Asian *Xanthorhysis* the M_3 and its hypoconulid are almost proportionately the same as in *Afrotarsius*. In both these genera, length of M_3 is only slightly less than M_2 but the crown area, as viewed from above, is distinctly smaller. Hence, the two features stressed by Ginsburg and Mein (1987) prove not to differ from fossil tarsiids. Despite this, their comments apparently generated the characterization of *Afrotarsius* as anthropoid. Finally, in a personal communication to the author, Ginsburg and Mein state that they now believe that *Afrotarsius* is a tarsiid.

In a discussion of the phyletic position of parapithecids, Fleagle and Kay (1987) comment on *Afrotarsius,* also citing the view of Ginsburg and Mein (1987). By building on the idea that *Afrotarsius* stands apart from *Tarsius* and that *Afrotarsius* shows anthropoid affinities, this interpretation developed a life of its own. Fleagle and Kay (1987) propose four *Afrotarsius* characters that diverge from tarsiers and are anthropoid-like: (1) the distal wall of M_{1-3} is smooth; (2) the molar cristid obliqua reaches only to the base of the trigonid, not significantly invading its distal wall; (3) small, midline, distal hypoconulids are present on M_{1-3}; and (4) a relatively small M_3. Fleagle and Kay

(1987) further state that "this is admittedly a rather weak list of features for determining the phyletic position of this important, but poorly known primate." The list is also weak because the first character has only a vague meaning that presents no distinction from the others. We now know that the remaining three characteristics are also seen in *Xanthorhysis,* and the second feature is similarly developed in *Tarsius* where the cristid obliqua reaches the trigonid in a manner similar to its location in *Afrotarsius.* The fact that the M_3 of *Afrotarsius* is relatively small, like that of *Xanthorhysis,* has already been mentioned; its length is about 90% of M_2 length in both genera. More importantly, the M_3 crown morphology of *Afrotarsius* resembles that of *Tarsius,* and existing differences do not resemble M_3 morphology of parapithecids or other anthropoids.

Ross et al. (1998) reiterate that *Afrotarsius* may be the sister taxon of Anthropoidea and cite Kay and Williams (1994) as evidence for this assignment with the help of four synapomorphies: (1) M_{1-2} cristid obliqua oriented mesiodistally toward the protoconid; (2) metacristid transverse to paraconid; (3) small lower molar hypoconulids; and (4) M_2 larger than M_3. Their wording of the second character is a lapsus; as in Kay and Williams (1994), it correctly reads "metaconid transverse to protoconid."

Comparisons reveal that, in fact, the cristid obliqua of *Afrotarsius* stands in the same position as in *Xanthorhysis. Xanthorhysis* and *Afrotarsius* show no greater extension or invasion of the trigonid posterior wall, suggesting a primitive condition. Kay and Williams (1994) state that this character (the first above) is "a synapomorphy of *Tarsius* and Anthropoidea." I see no difference that could separate *Afrotarsius* from tarsiids.

Both *Afrotarsius* and *Xanthorhysis* also share a larger M_3 than M_2. In addition, just as it is located in *Afrotarsius, Xanthorhysis* and *Tarsius thailandicus* both have metaconid transverse to protoconid, but in all three, the metaconid sits slightly more distal than the protoconid. In extant *Tarsius,* the metaconid lies somewhat more mesially. The M_3 hypoconulid, which is long in modern *Tarsius,* is comparatively smaller than in *Afrotarsius* and *Xanthorhysis.*

In consequence, none of the four weak dental features listed in Kay and Williams (1994), Kay et al. (1997), and Ross et al. (1998) can convincingly distinguish *Afrotarsius* from tarsiids or ally it with anthropoideans. Kay et al. (1997) conclude that "more complete material may demonstrate that eosimiids are actually Asian afrotarsiids." The statement suggests that both *Afrotarsius* and *Eosimias* may have shared derived similarities and can be characterized as taxonomically close to tarsiids.

Ross et al. (1998) have more confidence in the taxonomic placement of *Afrotarsius* (represented by three complete and two half teeth) than placement of *Rooneyia* (represented by a remarkably complete and undistorted cranium). The paper uses 182 dental characters and 49 cranial characters. Many of the dental characters, such as those cited in reference to *Afrotarsius,* are ambiguous or likely to form character complexes whose developmental basis may be linked (Jernvall and Jung, 2000; Sauther et al., 2001); thus, many of these characters should not be listed as separate features. For example, if "no inflation and vertical walls" is understood to refer to the pyramidal, angular, or trenchant cusps that characterize tarsiids, then it is actually another similarity between *Afrotarsius* and *Tarsius* shared in exclusion of anthropoideans.

In a paper on the historical biogeography of tarsiers, Beard (1998b) cites several papers that form the tradition of anthropoidean affinities for *Afrotarsius.* He then formulates for it three novel anthropoidean features: (1) apparent oblique orientation of P_4 (as occurs in *Eosimias* and most other basal anthropoids); (2) apparent large molar protoconids relative to metaconids (because lower molar trigonids on the *Afrotarsius* specimen are damaged, the existence of this character in *Afrotarsius* is not certain); and (3) reduced hypoconulid lobe on M_3 (as occurs in *Eosimias* and other basal anthropoids; cf. Ginsburg and Mein, 1987).

First, oblique orientation of P_4 in *Afrotarsius* cannot be confirmed because the crown of that tooth is broken away. This tooth may not sit more obliquely than the P_4 of *Xanthorhysis.* The oblique orientation of P_4 used by Beard (1988) as a derived feature of anthropoids is a particularly unconvincing character because tooth position and orientation can change during life. Furthermore, inferring tooth orientation requires unbroken specimens and more than one sample—neither of which is possible in this case. Using Beard's (1988) terminology and procedures, P_4 in another primate (*Plesiopithecus*— perhaps related to strepsirhines, lorisiforms, but definitely not anthropoids) is properly described as obliquely positioned. In addition, recently described postcrania of *Pondaungia* from the Late Middle Eocene of Myanmar (Ciochon, et al., 2001) strongly indicate a strepsirhine adapid, but the lower premolars of this animal, as with *Plesiopithecus,* show some transverse orientation. The broken P_4 of *Afrotarsius* cannot confirm oblique orientation. Second, the molar protoconids of *Afrotarsius, Tarsius,* and *Xanthorhysis* are to the same degree relatively large when compared to metaconids. Third, the hypoconulid lobe of *Afrotarsius* is of the same relative size as in *Xanthorhysis,* which Beard (1998a) describes as a tarsiid. The pattern of these subtle characters cannot form a strong basis for the "anthropoid affinity" argument.

Simons (1995, 1998) has illustrated the type specimen of *Afrotarsius* next to modern *Tarsius* (brought to the same size to show their near identity in molar structure), and this is reproduced here (fig. 1.3) relative to *Eosimias sinensis.* All three share a high, distinct trigonid with three separate and sharply delineated cusps and a talonid with posteriorly diminishing ento-

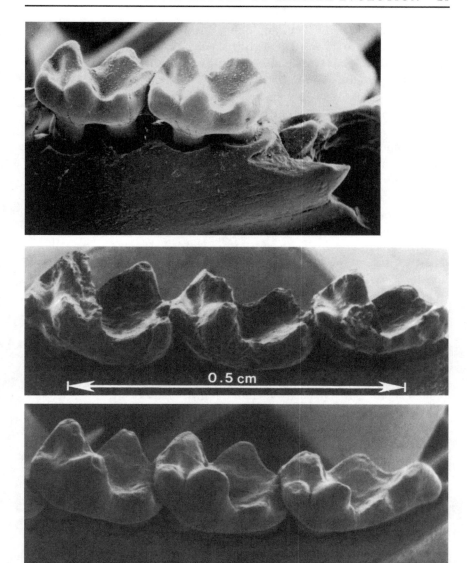

Figure 1.3. Mandibles of *Eosimias* (top), *Afrotarsius* (center), and extant *Tarsius* (bottom) demonstrating their molar similarities and differences.

conids. These similarities demonstrate no convincing morphological evidence on which to base a separate family or to characterize *Afrotarsius* outside Tarsiidae.

Rasmussen et al. (1998) describe a tibiofibula from Fayum Quarry M Early Oligocene of Egypt found in the same part of that quarry as the type mandible of *Afrotarsius*. They argue that this tibiofibula is morphologically similar to modern *Tarsius* but slightly smaller (as are the teeth of *Afrotarsius*). The discovery of this bone would provide an independent confirmation of the phyletic position for *Afrotarsius,* if the bone is tarsiid.

Dagosto et al. (1996) report tarsiid tibiofibulae attributed to *Tarsius eocaenus* from the Middle Eocene of Shanghuang, China, which may establish this as a tarsiid feature of great antiquity. However, fused tibiofibulae occur in some insectivores, macroscelideans, and rodents, and therefore the Quarry M tibiofibula must be firmly established as not pertaining to one of those groups. The cnemial crest of the tibia is anteroposteriorly (dorsoventrally) deep. It is not rounded as in *Tarsius* but instead is bladelike, like macroscelidean tibiofibulae—but no macroscelidean dentition of this size has yet been found at Quarry M. In addition, the lateral tibial condyle in all primates, including *Tarsius,* is more rounded than in the Quarry M bone. The back side of the dorsal surface of the proximal tibial shaft is squared and has very sharp medial and lateral sides, unlike typical primates. The phiomyid rodents, whose modern descendants are cane and rock rats, are the only Oligocene rodents found in abundance at Quarry M. My examination of modern cane rat (*Thryonomys*) hindlimb bones establish that tibia and fibula are fully separate bones, and hence they were not likely fused in any Fayum rodent. A small placental mammal, *Widanelfarasia,* recently described from the Late Eocene of Egypt (Seiffert and Simons, 2000) may be related to the Afro-Malagasy zalambdodonts, many of which have fused tibiofibulae. Nevertheless, none of these Fayum nonprimates have such a tibiofibula as that found in Quarry M. Presently, little alternative exists to the idea that the Quarry M tibiofibula is from a tarsier.

A final example of the manner in which current discussions of the taxonomic position of *Afrotarsius* can unintentionally be tilted occurs in Ross et al. (1998). Of the ten character state changes in their table 9, entitled "Unambiguous character state changes supporting an anthropoid clade that includes *Eosimias* and *Afrotarsius,*" only one can be observed in *Afrotarsius,* and this is the eighth character, which states that from a presumed ancestral M_3 trigonid and talonid of equal size, a shift toward a relatively large M_3 trigonid has occurred. Here, *Afrotarsius* is not different from *Xanthorhysis* (Beard, 1998a), an unquestioned tarsiid that was not published until after Ross et al. (1998). Additionally, both of the latter genera and species are documented by only a single specimen, and M_3 talonids in mammals are known to commonly vary in size between individuals of the same species.

Tarsius thailandicus

This species was described by Ginsburg and Mein in 1987 as *Tarsius thailandicus,* and curiously, these authors also place a Miocene species in a modern genus. Concerning the species name (DuCrocq et al., 1995; Beard, 1998), the ending should have a masculine generic ending according to the International Rules of Nomenclature. The type specimen comes from Early Miocene sediments at Li in Lampoon province, northwestern Thailand, and it consists of an isolated single right lower M_2, which resembles that of modern tarsiers. The site is dated by mastodont and rodent teeth (Ginsburg and Ukkakimapan, 1983) to the lower Kamlial provincial age, considered an equivalent of Middle Orleanien—16–18 million years old. The single tooth is slightly more than 75% the size of M_2 in the modern *T. spectrum*. In the discussion of their new species, these authors comment on its relationship to *Afrotarsius*. They propose that the latter represents a new family, Afrotarsiidae. They also recommend the placement of *Afrotarsius* as a branch of anthropoidean lineage that split before the separation of the catarrhine and platyrrhine stocks. Numerous Eocene/Oligocene primates have been called tarsioids, but in most cases they cannot be demonstrated as such. Many affinities exist between omomyoids and tarsioids, but here I discuss only two further forms that might have ties to tarsiers.

Fossil Tarsioids

A long controversy concerns whether many early Tertiary primates of the omomyiform sort belong in the same radiation as *Tarsius*. Whether or not this is the case for any omomyid has been discussed at length by Beard and MacPhee (1994). Since this book concerns living and fossil tarsiers, I do not include a discussion of most of these Paleogene genera. Nevertheless, two genera—*Pseudoloris* and *Eosimias*—concern the history of fossil tarsiers and will be discussed in the following pages. See also references elsewhere in this book and discussion of early tarsioids in the contribution by Schwartz (chapter 3).

Pseudoloris parvulus

This species, from the Quercy Phosphorites, French Middle Eocene, was described by Filhol (1889–90) as a species of *Necrolemur*. The species was placed in the new genus, *Pseudoloris*, by Stehlin (1916). Later, Teilhard de Chardin (1916–21) illustrated and discussed it in detail. He noted its resemblance to modern *Tarsius* and published comparative stereophotographs of the rostrum of *Pseudoloris* and skull of *Tarsius*, clearly illustrating their similarities. The rostrum of the former may be slightly longer relatively, and its absolute size is only about 60% of that of *Tarsius*.

Piveteau (1957) also discusses material of *Pseudoloris*, comparing it with *Tarsius*. Simons (1972) characterizes the species as a tarsiid while Szalay and Delson (1979) returned it to Microchoerinae among omomyid tarsiiforms.

Hence, the exact relationships of this species have been debated. The latter authors describe the resemblance as due to a strong dental convergence, but common characters extend beyond details of dentition. Piveteau (1957) considers the reduction of the mandibular coronoid process in *Pseudoloris* to resemble the reduced coronoid of tarsiers. The upper tooth rows in both are arranged almost identically with an outline in the shape of a bell. The dorsal aspect of two different rostra of *Pseudoloris* show that the orbits were enormously enlarged in a manner comparable only to modern *Tarsius*. The orbital fossae clearly reach to the midline so that the creature very likely, as in *Tarsius,* had a very narrow interorbital septum.

Considering the number and arrangement of upper teeth, Szalay and Delson (1979) note that the second tooth, presumably I^2, is more laterally located in tarsiers. Interestingly, in *Pseudoloris* the upper tooth row begins with a large I^1 followed by a small I^2, then the large canine, three premolars and three molars, just as in *Tarsius*. Functionally, the centrally shifted large incisor pair of *Tarsius* is most interesting, because with the dry nose, no interincisal diastema exists for a philtrum as in primates with rhinaria. The one difference in upper tooth spacing between *Pseudoloris* and *Tarsius* lies at the point in *Pseudoloris,* where the anteriormost teeth are spaced apart. This implies possession of a philtum and rhinarium rather than appressed central incisors as in *Tarsius*. If the relationship between tarsiers and *Pseudoloris* is real, in the Middle Eocene the tarsier ancestor might not have had a haplorhine nose (Beard, 1988).

Ranking *Pseudoloris* with the microchoerines is difficult because of the great difference in molar anatomy. The main difference, as Szalay and Delson (1979) remark, is "the acuity of the cusps and the relatively high crests" seen in both *Pseudoloris* and *Tarsius*. In *Microchoerus* and *Necrolemur,* the upper molars are more cuspidate; they have more mesiodistally elongated molar cusps and molar outlines. Nevertheless, Szalay and Delson (1979) wished to separate *Pseudoloris* and *Tarsius* because of differences in the relative size of the antecanine lower teeth. In the lower molars of *Pseudoloris,* the paraconid is situated more lingually. *Nannopithex,* another microchoerine, is possibly the ancestor of *Pseudoloris* (Godinot, pers. com.). In short, the range of dental variation in the European microchoerines is very great and, for instance, if Tarsiidae were allowed such a high degree of variation between individual species, including even such a supposedly different genus as *Eosimias* within, it would not be difficult.

The anterior lower teeth of *Pseudoloris* resemble those of other microchoerines, but the question is not how did this animal converge on a tarsierlike pattern but how did they diverge. The fossil record provides no information. Also, whether lower anterior teeth were arranged in tarsiers as in microchoerines is unknown. Clearly, in *Pseudoloris* only two incisors exist above, contra both Teilhard de Chardin (1916–21) and Piveteau (1957). Szalay and Delson (1979) are troubled by the fact that the premaxillary/maxillary suture cannot be determined in *Necrolemur* and *Microchoerus*. However, identifying them as the same teeth as in *Tarsius* seems reasonable, and as far as *Pseudoloris* is concerned, the only difference is that the central (most mesial) pair are more widely spaced apart. Since *Pseudoloris* and newly found *Tarsius eocaenus* and *Xanthorhysis* come from the Middle Eocene, imagining the oldest, most unique morphology from which *Pseudoloris* converged on *Tarsius* in dental anatomy is difficult. Since ancestral basal primates are hypothesized to have been small insectivores—similar to tarsiers (Kay et al., 1997)—the oldest, most unique morphology is even more difficult to imagine.

Eosimias sinensis and *E. centennicus*

These species have been described as belonging to a new family, Eosimiidae. The first species described (Beard et al., 1994; Beard and MacPhee, 1994) was *Eosimias sinensis* and the second was *E. centennicus* (Beard et al., 1996). The type and material of *E. sinensis* were collected from the Middle Eocene Shanghuang fissure fillings of southern Jiangsu Province, China. *E. centennicus* is based on material collected from the Eocene Heti Formation, Yuangqu Basin, southern Shanxi Province, China, which is believed to be younger than the Shanghuang fissures. Neither site has been conclusively dated to the Middle Eocene by independent means.

These two species constitute a new primate family placed into suborder Anthropoidea. In so doing, Beard et al. (1994a,b; Beard, 1998a,b) and Gebo et al. (2000a,b) attempted to change the very definition of anthropoid or anthropoidean. Furthermore, *Eosimias,* which literally means "dawn simian" or "dawn anthropoidean," reinforces the erroneous idea that these species are anthropoids. By defining stem anthropoideans around *Eosimias,* these authors suggest that this genus provides significant evidence concerning the nature of Eocene African Anthropoidea and "crown" anthropoids. Reported resemblances between *Eosimias* and Tarsiidae have been used to support the concept of Haplorhini, despite the fact that the evidence for this clade is primarily neontological. The articles, together with Beard (1998a), downplay Africa as the place of anthropoid origin and emphasize Asia as the point of origin for Tarsiidae. In this process, *Afrotarsius* has been marginalized as a tarsier relative, but in fact *Eosimias* has a very primitive and generalized dental morphology. *Eosimias*'s dental morphology resembles that of extant *Tarsius* in its trenchant, high, and sectorial molar cusps. However, many *Eosimias* peculiarities are not very consistent with the anthropoid radiation.

Although perhaps not relevant, known *Eosimias* species exist too late in time—*Algeripithecus* and *Tabelia* from the Middle Eocene of Algeria are both contemporary with *Eosimias,* if stratigraphic dating in both regions is correct.

Both Algerian forms have tooth morphologies resembling specific Late Eocene Egyptian and Oligocene Fayum anthropoideans (Godinot, 1994). For instance, unlike *Eosimias,* these Algerian species have distinctly developed hypocones and entoconids. *Eosimias* upper teeth do not conform with expectations of a basal anthropoid. Tong (1997) figures two upper molars of *Eosimias centennicus.* They show a nonanthropoid morphology with a huge M^2 parastylar cusp (larger than that seen in some omomyoids), and a buccolingually broad upper M^{2-3} that both narrow anteroposteriorly and are waisted in the middle. Unlike the case in *Catopithecus* and *Proteopithecus* where hypocones are small, there is no hypocone in the teeth illustrated.

Since the molars of both *Eosimias* and *Tarsius* are so primitive and insectivore-like, a valid morphological argument existed at first—when it was only known from anterior alveoli and P_3 to M_2 (type and paratype of *E. sinensis*) that the genus might not even have been a primate. To date, the arguments for anthropoid status of *Eosimias,* based on mandibular and lower tooth characters, are not very convincing. The partially published examples of upper dentition is even more unlike that of anthropoideans than is the lower dentition. The genus shows a combination of archaic primate features, features of both omomyids and tarsiids as well as some features not seen in any other primate group. (See also Gunnell and Miller, 2001.)

Szalay (2000) categorized *Eosimias* within tarsioids for a number of reasons. First, he is unimpressed with the erectness of the lower incisors as an anthropoid character, stating rather that this orientation "is also particularly similar to *Tarsius,* but omomyids such as *Loveina, Washakius,* or *Ourayia* also display vertical incisors." Second, he concludes that the shape and orientation of the canines could be structurally ancestral to the condition seen in tarsiids. He also believes that *Eosimias* lower premolars are tarsiidlike, which nevertheless lacks the hypertrophied relative size of the molars of modern *Tarsius.* Third, Szalay emphasizes the similarity between the two in possessing shearing, hypertrophied molar trigonids that lack the "distally progressive reduction of the paraconids" more typical of anthropoideans. Fourth, both families share mesially and distally, strong molar shearing crests. Finally, Szalay (2000) states that in both groups the talonid construction is unlike that of basal anthropoideans where the hypoconulid tends to be close to the entoconid.

Beard et al. (1996) itemize a series of *Eosimias* characters that resemble characters of anthropoideans or that are unlike those parts of adapids or omomyids. A unique combination of characters is the basis for justification of a new, separate family (Beard et al., 1994; Beard and MacPhee, 1994). Nevertheless, the long-continued delay in publishing details of the upper dentition with its nonanthropoid or unanthropoidean tooth crown anatomy has confused scholarly assessment.

Beard et al. (1996) discuss a number of features that resemble undoubted anthropoids. Six of these refer to size or orientation of teeth, not to anatomical or structural similarities. In turn, many of these orientational features may relate to a relatively anteroposteriorly crowded tooth row (Gunnell and Miller, 2001). Other analogues with higher primates appear meaningless. These include features such as *Eosimias* and *Catopithecus* sharing an unfused symphysis, a primitive character and likely coincidental. Supposed similarities based on size or orientation—not morphology—include the large canines, the vertical implantation of lower incisors, and the anteroposteriorly abbreviated and dorsally deep symphysis, said to resemble that of anthropoids. According to Gunnell and Miller (2001), the incisor orientation of omomyids and adapiforms show lower incisors that are "at least slightly more procumbent." One should be cautious about interpretation when there is little material to study. These conclusions about *Eosimias* are based on one tiny specimen only (IVPP V11000).

Another feature of "anthropoid" resemblance is the oblique orientation and exodaenodonty of the lower premolars, but this condition is seen in lower premolars of the undoubted Fayum prosimian *Plesiopithecus,* other primates, and nonprimates (Gunnell and Miller, 2001). Hence, orientation and exodaenodonty cannot be considered exclusive anthropoidean features. Other cited similarities do not exist.

In *Eosimias* two features, said to resemble Fayum early anthropoids, are incorrect. One incorrect feature is that the lower incisors of *Eosimias* show "only subtle differences" from those of *Arsinoea* (Beard et al., 1996). In fact, *Arsinoea* lower incisors are quite unlike those of *Eosimias*. The lateral incisor of *Arsinoea* has a more flattened, less pointed leading edge, and the central incisor (although broken) appears similar in size and shape to the lateral. In *Eosimias* these two teeth are highly dimorphic.

The second incorrect feature from Beard et al. (1996) is that the (broken) mandible angle of *E. centennicus* resembles mandibular angles of Fayum Eocene higher primates. Actually, the angle in *Eosimias* looks more like the mandible angles in *Tarsius* (figs. 1.1 and 1.4) than that of any of the several Fayum late Eocene anthropoideans. Two other features of *Eosimias* are not expected in Middle Eocene anthropoideans: (1) the relatively small lower P_2 (fig. 1.4), and (2) the deep mandible characteristic of Oligocene, not late Eocene Fayum anthropoids.

Gebo et al. (2000a,b) recovered tali and calcanea from the Shanghuang Middle Eocene fissures in China that they allocated to *Eosimias*. These authors argue that the bones exhibit several anthropoid-like features. On the talus, the medial facet for the tibial malleolus is said to resemble that of anthropoideans. However, Covert (pers. com.) found this facet in several *Omomys* specimens—it can occur in a nonanthropoid. On the *Eosimias*

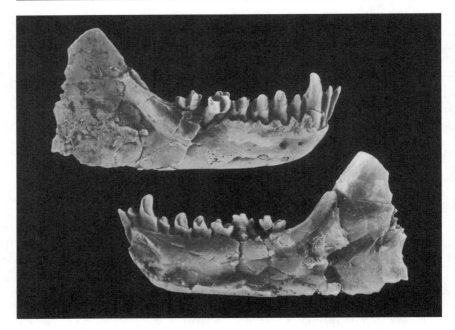

Figure 1.4. Views of the mandible of *Eosimias centennicus*. Lateral view on top and internal view on bottom.

calcaneum, a nonarticular wedge is similar to that seen in anthropoideans. However, similar wedges are found in *Adapis, Pondaungia,* and some modern strepsirhines. In addition, the distal segment of attributed *Eosimias* calcanea is shortened, unlike the almost certainly derived condition in *Tarsius.* Also, attributed *Eosimias* specimens lack a posterior astragalar shelf such as that found in most adapiformes, omomyiformes, and in many strepsirhines, but not in tarsiers or higher primates.

These shared features are not compelling unless one assumes Haplorhini validity. Other features are shared between tarsiers and strepsirhines to the exclusion of anthropoideans. If *Tarsius* is the sister of higher primates, one could deduce that the shared postcranial resemblances between tarsiers and strepsirhines, to the exclusion of anthropoideans, are primitive. Hence, shared features of *Eosimias* and anthropoideans could be considered shared-derived. These features need to be analyzed in relation to other groups outside Primates, such as tupaias or plesiadapiformes, to determine which characters are primitive and which are derived.

In sum, *Eosimias* anatomy is not convincingly anthropoidean. It is tarsierlike or omomyidlike, possibly only due to primitiveness. These characters, together with unique primate features (especially the upper teeth), create a creature that is somewhat of a chimera. Selection has combined its cranial, dental-gnathic, and postcranial material from a varied primate fauna where no associations exist. The genus occurs too late in time and, in some respects, is too derived to represent anything like the morphology expected to precede that of the Fayum anthropoideans (Gunnell and Miller, 2001). For instance, the deep mandible, large, tall canines, and apparently orthally oriented incisors could all be convergent on similar conditions seen in Oligocene Fayum anthropoideans. The possible affinities with tarsiids should be further analyzed to ascertain whether it is real or due to shared primitive features.

The presence of tree shrews, flying lemurs, and tarsiers today in Asia combined with an *Eosimias* characterization as a stem anthropoid may suggest primate, haplorhine, or anthropoidean origins in Asia. Nevertheless, Godinot (1994), in a detailed, tooth-by-tooth analysis of *Altiatlasius* from the late Paleocene of Morocco, concluded that it might lie closer to early simians than to oldest prosimians. In addition, a cladogram in Beard (1998b) characterizes *Altiatlasius* as a stem anthropoid. Inasmuch as this genus is thought to be of latest Paleocene age, such a conclusion establishes anthropoideans in Africa long before the Middle Eocene Asian occurrence of eosimiids or amphipithecids. Postcranial evidence that amphipithecids are not anthropoidians but instead have close affinities with the adapoids (Ciochon et al., 2001), combined with the unlikelihood that eosimiids are stem anthropoideans, seems to have left Asia untenanted by few if any higher primates until the Late Miocene appearance of apes, and a little later, monkeys.

Conclusions

Normal evolutionary change distinguishes primitive members of a clade from the clade of their descendants because the oldest members have no derived characters seen later in the crown group. The family Tarsiidae provides a good example because modern "crown group" tarsiers have many derived features connected with their visual, olfactory, and locomotor systems and anterior dentition, none of which are found in other living primates. Although fossil tarsiers have been traced back to the Eocene, none of these early specimens preserve the areas where the derived features of present-day species can be observed, with one possible exception: presence of fused tibiofibulae in *Tarsius eocaenus* (Gebo, pers. com.; Dagosto et al., 1996).

Parsimony trees by Kay et al. (1997) and Ross et al. (1998) link tarsiids, anthropoideans and eosimiids. Their analysis is based, in part, on dental features whose biological bases are currently not fully understood. The database is loaded with dental features, and these are grouped with equal weight with postcranial features and a few phyletically significant cranial characters (such as postorbital or metopic closure). For example, the authors take the orientation of the cristid obliqua on M_{1-2} as evidence for an *Afrotarsius*-anthropoidean clade. However, this feature cannot unite *Eosimias* with anthropoideans as the crest is relatively more oblique than in *Afrotarsius* and is located just as it is in tarsiers. The standard approach to sorting phyletic relationships has been to place greater importance on the more evolutionarily meaningful cranial characters, such as the position of the ectotympanic ring, carotid foramen, or postorbital bar (all of which are less prone to homoplasy than are dental characters). The use of numerous, possibly non-independent, dental features could, for example, lead to a grouping of the Microsyopidae (which lack both petrosal bulla and postorbital bar) within Anthropoidea.

For hypothetical reasons, those who believe that early anthropoideans should fall close to tarsiids generally oppose the scenario that Tarsiidae arose from or is nested in the Omomyidae. Because tarsiers possess a series of uniquely derived characters (such as loss of a pair of lower incisors, development of large circumorbital flanges or periorbital cone, tibiofibular fusion), they will conveniently fall outside the omomyid clade.

Some scientists believe, however, that the tarsier/omomyid link is stronger than the tarsier/anthropoid link. The principal consequence of accepting the former link is to invalidate, as being convergent, many of the most important features hypothesized to be shared-derived anthropoid-tarsiid features. Tarsiidae/Omomyidae have many shared cranial and postcranial similarities seen throughout the anatomy of these groups. In the minds of some, such characters (e.g., postorbital plates, zygomatico-alisphenoid contact, dry rhinarium, and possession of an auditory anterior accessory cavity—all of which do not, or are inferred not to have existed in omomyids) are invalidated as being shared-derived with anthropoideans.

In sum, two principal points are apparent: (1) the fossil record of tarsiers is presently very poorly known, and (2) most conclusions about it have been influenced by conflicting hypotheses about extant tarsiers.

Acknowledgments

I thank the staff of the Egyptian Geological Survey and Mining Authority and especially the hard-working staff of the Cairo Geological Museum for help and support in Egypt, where a fossil tarsier has been found. Thanks go also to Prithijit Chatrath for field management and as the collector who found the type of *Afrotarsius*. I am grateful to our extensive team of other collectors working during many seasons, among them, especially, Tom Bown, Don DeBlieux, John Fleagle, Pat Holroyd. K. C. McKinney, Ellen Miller, Tab

Rasmussen, and Verne Simons. Gregg Gunnell, Jukka Jernvall, Friderun Ankel-Simons, Erik Seiffert, Cornelia Simons, Tab Rasmussen, and Ian Tattersall have reviewed and contributed suggestions to the manuscript. I thank Chris Beard for sending the photographs in figure 1.4. Photos for figures 1.1 and 1.2 were taken by R. L. Usery, and the SEM work of figure 1.3 is by E. M. Eibest. This research has been funded by the National Science Foundation, Anthropology Division, SBR 98-07770. This is Duke Primate Center publication number 693.

References

Ankel F. 1967. Morphologie der Wirbelsäule und Brustkorb. Primatologia, Handbook of Primatology IV, 4, 1–120. Basel: Karger.

Beard KC. 1988. The phylogenetic significance of strepsirrhinism in Paleogene primates. Int J Primatol 9: 83–96.

Beard KC. 1998a. A new genus of Tarsiidae (Mammalia: Primates) from the Middle Eocene of Shanxi Province, China, with notes on the historical biogeography of tarsiers. Bull Carnegie Mus Nat Hist 34: 260–277.

Beard KC. 1998b. East of Eden: Asia as an important center of taxonomic origination in mammalian evolution. Bull Carnegie Mus Nat Hist 34: 5–39.

Beard KC, MacPhee RDE. 1994. Cranial anatomy of *Shoshonius* and the antiquity of Anthropoidea. In Fleagle JG, Kay RF, editors, Anthropoid origins, 55–97. New York and London: Plenum Press.

Beard KC, Krishtalka L, Stucky RK. 1991. First skulls of the early Eocene primate *Shoshonius cooperi* and the anthropoid-tarsier dichotomy. Nature 349: 64–67.

Beard KC, Qi T, Dawson MR, Wang B, Li C. 1994. A diverse new primate fauna from middle Eocene fissure-fillings in southeastern China. Nature 368: 604–609.

Beard KC, Tong Y, Dawson MR, Wang J, Huang X. 1996. Earliest complete dentition of an Anthropoid primate from the Late Middle Eocene of Shanxi Province, China. Science 272: 82–85.

Cartmill M. 1980. Morphology, function and evolution of the anthropoid postorbital septum. In Ciochon RL, Chiarelli AB, editors, Evolutionary biology of the new world monkeys and continental drift, 243–274. New York: Plenum Press.

Ciochon RL, Gingerich PD, Gunnell GF, Simons EL. 2001. Primate postcrania from the late Middle Eocene of Myanmar. Proc Natl Acad Sci USA, 7672–7677.

Dagosto M, Gebo DL, Beard C, Qi T. 1996. New primate postcranial remains from the middle Eocene Shanghuang fissures, southeastern China. Am J Phys Anthro Suppl 22: 92–93.

Ducrocq S, Jaeger JJ, Chaimanee Y, Suteethorn V. 1995. New primate from the Palaeogene of Thailand, and the biogeographical origin of anthropoids. J Hum Evol 28: 477–485.

Filhol H. 1889-90. Description d'une nouvelle espèce de Lémurien fossile (*Necrolemur parvulus*). Bull Soc Philo Paris 8 (2): 39–40.

Fleagle JG, Kay RF. 1987. The phyletic position of the Parapithecidae. J Hum Evol 16: 483–532.

Gebo DL, Dagosto M, Beard KC, Qi T, Wang J. 2000a. The oldest known anthropoid postcranial fossils and the early evolution of higher primates. Nature 404: 276–278.

Gebo DL, Dagosto M, Beard KC, Qi T. 2000b. The smallest primates. J Hum Evol 38: 585–594.

Ginsburg L, Mein P. 1987. *Tarsius thailandica* nov sp, premier Tarsiidae (Primates, Mammalia) fossile d'Asie. C R Acad Sci Paris, Ser II 304: 1213–1215.

Ginsburg L, Ukkakimapan Y. 1983. Un Cervidé nouveau du Miocène d'Asie et l'âge des lignites des bassins intramontagneux du Nord de la Thaïlande. C R Acad Sci Paris, Ser II 297: 297–300.

Godinot M. 1994. Early North African primates and their significance for the origin of Simiiformes (Anthropoidea). In Fleagle JG, Kay RF, editors, Anthropoid origins, 235–296. New York: Plenum.

Gunnell GF, Miller ER. 2001. Origin of Anthropoidea: dental evidence and recognition of early anthropoids in the fossil record, with comments on the Asian anthropoid radiation. Am J Phys Anthro 114: 177–191.

Hofer HO. 1977. The anatomical relations of the *Ductus vomeronasalis* and the occurrence of taste buds in the papilla palatina of *Nycticebus coucang* (Primates, Prosimiae) with remarks on strepsirrhinism. Gegenbauers morph Jb 123: 836–856.

Hofer H. 1980. The external anatomy of the oro-nasal region of primates. Z Morph Anthro 71: 233–249.

Hunter JP, Jernvall, J. 1995. The hypocone as a key innovation in mammalian evolution. Proc Natl Acad Sci USA 92: 10718–10722.

Jernvall J. 2000. Linking development with generation of novelty in mammalian teeth. Proc Natl Acad Sci USA 97: 2641–2645.

Jernvall J, Jung HS. 2000. Genotype, phenotype, and developmental biology of molar tooth characters. Yrbk Phys Anthro 43: 171–190.

Kay RF, Williams BA. 1994. Dental evidence for anthropoid origins. In Fleagle JG, Kay RF, editors, Anthropoid origins, 361–446. New York: Plenum Press.

Kay RF, Ross C, Williams CA. 1997. Anthropoid origins. Science 275: 797–804.

Luckett WP. 1971. A comparison of the early development of the fetal membranes of Tupaiidae, Lorisidae and *Tarsius,* and its bearing on the evolutionary relationships of the prosimian primates. Proc 3rd Int Congr Primatol Zurich 1: 238–245.

Luckett WP. 1974. Comparative development and evolution of the placenta in primates. Contrib Primatol 3: 143–234.

Luckett WP. 1976. Cladistic relationships among primate higher categories: evidence of the fetal membranes and placenta. Folia Primatol 25: 245–276.

Luckett WP. 1982. The uses and limitations of embryological data in assessing the phylogenetic relationships of *Tarsius* (Primates, Haplorhini). Geobios mém spécial 6: 289–304.

Luckett WP. 1993. Developmental evidence from the fetal membranes for assessing archontan relationships. In MacPhee RDE, editor, Primates and their relatives in phylogenetic perspective, 149–186. New York: Plenum Press.

MacPhee RDE, Cartmill M. 1986. Basicranial structures and primate systematics. Swindler D, Erwin J, editors, Comparative primate biology, vol. 1, Systematics, evolution and anatomy, 219–275. New York: Alan R. Liss.

Maier W. 1980. Konstruktionsmorphologische Untersuchungen am Gebiss der rezenten Prosimier (Primates). Abh Senckenb Naturforsch Ges 538: 1–158.

Murphy WJ, Elzirik E, Johnson WE, Zhang YP, Ryder O, O'Brien S. 2001. Molecular phylogenetics and the origins of placental mammals. Nature 409: 614–618.

Piveteau J. 1957. Primates: paléontologie humaine. Traité de Paleontologie, Vol. 7a. Paris: Masson et Cie.

Pocock RI. 1918. On the external characters of the lemurs and of *Tarsius*. Proc Zool Soc London 1918: 19–53.

Polly PD. 1998. Variability, selection, and constraints: development and evolution in viverid (Carnivora, Mammalia) molar morphology. Paleobiology 24: 409–429.

Rasmussen DT, Conroy GC, Simons EL. 1998. Tarsier-like locomotor specializations in the Oligocene primate *Afrotarsius*. Proc Natl Acad Sci USA 95: 14848–14850.

Rasmussen DT, Shekelle M, Walsh SL, Reiney BO. 1995. The dentition of *Dyseolemur* and comments on the use of the anterior teeth in primate systematics. J Hum Evol 29: 301–320.

Rosa MGP, Pettigrew JD, Cooper HM. 1996. Unusual pattern of retinogeniculate projections in the controversial primate *Tarsius*. Brain, Behavior and Evolution 48: 121–129.

Rosenberger AL. 1985. In favor of the *Necrolemur*-tarsier hypothesis. Folia Primatol 45: 179–194.

Rosenberger AL, Szalay F. 1980. On the tarsiiform origins of Anthropoidea. In Ciochon RL, Chiarelli AB, editors, Evolutionary biology of the New World monkeys and continental drift, 139–157. New York: Plenum Press.

Ross C. 1994. The craniofacial evidence for anthropoid and tarsier relationships. In Fleagle JG, Kay RF, Anthropoid origins, 469–548. New York: Plenum Press.

Ross C, Covert H. 2000. The petrosal of *Omomys carteri* and the evolution of the primate basicranium. J Hum Evol 39: 225–251.

Ross C, Williams B, Kay RF. 1998. Phylogenetic analysis of anthropoid relationships. J Hum Evol 35: 221–306.

Sánchez-Villagra MR, Kay RF. 1996. Do Phalangeriforms (Marsupialia: Diprodontia) have a "hypocone"? Aust J Zool 44: 461–467.

Sauther ML, Cuozzo FP, Sussman RW. 2001. Analysis of dentition of a living wild population of ring-tailed lemurs (*Lemur catta*) from Beza Mahafaly, Madagascar. Am J Phys Anthro 114: 215–223.

Schwartz JH, Tattersall I. 1987. Tarsiers, adapids and the integrity of Strepsirrhini. J Hum Evol 16: 23–40.

Seiffert ER, Simons EL. 2000. *Widanelfarasia*, a diminutive placental from the late Eocene of Egypt. Proc Natl Acad Sci USA 96: 2646–2651.

Simons EL. 1961. Notes on Eocene Tarsioids and a revision of some Necrolemurinae. Bull Brit Mus Nat Hist Geol 5(3): 43–69.

Simons EL. 1972. Primate evolution: an introduction to man's place in nature. New York: Macmillan.

Simons EL. 1974. Notes on early Tertiary prosimians. In Martin RD, Doyle GA, Walker A, editors, Prosimian biology, 415–433. London: Duckworth.

Simons EL. 1987. New faces of *Aegyptopithecus* from the Oligocene of Egypt. J Hum Evol 16: 273–289.

Simons EL. 1995. Egyptian Oligocene primates: a review. Yrbk of Phys Anthro 38: 199–238.

Simons EL. 1997a. Discovery of the smallest Fayum Egyptian primates (Anchomomyini, Adapidae). Proc Natl Acad Sci USA 94: 180–184.

Simons EL. 1997b. Preliminary description of the cranium of *Proteopithecus sylviae*, an Egyptian late Eocene anthropoidean primate. Proc Natl Acad Sci USA 94: 14970–14975.

Simons EL. 1998. The prosimian fauna of the Fayum Eocene/Oligocene deposits of Egypt. Folia Primatol 69 (suppl. 1): 286–294.

Simons EL, Bown TM. 1985. *Afrotarsius chatrathi,* first tarsiiform primate (Tarsiidae) from Africa. Nature 315: 477.

Simons EL, Rasmussen DT. 1989. Cranial morphology of *Aegyptopithecus* and *Tarsius* and the question of the tarsier-anthropoidean clade. Amer J Phys Anthro 79: 1–23.

Simons EL, Russell DE. 1960. Notes on the cranial anatomy of *Necrolemur*. Breviora MCZ, Harvard University 127: 1–14.

Simpson GC. 1953. The major features of evolution, 434. New York: Columbia University Press.

Sprankel H. 1965. Untersuchungen an *Tarsius*. I. Morphologie des Schwanzes nebst ethologischen Bemerkungen. Folia Primatol 3: 153–188.

Starck D. 1955. Embryologie. Stuttgart: Thieme-Verlag

Starck D. 1984. The nasal cavity and nasal skeleton of *Tarsius*. In Niemitz C, editor, Biology of tarsiers, 275–290. New York: Gustav-Fischer-Verlag.

Stehlin HG. 1916. Die Säugetiere des schweizerischen Eozäns. Kritischer catalog der materialien, part 7, second half. Abh Schweiz Pal Ges 41: 1299–1552.

Szalay FS. 1976. Systematics of the Omomyidae (Tarsiiformes, Primates): taxonomy, phylogeny, and adaptations. Bull Am Mus Nat Hist 156: 157–450.

Szalay FS. 2000. Eosimiidae. In Delson E, Tattersall I, van Couvering JA, Brooks AS, editors, Encyclopedia of human evolution and prehistory, 2d ed., 235. New York, London: Garland.

Szalay FS, Delson E. 1979. Evolutionary history of the primates. New York: Academic Press.

Teilhard de Chardin P. 1916-21. Sur quelques primates des phosphorites du Quercy. Ann Paléontol 10: 1–20.

Tong Y. 1997. Middle Eocene small mammals from Liguanqiao Basin of Henana Province and Yuanqu Basin of Shanxi Province, central China. Palaeontol Sinica Whole no. 18, n.s. C, 20: 41–49, 198–201, 251–252.

CHAPTER 2

The Evolution of the Tarsiid Niche

Nina G. Jablonski

Long before indisputable fossil evidence of tarsiers was discovered, tarsiers were considered "living fossils." This assignment was based as much on their primitive *gestalt* as on the morphological affinities of tarsiers to members of various Paleogene primate and plesiadapiform taxa, and on predictions based on cladistic reconstructions of the tarsier's phylogenetic position. Thus, through many years of learned discourse on tarsiers, "the 'living fossil' [had] no fossil record!" (Schwartz 1984, 47). In recent years, this situation has improved, although the fossil record of tarsiers is still more gap than record. The fragmentary remains of fossil tarsiids recovered from deposits of middle Eocene age onward from Egypt, China, and Thailand indicate that the tarsier's "living fossil" moniker is well deserved. The morphology of these fragments is remarkably modern, or perhaps better said, the body plan of modern tarsiers is remarkably ancient and conservative. This leads to the inevitable question: What made tarsiers successful in the first place, and why have they persisted, little changed, through most of the Tertiary through to the present day? These questions are at the heart of this paper.

Reconstructing the niche of an extinct species or the successive niches of the members of an ancient lineage is difficult because missing or fragmentary data are many and unassailable facts are few. The exact nature of the forest environments inhabited by the fossil tarsiids of the Eocene and Miocene, for instance, is not known, yet a great deal is known about the character of Paleogene and early Neogene forest environments in general (e.g., Upchurch and Wolfe, 1987) and the forests in which tarsiers currently dwell (e.g., Davis, 1962; Crompton and Andau, 1986; Musser and Dagosto, 1987; Whitten et al., 1987; MacKinnon et al., 1996). Similarly, relatively little is known of anatomy, ecomorphology, and life-history parameters of ancient tarsiids, but much is known in all of these areas in connection with the living species of tarsiers (e.g., Niemitz, 1977; Crompton and Andau, 1986; Jablonski and Crompton, 1994; Roberts, 1994; Gursky, 1997). The similarities in habitat preference between extinct and extant tarsiids, and the strong morphological (especially dental) resemblances between these forms, thus mean that the tarsiers of today likely appear and behave similarly to those of the distant past. In this chapter, the ecological role of tarsiers is examined, and the reasons for their success and persistence through 45 million dynamic years of mammalian evolution are explored.

Tarsiids Past and Present

The fossil record of the Tarsiidae comprises mostly isolated teeth and lower jaws. These elements have proven useful for determining the phyletic relationships and the putative diets of the animals they represent, but they leave us begging for more. Tarsiers today are distinguished by their enormous eyes and greatly elongated hindlimbs—the seeming anatomical underpinnings of their success. Thus, we must recognize that the absence of crania preserving any portions of the orbits and the near absence of postcranial remains in the tarsiid fossil record are impediments to understanding the evolution of the tarsiid niche.

The earliest recognized Tarsiidae species have been recognized from Middle Eocene deposits of China. *Tarsius eocaenus* has been recovered from fissure-filling deposits at Shanghuang in Jiangsu Province, China (Beard and Qi, 1994; Qi, Beard, et al., 1996; Beard, 1998). The species is represented by isolated teeth and was almost certainly sympatric with the basal anthropoid, *Eosimias sinensis,* the omomyid, *Macrotarsius macrorhysis,* and at least two members of the adapiform clade Adapina recovered from the same locality (Beard and Qi, 1994; Dagosto and Gebo, 1996; Gebo and Dagosto, 1996; Qi et al., 1996; Beard, 1998). The Shanghuang fossil mammal assemblage is rich in micromammals, including a primitive lagomorph, two species of microchiropterans, several species of primitive cricetid rodents, and numerous insectivores (Qi, Beard, et al., 1996). Larger mammals from Shanghuang include other species known to favor closed, tropical or paratropical forest environments such as miacid carnivores, representatives of several families of browsing perissodactyls, and a small, primitive anthracothere (Qi, Zong, et al., 1991; Beard and Qi, 1994; Qi et al., 1996). Biostratigraphic correlation indicates that the mammalian fauna of the Shanghuang fissures is likely to be about 45 myr old (Qi, Beard, et al., 1996).

Xanthrorhysis tabrumi is another Paleogene ancient tarsiid that has been recovered from China. This species was recovered from a small drainage in Yuanqu County of southern Shanxi Province and is estimated to be of Late Middle Eocene age. Based on the morphology of the dentary and its contained teeth, *X. tabrumi* has been described and interpreted by Beard as being a sister group of all living and fossil tarsiers, including *Tarsius eocaenus* (Beard, 1998).

Xanthrorhysis tabrumi and *Tarsius eocaenus* are the oldest known tarsiers (*sensu lato*), but how close they are to the phyletic and biogeographic origin of the Tarsiidae is still a matter of speculation. Beard has presented a convincing case that these fossil species are ancient, but are probably not the oldest tarsiers (Beard, 1998), and that the origin of the Tarsiidae can be traced to the early Paleogene of the Asian mainland. The Eocene tarsiids of China are smaller than any of the living species of *Tarsius,* but the close similarity in molar morphology between *X. tabrumi, T. eocaenus,* and living tarsiers is so clear as to leave no doubt that the fossil species truly are ancient tarsiers, and that the molar morphology of modern tarsiers is primitive (Martin, 1994). The important dietary implications of this molar morphology are discussed below.

The next oldest fossil tarsiid is also the only one, living or fossil, of African origin. *Afrotarsius chatrathi* was first described on the basis of a fragmentary mandible, with molar teeth very similar to those of living *Tarsius,* derived from early Oligocene sediments of Quarry M of the Jebel Qatrani Formation of the Fayum Province of Egypt (Simons and Bown, 1985). Since the recovery of the type mandible, a nearly complete tibiofibula was found in the same quarry and was assigned to the same species on the basis of its near identity to the comparable element in living tarsiers (Rasmussen et al., 1998). As is discussed below, the morphology of the fused tibiofibula in *Afrotarsius* is a strong indication that it exhibited a mode of leaping locomotion very similar to that of modern *Tarsius* (Rasmussen et al., 1998) and implies a similarity of ecological role between fossil and living forms that previously was only suspected.

The early Miocene form, *Tarsius thailandicus,* is represented by only a single isolated lower molar found at the site of Li in northwestern Thailand, as part of a forest-associated fauna dominated by rodents and insectivores (Ginsburg and Mein, 1987).

The five recognized species of living tarsiers inhabit rain-forest environments in island southeast Asia. (See fig. 8 in Musser and Dagosto [1987] for a summary distribution map.) The three most widely distributed species, *Tarsius bancanus, T. syrichta,* and *T. spectrum,* live in a zoogeographic regions distinct and different from one another. *T. bancanus* is native to islands on the Sunda Shelf, namely mainland Borneo, Pulau Serasan in the South Natuna islands, Pulau Belitung, Pulau Bangka, and southern Sumatra (Musser and Dagosto, 1987). *T. syrichta* occurs in several islands of the southern Philippines once united as a land mass called Greater Mindanao during the Late Pleistocene (Musser and Dagosto, 1987). *T. spectrum* is known only from the Sulawesi region, including mainland Sulawesi, the Sangihe Islands, Pulau Peleng, and Pulau Selajar (Musser and Dagosto, 1987). The two remaining tarsier species, *T. pumilis* and *T. dianae,* enjoy highly geographically restricted distributions within Sulawesi. The former inhabits the montane mossy forest of central Sulawesi while the latter inhabits the lowland forest of the same area (Musser and Dagosto, 1987; Niemitz, 1991). The degree to which these species are actually sympatric with one another or with *T. spectrum* is unclear (Musser and Dagosto, 1987; Niemitz, 1991).

Tarsiid Environments Past and Present

The establishment of *T. eocaenus* as the oldest tarsiid establishes a minimum age for the lineage of about 45 myr (Beard and Qi, 1994; Qi, Beard, et al., 1996). The period spanning the Late Cretaceous through the Eocene, which no doubt witnessed the origin of the group, was a time of climatic quiescence and dramatically increasing mammalian diversity. It was one of the warmest intervals in the history of the Earth, and climatic patterns followed closely the zonal component of atmospheric circulation (Parrish, 1987). It was in this environment of stability that angiosperms underwent their greatest diversification (Parrish, 1987). A conspicuous restructuring of vegetational communities appears to have occurred after the mass kill of vegetation at the Cretaceous-Tertiary boundary (Wolfe and Upchurch, 1986; Upchurch and Wolfe, 1987). The vegetational recovery that occurred at this time resembled a modern secondary succession in many respects, although over a longer time course (Wolfe and Upchurch, 1986; Upchurch and Wolfe, 1987). At low and middle latitudes, precipitation increased greatly, initiating the expansion of humid multistratal rain forests with closed canopies (Wolfe, 1985; Upchurch and Wolfe, 1987; Wing and Tiffney, 1987). It is interesting to note, in the broader context of mammalian evolution, that the earliest extensive closed-canopy multistoryed rain forests appeared only after the disappearance of the dinosaurs at the end of the Cretaceous (Morley, 2000). Their proliferation may, in fact, be directly linked to the removal of large generalist herbivores and their replacement by seed-dispersing small mammals (Morley, 2000).

By the early Eocene, angiosperms had come to dominate late successional communities, for the first time forming a closed forest vegetation of modern aspect (Wing and Tiffney, 1987). Such forests, characterized by tall trees with buttressed trunks, lianas, epiphytes, and a distinct understory, would have provided many new or greatly expanded niches for plant and animal adaptation (Upchurch and Wolfe, 1987). Of particular relevance to the discussion in this chapter is the fact that the early Tertiary saw great increases in the abundance and diversity of lianas and understory plants (Upchurch and Wolfe, 1987). The acme of global temperatures, tropicality, and angiosperm-dominated closed forests was achieved at this time (Behrensmeyer and Damuth, 1992; Janis, 1993), with subtropical vegetation probably having extended to 60 degrees north latitude and full tropical multistratal rain forests occurring as far north as 30 degrees north latitude (Wolfe, 1985).

The closed angiosperm forests of the early Tertiary were dominated by large trees with large diaspores, and were populated mostly by small vertebrate herbivores such as multituberculates, primates, and rodents who concentrated on fruit and seeds (Rose, 1981; Wing and Tiffney, 1987; Collinson and Hooker, 1991; Janis, 1993). These "little herbivore/big angiosperm" systems (Wing and Tiffney, 1987) were also the homes of many insect pollinators and phytophagous insects (Crepet and Friis, 1987) and the vertebrates that preyed on them. It is significant that most of the nonflying vertebrates that evolved to occupy rain-forest environments were consumers of the fruits, seeds, and leaves that were its primary products (Collinson and Hooker, 1991; Emmons, 1995). Only a small fraction of these animals concentrated on invertebrates, and of these the only groups to achieve significant diversity were bats and birds (Emmons, 1995). The mammals inhabiting Early and Middle Eocene forests were broadly analogous to modern ones, but their community structures emphasized terrestrial frugivores and arboreal insectivores to a much greater extent than do the tropical and subtropical forests of the present day.

The emergence of the understory as a significant component of the humid multistratal rain forests of the Eocene provided important ecological opportunities for several evolving lineages of Paleogene mammals (Morley, 2000). At the forest floor, rain forests have low relative illuminance of visible light, often less than 1% (Grubb and Whitmore, 1967). The high moisture and low light conditions that characterized the understory selected for plants with large leaf size and the ability to photosynthesize under suboptimal light conditions (Upchurch and Wolfe, 1987). The leaves of understory plants are also characterized by softer textures and lower toxicities (Lowman, 1995), features that would have facilitated their exploitation by arthropods.

The arthropod fauna of the understory, like those elsewhere in the rain-forest canopy, shows greater numbers of small-bodied rather than large-bodied species because there is more usable space for smaller animals living on vegetation than for larger animals (Morse et al., 1985). The rain-forest understory with its larger average leaf size, however, supports a greater proportion of larger-bodied arthropods than do other forest environments (Morse et al., 1985). There is also some evidence that rain-forest arthropods are less host-plant specific than are the arthropods of other forest ecosystems, possibly because of the greater heterogeneity of plant species in rain forests and the reduced number of seasonal changes in tree condition that is found there (Stork, 1987).

Since the Middle Eocene, the distribution and composition of humid, multistratal rain forests have undergone significant changes as the result of a major cooling at the end of the Eocene (the terminal Eocene event) and the origin and spread of low-biomass vegetation types (e.g., savanna, steppe, and tundra) during the Neogene (Wolfe, 1985; Collinson and Hooker, 1991; Leopold et al., 1992). Of all the areas of the Old World once covered by tropical or paratropical rain forest in the Eocene, it is only small, low-latitude areas of southeast Asia that have retained such forests through the Neogene (Collinson and Hooker, 1991). Although the rain forests of southeast Asia have undergone some floristic changes since the late Eocene, the physical characteristics of the forest today appear to be essentially the same as they were 55 million years ago, as judged by similarities in foliar physiognomy and wood morphology (Wolfe, 1985; Upchurch and Wolfe, 1987; Collinson and Hooker, 1991; Morley, 2000).

The nature of the arthropod communities associated with the tropical and paratropical rain forests of the Eocene is not well known. The rise of the insect pollinators associated with early angiosperms is reasonably well understood, largely through inferences based on pollen and leaf remains (Crepet and Friis, 1987), but little is known of the details of what was no doubt an explosive diversification of phytophagous arthropods in the humid multistratal forests of the early Tertiary. The great antiquity of the Coleoptera (beetles) and the orthopteroids (including crickets and cockroaches) (Carpenter, 1976) and their prominence in the communities of leaf-eating arthropods in modern Asian rain forests suggest that their evolution and diversification were closely linked to that of the plants they exploited. In modern rain forests, many larger-bodied beetles and orthopteroid insects are nocturnal, presumably to avoid predation by birds and reptiles, and it is likely that the activity patterns of their early Tertiary relatives were the same. The lower level of host-plant specificity observed in tropical forest arthropods suggests, further, that the arthropods that preyed on the plants of the early Tertiary rain-forest plants may not have changed appreciably over time. This is also supported by the recognized functional comparability of rain-forest vegetation through time and space, in terms of leaf structure and toxicity (Morley, 2000).

This evidence leads us to conclude that the humid forests of eastern Asia occupied by *Tarsius eocaenus* and its putative ancestors were closely comparable to the tropical rain forests of southeast Asia today. This comparability extends from canopy structure through foliar physiognomy and physiology to the composition of the arthropod guilds that inhabited the forests.

The five recognized species of living tarsiers inhabit parts of southeast Asia that have been covered with tropical rainforests continuously since the Middle Eocene (Morley, 2000). These forests have a dynamic history of their own, having undergone considerable floristic changes as a result of climatic oscillations and migrations resulting in fluctuating land connections. Continuous land connections between southeast Asia and mid-latitude Asia since the earliest Tertiary allowed elements of Paleogene Northern Hemisphere rain forests to find refuge in the lower montane forests of southeast Asia following the mid-Tertiary global climatic deterioration (Morley, 2000). This event has no parallel in other regions (Morley, 2000) and accounts for the fact the rain forests of southeast Asia, despite fluctuations in composition and diversity, have a primitive character and more closely resemble the rain forests of the Paleogene than do rain forests elsewhere.

When the fossil evidence of Eocene tarsiers is considered together with this botanical information, the inescapable conclusion is that tarsiers as a group originated in the tropical forests of the early Paleogene in the Asian mainland, as argued by Beard (Beard, 1998), and that their range shift into Sundaland probably coincided with the southward migration of some elements of Asian mainland forests at the end of the Eocene. Tarsiers thus can be said to have inhabited more or less the same forests for the last 45 million years.

The Bornean tarsier, *Tarsius bancanus,* and its Philippine congener, *T. syrichta,* are restricted in their distributions to lowland evergreen rain forests (both primary and secondary). The distribution of the spectral tarsier of Sulawesi, *T. spectrum,* spans a greater altitudinal range from lowland evergreen rain forest near sea level through lower montane rain forest at 1500 m (Musser and Dagosto, 1987). *T. dianae* occurs in primary rain forest within the altitudinal range of *T. spectrum.* The pygmy tarsier, *T. pumilis,* of central Sulawesi, although still poorly known, appears to occupy the mossy upper montane rain forest at altitudes of approximately 1700 to 2200 m.

The subcanopies of lowland rain forests that are the homes to most tarsiers support an amazing density and diversity of arthropods, especially termites (Isoptera) and beetles (Coleoptera), that form the core of the decomposer community (Whitten et al., 1987). These animals are, in turn, the prey of frogs, toads, and skinks, which comprise the most numerous and diverse elements of the subcanopy vertebrate fauna (Whitten et al., 1987). Larger leaf-eating arthropods such as cockroaches (Blatteroptoidea) and grasshoppers (Orthopteroidea) occur at lower densities in subcanopy environments and represent some of the most ancient elements of the rain-forest invertebrate fauna (Rohdendorf, 1991). Much has been said about Paleogene mammalian evolution being dominated by "little herbivore/big angiosperm" systems (Wing and Tiffney, 1987), but this was not where the tarsiids succeeded. From the Paleogene onward, tarsiers have occupied a stable niche little affected by shifts in the composition or diversity of rain-forest plant communities, in which their primary focus has been the exploitation of relatively large, nutrient-dense arthropods and invertebrates.

The Tarsiid Niche

Modern tarsiers inhabit what has been termed the sapling trunk and ground zone of Asian rain forests, consisting of the lower understory and adjacent forest floor (Davis, 1962; Crompton and Andau, 1986; Crompton, 1989). Tarsiers spend most of their time foraging, specifically in the active scanning for food from perches near ground level (Crompton and Andau, 1986; Crompton, 1989). Clinging to vertical substrates is the dominant postural mode for living *Tarsius,* while leaping between such substrates, or between vertical supports and the ground, is the primary mode of locomotion (Crompton and Andau, 1986; Crompton, 1989). Tarsiers do not require undisturbed, primary rain-forest habitats, and in fact appear to favor secondary forests, at least in Borneo (Davis, 1962; Crompton and Andau, 1986; Crompton, 1989), perhaps because of greater densities of favored prey items found there. Although primary and secondary forests differ in floristics, their physical structures are similar. For tarsiers, the essential requirement of a rain forest is that it contain an understory with large arthropod and small invertebrate prey.

The nonflying mammals that evolved in early modern rain-forest environments were faced with many locomotor challenges, including how to move up and down large vertical tree trunks, how to balance on thin, flexible branches and lianas, and how to cross gaps (Emmons, 1995). Leaping has been a common solution evolved by several lineages of small, forest-dwelling mammals to the problem of crossing gaps and moving between widely dispersed resources (Emmons, 1995).

The prodigious leaping abilities of modern tarsiers can be attributed to the great elongation and increase in mass of its hindlimb elements (Covert, 1995; Rasmussen et al., 1998; Niemitz, 1984; Crompton and Andau, 1986; Anemone and Nachman, chapter 4). The calcaneus and navicular bones of the ankle are greatly elongated, and the fibula is reduced and fused to the tibia. Until recently, the postcranial morphology of extinct tarsiers was a matter of speculation. Thanks, however, to the recent recovery of a diagnostic tibiofibula from the early Oligocene *Afrotarsius chatrathi* (Rasmussen et al., 1998), we now have some direct and tantalizing evidence about the nature of posture and locomotion in ancient tarsiids. The fused tibiofibula of *A. chatrathi* closely resembles that of modern tarsiers in size and morphology and provides strong evidence that the species was engaged in a style of leaping very similar to that of modern tarsiers. The shaft of the fossil is gracile, and its proximal end is compressed mediolaterally as an adaptation to resist bending moments in the sagittal plane during leaping (Rasmussen et al., 1998). The anatomy of this important element thus demonstrates not only that *A. chatrathi* was a "classic tarsier" in its locomotor adaptation, but also suggests that leaping locomotion was part of the original tarsier adaptation that defined the tarsiid niche.

The strong similarities in molar structure between ancient and modern tarsiers denote strong similarities in molar function and, therefore, diet, between extant and extinct tarsiids. Modern tarsiers have been observed to eat a variety of arthropods and occasionally vertebrates, but large-bodied coleopterans and orthopteroids are the most common components of their diets (Davis, 1962; Crompton, 1989). These items are seized with the hands and then ingested whole in a process made possible by an extraordinary gape (Jablonski and Crompton, 1994). The masticatory apparatus of modern tarsiers is capable of extremely wide jaw opening and powerful jaw closure. The temporomandibular joint permits extensive anterior translation of the mandible, powered by strong mandibular depressor muscles (the *M. digastricus*), when the mouth needs to be opened (Jablonski and Crompton, 1994). Large, fleshy, and long-fibered jaw adductors (*M. temporalis* and *M. massetericus*) permit extremely wide jaw opening and the ability to initiate forceful jaw closure to initiate ingestion and chewing (Jablonski and Crompton, 1994). The molars of modern tarsiers make possible the rapid, forceful cracking of brittle insect exoskeletons as well as the skeletons of small vertebrates (Crompton, 1989; Jablonski and Crompton, 1994). This process quickly liberates the protein-, fat- and readily digestible carbohydrate-rich tissues of insect and small vertebrate bodies for digestion in the tarsier's simple gut (Crompton, 1989; Jablonski and Crompton, 1994). Although the smaller size of Eocene tarsiids compared to modern tarsiers suggests that they focused on arthropod prey items of smaller average size, the inference from their molar morphology was that their diet was comparable in structure and content to that of modern tarsiers. The molars of *Afrotarsius chatrathi* and *Tarsius thailandicus* are comparable in size and cusp morphology to modern tarsiers (Simons and Bown, 1985; Ginsburg and Mein, 1987), supporting the hypothesis of comparability of diet and dental function in all known tarsiids.

If ancient tarsiids concentrated on the same kinds of arthropod prey that modern tarsiers do, some cautious inferences about foraging behavior in the fossil species are possible. The large-bodied coleopterans and orthopteroids favored by modern tarsiers are relatively rare elements of the arthropod faunas of Asian rain forests and are widely scattered (Davis, 1962; Crompton and Andau, 1986; Crompton, 1989). Also, many are only active and visible at night. Concentration on such prey requires, among other things, the ability to travel quickly at night within the forest so as to locate and seize prey. This is one of many reasons why the niche of the tarsier has been likened to that of the owl (Niemitz, 1983). Ancient tarsiids, exploiting large, nocturnal arthropods and small vertebrates as prey items, almost certainly employed comparable, if not identical, foraging strategies as modern tarsiers to obtain those items. The preserved anatomy of Paleogene and early Neogene tarsiers provides few details here, and our reconstruction of ancient tarsiid foraging strategies and the nature of the tarsiid niche relies both on careful speculation and interpretation of stable form-function relationships through time.

The prey items favored by tarsiids today (and probably in the past) are more active at night than during the day, adding credence to the proposition that tarsiids throughout most or all of their history have been nocturnal. A nocturnal habitus makes possible the exploitation of uniquely nocturnal food resources and avoidance of diurnal predators, but is limiting in many other respects (Wright, 1989; Wright, 1994). Major adaptations in the realms of vision, energy expenditure, communication, ranging behavior, and social structure distinguish nocturnal primates (Charles-Dominique, 1975; Wright, 1989). For highly visually oriented mammals like primates, adaptation to reduced light levels has been the most important aspect of their success in nocturnal niches. The tarsier, like all nocturnal primates, has evolved a highly enlarged eyeball and a large retina capable of capturing and focusing available light (Castenholz, 1984). Importantly, modern tarsiers lack a tapetum lucidum, the highly reflective layer behind the retina that assists vision at reduced light levels in many nocturnal mammals but reduces visual acuity (Wright, 1989). Tarsiers instead have a well-developed fovea in the center of an all-rod retina, where visual acuity is concentrated and where the visual image is intensified because of the dense arrangement of visual receptor cells (Castenholz, 1984). This provides the tarsier with the most acute scotopic vision of all primates (Castenholz, 1984).

Among the most profound adaptations to nocturnality that have evolved in tarsiers and other nocturnal primates are those relating to metabolism, energy expenditure, and reproduction. Small animals facing cool nighttime temperatures could, potentially, exhaust their energy reserves if they attempted to maintain high metabolic rates around the clock and throughout the year. It is, therefore, not surprising that prosimians and tarsiers have low basal metabolic rates and low body temperatures (McNab and Wright, 1987). Tarsiers do not exhibit torpor, however, and in this way are distinct from several nocturnal prosimians who enter a lengthy period of physiological quiescence as an adaptation to seasonal food shortages (Wright and Martin, 1995). In tarsiers, a host of reproductive parameters appear to be related to their slower metabolism and reduced body temperatures (McNab and Wright, 1987). Tarsiers have single offspring and long gestation periods (Izard et al., 1985), and their rates of fetal and postnatal growth are among the slowest recorded for any mammal (Roberts, 1994). Tarsier infants exhibit the largest neonatal sizes of all nonanthropoid primates, and a large proportion of their neonatal mass is invested in brain mass, eyes and cranium (Roberts, 1994). Because of the precocious development of the brain, maturation of foraging and locomotor behaviors in tarsiers is extremely rapid (Roberts, 1994). This "slow" life-history pattern is characteristic of anthropoid primates that have evolved in the stable ecosystems of equable forests (Jablonski et al., 2000).

Tarsiers teeter on an energetic knife-edge. They must not only leap to eat and eat to leap (Jablonski and Crompton, 1994), they must stay warm and reproduce. Growing evidence from studies of maternal time budgets in tarsiers suggests that the burden of reproduction in female tarsiers is enormous. Observations by Gursky (1997) indicate that pregnant female tarsiers exhibit lower mobility, impaired foraging abilities, and maintain smaller home ranges than do nonpregnant females. Postpartum females, further, appear to be unable to sustain the energetic costs of lactation and continual infant transport (Gursky, 1997).

The niche or ecological role of tarsiids through time has been that of the small-bodied nocturnal insectivore and carnivore of the tropical rain-forest understory. Clearly, the survival of tarsiers in the short and very long term is related to their focus on arthropod and small vertebrate prey that are less subject to dramatic seasonal fluctuations than are plant resources. Once filled, the tarsiid niche was not subject to high levels of competition because flying (by bats or birds) is not easily accomplished in the dense understory at night. Poikilothermic vertebrates would also have posed little competitive threat to tarsiers because of their inability to chase large, highly mobile arthropods at night.

Conclusions

Dwight Davis (1962, 55) wrote that tarsiers "are astonishingly deliberate and stupid-appearing in behavior, so much so that it seems a miracle that they can survive." But survive they have. Many facets contribute to the original success and persistence of tarsiids. Certainly their ability to detect, hunt, and quickly ingest high-quality food items has been one of the most important keys to their success and survival. Leaping is an energy-intensive mode of locomotion, but it appears that the costs of leaping in tarsiids have never exceeded the benefits because of the high-quality nature of the food rewards that occasion the leaps. Are tarsiids in the jaws of a perilous evolutionary trap, or are they simply occupying a stable, if unusual, niche made possible by long-term environmental stability? Well, probably both, but the ecological and morphological evidence available suggests "so far, so good."

The tarsiid niche is unique among primates in many respects, but is perhaps most remarkable because of its great antiquity. Tarsiid primates, as well as sivaladapine adapiforms, maintained relictual distributions in southern and southeastern Asia long after their close relatives on other Holarctic continents became extinct (Qi and Beard, 1998). This was made possible by the persistence in parts of southern and southeast Asia of humid multi-stratal rain-forest ecosystems from the earliest Tertiary to the present day (Morley, 2000). Although the composition of these forests has changed subtly through time, the physical structure of the forests and the nature of the insect guilds inhabiting the forests has not. The tarsiid niche has persisted for at least 45 million years because of the environmental stability of the tropical forest refuges of southeastern Asia and because of the trophic level (of the insectivore/carnivore) at which tarsiers exist.

References

Beard KC. 1998. A new genus of Tarsiidae (Mammalia: Primates) from the middle Eocene of Shanxi Province, China, with notes on the historical biogeography of tarsiers. In Beard KC, Dawson MR, editors, Dawn of the age of mammals in Asia, no. 34, 260–277. Pittsburgh: Carnegie Museum of Natural History.

Beard KC, Qi T. 1994. A diverse new primate fauna from middle Eocene fissure-fillings in southeastern China. Nature 368: 604–609.

Behrensmeyer AK, Damuth JD. 1992. Terrestrial ecosystems through time. Chicago: University of Chicago Press.

Carpenter FH. 1976. Geological history and the evolution of the insects. In D. White, editor, Proceedings of the 15th Congress of Entomology, 63–70. Washington, DC: American Entomological Society.

Castenholz A. 1984. The eye of *Tarsius*. In Niemitz C, editor, Biology of tarsiers. Stuttgart: Gustav-Fischer-Verlag.

Charles-Dominique P. 1975. Nocturnal primates and diurnal primates: an ecological interpretation of these two modes of life by analysis of the higher vertebrate fauna in tropical forest ecosystems. In Luckett WP, Szalay FS, editors, Phylogeny of the primates, 69–88. New York: Plenum Press.

Collinson ME, Hooker JJ. 1991. Fossil evidence of interactions between plants and plant-eating mammals. Phil Trans R Soc Lond B 333: 197–208.

Covert HH. 1995. Locomotor adaptations of Eocene primates: adaptive diversity among the earliest prosimians. In Alterman L, Doyle GA, Izard MK, editors, Creatures of the dark: the nocturnal prosimians, 495–509. New York: Plenum Press.

Crepet WL, Friis EM. 1987. The evolution of insect pollination in angiosperms. In Friis EM, Chaloner WG, Crane PR, editors, The origins of angiosperms and their biological consequences, 181–201. Cambridge: Cambridge University Press.

Crompton RH. 1989. Mechanisms of speciation in *Galago* and *Tarsius*. J Hum Evol 4: 105–116.

Crompton RH, Andau PM. 1986. Locomotion and habitat utilization in free-ranging *Tarsius bancanus:* a preliminary report. Primates 27: 337–355.

Dagosto M, Gebo DL. 1996. New primate postcranial remains from the middle Eocene Shanghuang fissures, southeastern China. Am J Phys Anthro (suppl.) 22: 92–93.

Davis DD. 1962. Mammals of the lowland rain-forest of north Borneo. Bull Sing Nat Mus, no. 31: 1–129.

Emmons LH. 1995. Mammals of rain forest canopies. In Lowman MD, Nadkarni NM, editors, Forest canopies, 199–223. San Diego: Academic Press.

Gebo DL, Dagosto M. 1996. New primate tarsal remains from the middle Eocene Shanghuang fissures, southeastern China. Am J Phys Anthro (suppl.) 22: 111.

Ginsburg L, Mein P. 1987. *Tarsius thailandica* nov. sp., premier Tarsiidea (Primates, Mammalia) fossile d'Asie. C R Acad Sci Paris, Ser II 304, no. 19: 1213–1215.

Grubb PJ, Whitmore TC. 1967. A comparison of montane and lowland forest in Ecuador. III. The light reaching the ground vegetation. J Ecol 47: 33–57.

Gursky SL. 1997. Modeling maternal time budgets: the impact of lactation and infant transport on the time budget of the spectral tarsier, *Tarsius spectrum*. Ph.D dissertation, Anthropological Sciences, State University of New York at Stony Brook.

Izard MK, Wright PC, Simons EL. 1985. Gestation length in *Tarsius bancanus*. Am J Primatol 9: 327–331.

Jablonski NG, Crompton RH. 1994. Feeding behavior, mastication, and tooth wear in the western tarsier, *Tarsius bancanus*. Int J Primatol 15: 29–59.

Jablonski NG, Whitfort MJ, Roberts-Smith N, Xu Q-Q. (2000). The influence of life history and diet on the distribution of catarrhine primates during the Pleistocene in eastern Asia. J Hum Evol 39: 131–157.

Janis CM 1993. Tertiary mammal evolution in the context of changing climates, vegetation, and tectonic events. Ann Rev Ecol Syst 24: 467–500.

Leopold EB, Liu G-W et al. 1992. Low-biomass vegetation in the Oligocene? In Prothero DR, Berggren WA, editors, Eocene-Oligocene climatic and biotic evolution, 399–420. Princeton: Princeton University Press.

Lowman MD. 1995. Herbivory as a canopy process in rain forest trees. In Lowman MD, Nadkarni NM, editors, Forest canopies, 431–455. San Diego: Academic Press.

MacKinnon K, Hatta G, Halim H, Mangalir A. 1996. The ecology of Kalimantan. The Ecology of Indonesia series, volume 3. Hong Kong: Periplus Editions.

Martin RD. 1994. Bonanza at Shanghuang. Nature 368: 586–587.

McNab BK, Wright PC. 1987. Temperature regulation and oxygen consumption in the Philippine tarsier *Tarsius syrichta*. Physiol Zool 60: 596–600.

Morley RJ. 2000. Origin and evolution of tropical rain forests. New York: John Wiley and Sons.

Morse DR, Lawton JH, Dodson MM, Williamson MH. 1985. Fractal dimension of vegetation and the distribution of arthropod body lengths. Nature 314: 731–733.

Musser GG, Dagosto M. 1987. The identity of *Tarsius pumilis*, a pygmy species endemic to the montane mossy forests of central Sulawesi. Am Mus Nov (2867): 1–53.

Niemitz C. 1977. Zur Funktionsmorphologie und Biometrie der Gattung *Tarsius* Storr, 1790. Herleitung von Evolutionsmechanismen be einem Primaten. Cour Forsch Senck 25: 1–161.

Niemitz C. 1983. Can a primate be an owl? 1st International Symposium on Vertebrate Morphology. Giessen, Germany: Gustav-Fischer-Verlag.

Niemitz C. 1984. Synecological relationships and feeding behaviour of the genus *Tarsius*. In Niemitz C, editor, Biology of tarsiers, 117–118. Stuttgart: Gustav-Fischer-Verlag.

Niemitz C. 1991. *Tarsius dianae:* a new primate species from central Sulawesi (Indonesia). Folia Primatol 56: 105–116.

Parrish JT. 1987. Global palaeogeography and palaeoclimate of the Late Cretaceous and Early Tertiary. In Friis EM, Chaloner WG, Crane PR, editors. The origins of angiosperms and their biological consequences, 51–73. Cambridge, UK: Cambridge University Press.

Qi T, Beard KC. 1998. Late Eocene sivaladapid primate from Guangxi Zhuang Autonomous Region, People's Republic of China. J Hum Evol 35: 211–220.

Qi T, Beard KC et al. 1996. The Shanghuang mammalian fauna, Middle Eocene of Jiangsu: history of discovery and significance. Vert Pal As 34: 202–214.

Qi T, Zong G-F et al. 1991. Discovery of *Lushilagus* and *Miacis* in Jiangsu and its zoogeographical significance. Vert Pal As 29: 39–63.

Rasmussen DT, Conroy GC, Simons EL. 1998. Tarsier-like locomotor specializations in the Oligocene primate *Afrotarsius*. Proc Natl Acad Sci USA 95: 14848–14850.

Roberts M. 1994. Growth, development, and parental care in the western tarsier (*Tarsius bancanus*) in captivity: evidence for a "slow" life-history and non-monogamous mating system. Int J Primatol 15: 1–28.

Rohdendorf BB. 1991. Fundamentals of paleontology. New Delhi: Amerind.

Rose KD. 1981. The Clarkforkian land-mammal age and mammalian faunal composition across the Paleocene-Eocene boundary. Univ Mich Pap Paleontol 26: 1–197.

Schwartz, JH. 1984. What is a tarsier? In Eldredge N, Stanley SM, editors, Living fossils, 38–49. New York: Springer-Verlag.

Simons EL, Bown TM. 1985. *Afrotarsius chatrathi*, first tarsiiform primate (Tarsiidae) from Africa. Nature 313: 475–477.

Stork NE. 1987. Arthropod faunal similarity of Bornean rain forest trees. Ecol Entomol 12: 219–226.

Upchurch GR, Wolfe JA. 1987. Mid-Cretaceous to Early Tertiary vegetation and climate: evidence from fossil leaves and woods. In Friis EM, Chaloner WG, Crane PR, editors, The origins of angiosperms and their biological consequences, 75–105. Cambridge, UK: Cambridge University Press.

Whitten AJ, Mustafa M, Henderson G. 1987. The ecology of Sulawesi. Yogyakarta: Gadjah Mada University Press.

Wing SL, Tiffney BH. 1987. Interactions of angiosperms and herbivorous tetrapods through time. In Friis EM, Chaloner WG, Crane PR, editors, The origins of angiosperms and their biological consequences, 203–224. Cambridge, UK: Cambridge University Press.

Wolfe JA. 1985. Distribution of major vegetational types during the Tertiary. Geophys Monogr 32: 357–375.

Wolfe JA, Upchurch GR. 1986. Vegetation, climatic and floral changes at the Cretaceous-Tertiary boundary. Nature 324: 148–152.

Wright PC. 1989. The nocturnal primate niche in the New World. J Hum Evol 18: 635–658.

Wright PC. 1994. The behavior and ecology of the owl monkey. In Baer JF, Weller RF, Kakoma I, editors, Aotus: the owl monkey, 380–392. San Diego: Academic Press.

Wright PC, Martin LB. 1995. Predation, pollination and torpor in two nocturnal prosimians: *Cheirogaleus major* and *Microcebus rufus* in the rain forest of Madagascar. In Alterman L, Doyle GA, Izard MK, editors, Creatures of the dark: the nocturnal prosimians, 45–60. New York: Plenum Press.

CHAPTER 3

How Close Are the Similarities between *Tarsius* and Other Primates?

Jeffrey H. Schwartz

For centuries, the tiny primate *Tarsius* has amazed and frustrated those who study it. Its bizarre attributes make its phylogeny difficult to establish. Still, systematists support tarsier phylogenies based on interpretations of data constrained by assumed phylogenies or reconstructed transformation series. Consequently, retrieving specific details of *Tarsius* anatomy and considering them in light of alternative interpretations is difficult. This situation is an unfortunate legacy of taxonomic practice, whereby the identity (diagnosis) of a new taxon is defined less by the features of the organism than by the ways in which it is thought to be similar to other taxa, which are also defined comparatively.

To rectify this, I present first, a morphological overview and second, a brief account of its systematic history. I use the development of the orbital region as an example of how we might try to delineate primitive retention from shared derivedness. I also discuss features that pertain to *Tarsius*'s potential phylogenetic relationships to other extant primates. Since hypotheses of phylogenetic relatedness of various fossils to lower or higher primates often rely on comparisons with *Tarsius,* I review fossils offered as *Tarsius,* sister taxa of *Tarsius,* and more basal anthropoids.

Extant *Tarsius:* An Overview

The tarsier (genus *Tarsius*)—a small, nocturnal, totally carnivorous, large and bug-eyed, long-ankled, scaly-tailed primate—has been known to the scientific community since Camel's 1706 publication of a new small mammal from the Philippines. The five currently recognized species of *Tarsius* (*T. bancanus, T. syrichta, T. spectrum, T. pumilis,* and *T. dianae*), and two additional unnamed species (Groves, pers. com.) are distributed throughout the tropical evergreen rain forests of much of the southeast Asian archipelago.

Tarsiers are nocturnal, extremely hindlimb dominated, arboreal mammals (Niemitz, 1984b), which, relative to their small body size, perform extraordinary leaps, covering wide spaces in kangaroo style. Some saltatory primates (e.g., *Microcebus* and *Hapalemur*) typically land hands first, but *T. bancanus* usually lands feet first, especially after traversing great distances (Peters and Preuschoft, 1984). *T. bancanus* uses both feet simultaneously in propelling itself, whereas *T. syrichta* takes a step prior to lift-off and upon landing. Only *T. bancanus* displays a preference for vertical rather than horizontal supports for perching and inclined branches for sleeping (Niemitz, 1984b). Like *Galago senegalensis, T. spectrum* exploits varied substrates while *T. dianae* tends toward walking quadrupedally on more horizontal supports.

Tarsius is the only totally carnivorous and insectivorous (animalivorous) primate. It preys on a diversity of insects (including retaliatory red ants), various crustaceans (shrimp, crabs), small vertebrates (bats, birds, fish, and flying frogs), and even snakes (including neurotoxic species) (Niemitz, 1979, 1984a; Jablonski and Crompton, 1994). Tarsiers typically consume the entire prey, even if the prey's mass exceeds its own. In hunting, tarsiers usually prowl noiselessly above ground level and then ambush their prey by pouncing on it, eyes closed, killing it with a series of powerful bites inflicted by their pointed antemolar teeth (Jablonski and Crompton, 1994). As noted in captive *T. bancanus,* for their small size (body weight: 115–150 g), tarsier proclivities toward relatively large prey (lizards of 3–5 g and mice of 5–8 g) results in marked fluctuations in daily body weight (Izard et al., 1985).

Typical of tropical insectivorous mammals of body weights in excess of 35 g, at least *Tarsius syrichta* and *T. bancanus* have low body temperatures and, as calculated for the Philippine tarsier, a low basal metabolic rate (65% of what is expected of an animal of similar body weight) (McNab and Wright, 1987). *T. bancanus* may be torpid during the day (Niemitz et al., 1984), but this condition has not been detected in *T. syrichta*—and would not be expected in mammals living in tropical forests within a 10 degree latitude of the equator (McNab and Wright, 1987). Interscapular brown body fat (the "interscapular hibernating gland"), which is thermoregulatory tissue in neonatal mammals (including humans) that remains active in adult hibernating mammals, also occurs in adult *Tarsius* (Niemitz et al., 1984).

Gestation in captive *Tarsius bancanus* is 178 days (Izard et al., 1985), whereas it is 157 days in *T. spectrum* and 180 in *T. syrichta* (Roberts, 1994). In a comparison to 26 prosimian species (Roberts, 1994), although tarsiers ranked among the smallest (adult *T. bancanus* weighs 130 g, *T. spectrum* weighs 200 g, *T. syrichta* weighs 113 g), tarsiers have long gestation periods (cf. *Microcebus murinus* [adults weigh 84 g and gestate 61 days]). *Galago demidovii* (adults weigh 63 g and gestate 111 days) is most similar to *Tarsius*. In *Loris tardigradus* (298 g) and *G. senegalensis* (201 g), gestation lengths are 166–169 and 141 days, respectively. Izard et al. (1985) suggest that *Tarsius*'s relatively long gestation period is due either to its restricted animalivorous diet and/or its apparently low metabolic rate, the latter of which Niemitz et al. (1984) think correlated with producing only one offspring per year. Also possibly correlated with gestation length, as seen in *T. bancanus* (Roberts, 1994), are (1) prenatal development slowest among mammals; (2) brain at birth largest among mammals relative to body size (60–70% adult size); and (3) neonate approximately 20% maternal body mass.

Tarsius is the only primate with a bicornuate uterus that also develops hemochorial placentation, which it achieves through a unique sequence of events: Rauber's layer is lost prior to implantation, as it is in prosimians, and amniogenesis begins by cavitation with the formation of a primordial amniotic cavity as in anthropoids, but then changes to folding for the duration (Luckett, 1974, 1976; but see Starck, 1984).

The pelage of the large tarsier species is generally grayish buff; the fur of *T. dianae* is somewhat woollier in appearance (Niemitz et al., 1991). The tail is sparsely covered with hair, but the tip bears a distinct tuft, which is densest in *T. bancanus*. *T. spectrum* and apparently *T. bancanus* develop edentate- and reptile-like scales along the tail and around the areolae (Niemitz, 1979). *T. dianae* is distinguished at least from *T. spectrum* in having a hairless patch at the base of its ear (Niemitz et al., 1991).

Orbital size increases from *T. pumilus* (smallest), *T. spectrum,* and perhaps *T. dianae, T. syrichta,* to *T. bancanus* (largest). In all tarsiers, the orbital enlargement dominates the skull so that the size relationships of the organs of the head are unique among mammals (Starck, 1984). For example, the average volume of a tarsier's eyeball (2.8 cm^3) equals its average cranial capacity (Castenholz, 1984). The unusually spheroidal/ tubular eyeball protrudes more than half its length beyond the margin of the shallow orbit cupping it; the eyeball/orbital volume ratio is 1.79 in *Tarsius,* roughly 1.0 for various lower primates, and 0.32 for *Homo.* With its eyes immobilized in their orbits, *Tarsius* makes visual adjustments not by eyeball movement, but by rotating the entire head through an arc of approximately 180 degrees to either side.

Tarsius also develops a small, triangular, hairless region at the medial border of each nostril that is continuous with the hairless skin of the inside its nostrils (Klauer, 1984). The rest of the oro-nasal region is profusely covered with different kinds of hairs, including thick vibrissal (i.e., sensory or sinus) hairs, which occur everywhere except, perhaps, in the internarial region (Hofer, 1979, 1980; Klauer, 1984). The perimeter of a tarsier's lips bears an unusual arrangement of interdigitating hairs and distinct lateral papillae that cover the gaps between the anterior and most of the cheek teeth. The upper and lower lips bear an abundance of sebaceous and apocrine glands throughout, creating *Tarsius*'s distinctive circum-oral organ.

Systematic History

Tarsius was first described by the missionary J. G. Camel (1706–8), who named it *Cercopithecus luzonis minimus* because he mistakenly thought it was a monkey from the island of Luzon in the Philippines. In the tenth edition of the *Systema Naturae* (in which he replaced Anthropomorpha with primates), Linnaeus (1758) used Camel's description for the binomial *Simia syrichta*. Buffon (1765) described a juvenile mammal of unknown origin; because of the extreme length of the hindlimb (especially the foot and upper tarsals), he coined the name "tarsier" for it. Since Buffon did not realize that his "tarsier" was the same animal as Linnaeus's *Simia syrichta,* he suggested that it might be a jerboa or an opossum, which Linnaeus unwittingly accepted. In 1777, Erxleben argued that the tarsier was a primate most similar to lemurs and coined *Lemur tarsier* to include both of Linnaeus's and Buffon's animals. Ultimately, Storr (1780) created the genus *Tarsius* in order to distinguish this primate from lemurs.

Since Linnaeus's *Simia syrichta* was based on a specimen from the Philippines, the correct designation for this particular primate is *Tarsius syrichta.* In 1778, Pallas recorded the first tarsier of possible Celebesian (i.e., Sulawesian) origin and referred to it as *Lemur spectrum.* Thus, this species name has priority over those subsequently proposed for Sulawesian tarsiers (*T. fuscus, T. fuscomanus,* and *T. fisheri*), and the correct binomial is *T. spectrum.* In 1821, Horsfield described tarsiers from Bangka, which he named *T. bancanus. T. pumilus* was first identified by Miller and Hollister (1921) and rediscovered by Musser and Dagosto (1987). *T. dianae* was recently described by Niemitz et al. (1991).

Because *Tarsius* and *Galago* have similarly distinctive hindlimb morphologies, taxonomists such as Geoffroy-Saint Hilaire and Cuvier (1795) and de Blainville (1839) grouped these taxa together. In 1811, Illiger placed them in their own family, Macrotarsi, and relegated lemurs and lorises to another, Prosimia (Prosimii). Fitzinger (1861) and Gray (1870) kept Illiger's Macrotarsi, as did Brehm (1868), who also added to it the long-footed *Microcebus.* In 1883, as the counterpart to Mivart's (1864) taxon for higher primates, Anthropoidea, Flower proposed the suborder Lemuroidea, which he divided into three families: Lemuridae (lemurs, lorises, bush babies), Chiromyidae (*Daubentonia,* which was then still called *Chiromys*), and Tarsiidae (*Tarsius*). Illiger's suborder Prosimii eventually replaced Flower's Lemuroidea as the taxonomic partner of Anthropoidea. With Hubrecht's (1898) description of a type of hemochorial placentation occurring in *Tarsius* (as well as in *Tupaia*), which he contrasted with the epitheliochorial placentation of bush babies, support for a *Tarsius-Galago* relationship began to erode. In 1918, on the basis of an incorrect assessment of narial morphological types, Pocock (1918) suggested that *Tarsius* should be dissociated from other prosimians—which he relegated to Geoffroy's (1812) suborder Strepsirrhini—and grouped with anthropoids in the suborder Haplorhini.

Pocock's (1918) taxonomic suggestion lay virtually unnoticed until the 1950s, when Hill (1953) devoted one volume of his treatise on comparative primate anatomy to Strepsirrhini and the rest to Haplorhini, the first volume of which dealt solely with *Tarsius* (Hill, 1955). Nevertheless, the suborders Prosimii and Anthropoidea continued to be the favored divisions of Primates until the 1970s and '80s, when studies, especially on placentation, the auditory bulla, carotid circulation, and craniofacial morphology, were seen as indicating a closer relationship between *Tarsius* and anthropoids than between *Tarsius* and lemurs and lorises. There are, however, a number of features that appear to link *Tarsius* closely with the lorisiform group of strepsirhine primates (e.g., Schwartz, 1984, 1986, in press). Recently, the pendulum has begun to swing back toward recognizing the groups Prosimii and Anthropoidea (e.g., see contributions in Fleagle and Kay, 1994; Ankel-Simons, 2000; and Delanty and Ross, 2000).

While eighteenth-century comparative anatomists grouped *Tarsius* with *Galago* and, eventually, also with *Microcebus* on the basis of pedal anatomy, the picture of Prosimii inherited by the twentieth century was not that *Tarsius*'s relationships were within that group, but that it was a lower primate or an evolutionarily intermediate between the lower and higher primates. Superficially, at least, *Tarsius,* with its large eyes, somewhat globular cranium, seemingly short snout, vertical torso, and primitive molar morphology, made a convenient model from which to derive anthropoids, who were seen as progressively evolving these attributes, culminating in *Homo sapiens*. (Jones Wood [1920] even argued that especially *Tarsius*'s upright posture made it a perfect model for the ancestor of the human lineage alone.) *Tarsius* was kept with lemurs and lorises largely because it was seen as a rather primitive primate.

The notion that *Tarsius* is a lower primate, foreshadowing higher primate evolution, was bolstered by interpretations of a diversity of small Eocene fossils. European forms (such as *Necrolemur* and other microchoerines) were taken as the closest potential relatives of tarsiers. North American Eocene fossils were considered both sufficiently similar to modern tarsiers in size, cheektooth morphology, cranial shape, and development of moderate tarsal elongation that they could serve as tarsier predecessors. They were also primitive enough in molar cusp pattern to provide convincing anthropoid ancestors.

Thus, fossil tarsioids or tarsiiforms, as they were variably called, became the link between *Tarsius* and anthropoids. Because fossils could be linked to living tarsiers, which, in turn, were supposedly intermediate between higher and lower primates, primate taxonomists could choose to embrace the subordinal divisions Strepsirrhini and Haplorhini. Whereas in the scheme of dividing primates into the suborders Prosimii and Anthropoidea, fossil tarsioids could be accepted as prosimian because of their primitive molar morphology (in spite of the fact that an animal's primitiveness does not indicate phylogenetic relationship), in the divisions Strepsirrhini-Haplorhini these fossils became de facto haplorhines only because of their association with *Tarsius.* Haplorhini was, and still is, defined largely on the basis of unfossilizable anatomies (e.g., see Beard's [1988] argument for a rhinarium in extinct tarsiiforms).

Before we proceed with evaluating the alternatives—Prosimii and Anthropoidea versus Strepsirrhini and Haplorhini—we must clarify the features that support the competing hypotheses. With this goal in mind, I review the salient morphologies.

Morphology: Comparisons and Interpretation

The Eye and Orbital Region

Various features of the eye—a well-developed central retinal fovea, a macula lutea, and the lack of a tapetum—have been taken as being similar in *Tarsius* and anthropoids. Although *Tarsius* does not develop a tapetum, it also does not consistently develop a retinal fovea (Castenholz, 1984). When a retinal fovea is present, it is relatively quite small and composed entirely of rods. *Galago* also develops a small central fovea (ibid.). The structure identified in *Tarsius* as a "macula lutea" (Woollard, 1926) was an artifact of fixation (Kolmer, 1930). Anthropoids lack a tapetum and develop a large central foveae; only the nocturnal *Aotus*'s retina is composed entirely of rods. Wolin (1974) suggested that, in retinal organization, *Aotus* may be less like other anthropoids than *Tarsius,* and Castenholz (1984: 317) concluded that, because "the eye of *Tarsius* is the most specialized eye found in primates," it is impossible to "decide which one of the specializations [is] of phylogenetic significance."

The perimeter of the enormous bony orbit of *Tarsius* is somewhat lipped; it is most distended superiorly and laterally. The superior orbital rim approaches (and, as in *T. bancanus,* may exceed) the height of the neurocranium; the inferior margin is level with the tooth row. The enormous orbits also add to the illusion that the facial skeleton is hafted low on the cranium (cf. Rosenberger and Szalay, 1980; Beard and MacPhee, 1994; Ross, 1994), as it is in anthropoids (ibid.), but radiographic analysis demonstrates that this is not the case (fig. 3.1). Near-frontal alignment of *Tarsius*'s eyeballs (not the orbits, which, as in *Galago,* retain the primitive orientation

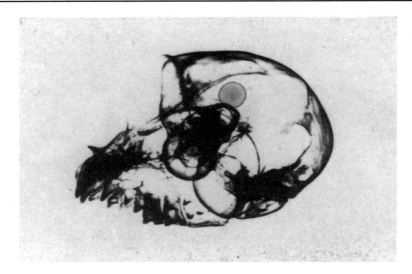

Figure 3.1. Radiograph of a skull of *Tarsius spectrum* (Rijksmuseum van Natuurlijke Historie [RMNH] d, Leiden). Note that the auditory region is vacuous, that the facial skeleton is not hafted low upon the cranium, and that the huge orbit encroaches upon the elongated snout. Arrows point to superior and inferior orbital margins. (© J. Schwartz)

[Schwartz, 1996]) is associated with extreme medial orbital convergence. This is the most pronounced among primates and both compresses the entire nasal region dorsally and reduces the nasal cavity and paranasal sinuses (Starck, 1975, 1984). Anthropoids, however, with their relatively smaller orbits, have absolutely reduced snouts and nasal cavities.

Orbital size and convergence in *Tarsius* appears correlated with its development of a large and extensive (apical) interorbital septum, which is formed by contributions not only from the vomer and an ossified nasal septum, but also and uniquely from the frontal bones and the interorbital lamina of the presphenoid (Starck, 1984). Severe orbital convergence in some specimens results in incomplete ossification of the interorbital septum. *Tarsius* is further unique among mammals in general in that the large rostral portion of its cranial cavity, together with the bulbous chamber of the roof and the small lamina cribrosa of the nasal capsule, is associated with a long olfactory tube formed entirely from the frontal bone (Starck, 1984: 287).

Various authors (e.g., Cartmill, 1978, 1980, 1994; Ross, 1994) have emphasized the orbital region in their arguments for *Tarsius*'s being either intermediate between lower and higher primates or specifically related to anthropoids, because, while all living primates possess a postorbital bar, only *Tarsius* and anthropoids develop some degree of postorbital closure.

However, the bony flanges that contribute to postorbital closure in *Tarsius* hardly impinge upon this region as much as they do in various New World monkeys (even those that achieve only incomplete postorbital closure). Catarrhines essentially have complete postorbital closure. If those extinct "tarsiiforms" for which crania are known are indeed related to *Tarsius* (see below), then postorbital closure would have occurred independently in *Tarsius* and anthropoids because the fossils have only postorbital bars. Indeed, independent attainment of postorbital closure in *Tarsius* and anthropoids through different growth patterns best explains the differences in the relative bony contributions to their respective orbital regions.

Cartmill (e.g., 1978, 1980) suggested that *Tarsius* and anthropoids achieve postorbital closure from the primitive condition of a postorbital bar via varying degrees of enlargement of the same three bones: the zygoma, frontal, and maxilla. As he portrayed it, in *Tarsius* and in anthropoids, the alisphenoid becomes enlarged and the frontal either displaces or overrides the parietal and squamosal. In support of the supposition that *Tarsius* is intermediate between prosimians (with postorbital bars) and anthropoids (with postorbital closure), Cartmill (1980) proposed that a transformation series from *Galago senegalensis* to *Tarsius* to *Saimiri sciureus* illustrates how expansion of the alisphenoid and frontal could have occurred. If the eyeballs of these three primates were the same size, the suggested transformation series from *Galago* to *Tarsius* might be phylogenetically relevant. However, *Tarsius*'s eyeball is huge, which has implications for interpreting orbital morphology.

In spite of *Tarsius*'s extreme degree of orbital hypertrophy, the circumferentially enormous but shallow orbits barely cup its eyeballs (fig. 3.2). In other prosimians and especially in anthropoids, the orbits are deep at birth and remain so in the adult (fig. 3.3). In juvenile and adult anthropoids, the frontal process of the zygoma tapers superiorly in frontal view; it is also deep anteroposteriorly, as is reflected in the long zygomaticofrontal suture (fig. 3.4). Within the orbit of juvenile and adult anthropoids, the sphenoid is generally rectangular in outline and its lateralmost extremity, delineated by the zygomaticosphenoid suture, is contained well within the orbital cone (fig. 3.3). Externally in juvenile and adult anthropoids, the (ali)sphenoid, which lies behind and well away from the lateral orbital rim, faces laterally and contributes to the wall of the temporal fossa (fig. 3.4). The shape of the (ali)sphenoid may differ between anthropoid subclades, but whatever the configuration in the juvenile, it is retained in the adult of that taxon (e.g., fig. 3.4). Thus, the essential features of adult anthropoid postorbital closure are established early in life.

In the reverse of the anthropoid configuration, the zygomatic frontal process of juvenile and adult tarsiers broadens superiorly. Instead of closing

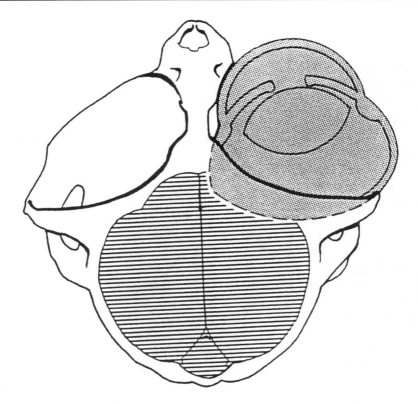

Figure 3.2. Superior view of skull of *Tarsius* illustrating the unusual morphology of the eye, its protrusion from the bony orbit, and its size relative to the brain. (Modified from Castenholz, 1984, and Stephan, 1984. Drawing by J. Anderton.)

off the back of the orbit, this strut of bone is well separated from the lateral wall of the brain case in the juvenile and remains so in the adult. Thus, *Tarsius*'s zygomatic frontal process is always differentiated as part of a postorbital bar. Within the orbit, the medially expanded sphenoid tapers laterally to a thin and short "tail" that nestles in the notch formed by the contact of the zygomatic frontal process and the frontal bone. Externally, the laterally extended sphenoid thus protrudes into the space of the temporal fossa. Also externally, as seen in juveniles, *Tarsius*'s (ali)sphenoid has two surfaces essentially at right angles to each other: one surface faces back upon the temporal fossa and the other faces laterally outward.

It is often difficult to delineate sutures in adults, but it is obvious that the configuration established in juveniles is maintained, if not exaggerated, in adults. *Tarsius* thus differs significantly from anthropoids in that the sphenoid extends from within the orbit to protrude laterally beyond the plane

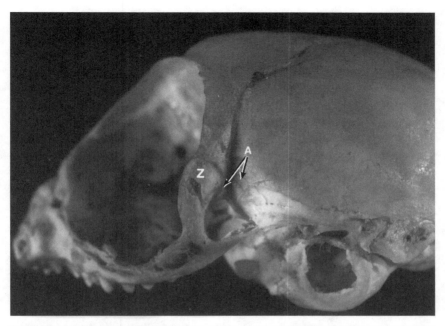

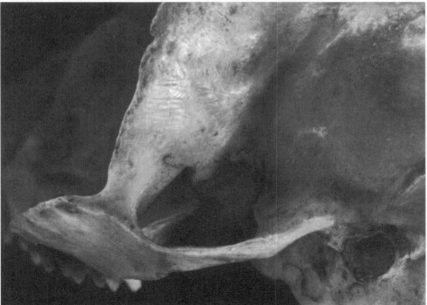

Figure 3.3. Closeup in lateral view of the bony orbital region of two *Tarsius spectrum:* (above RMNH n) a juvenile individual (antemolar deciduous teeth in place, with M2s beginning to erupt) and (below) a fully adult individual (RMNH d). In the juvenile, the superiorly expanded zygoma (Z), which is separated from the cranial vault as is a postorbital bar, is clearly delineated, as is the alisphenoid (A) with its posteriorly and laterally facing surfaces, the former of which contributes to the orbital lip. Although the sutural distinctions are obscured in the adult, it is obvious that the relative dispositions of the zygoma and alisphernoid are not appreciably different from the juvenile state. Not to scale. (© J. Schwartz)

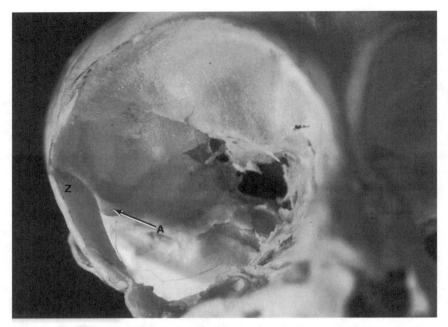

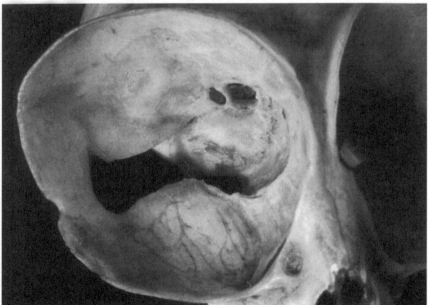

Figure 3.4. Frontal view of same specimens as in fig. 3.3. In the juvenile (above), the superiorly expanded zygoma (Z) is clearly separated from the cranial wall and a thin lateral extension of the otherwise bulbous alisphenoid (A) extends laterally beyond the cranial wall to insert in the notch between the zygoma and frontal. In the adult (below), the outline of the superiorly expanded zygoma is faintly visible and it appears that the lateral extension of the alisphenoid has expanded somewhat in its vertical dimension. The lateral extremity of the alisphenoid has both an orbital and a temporal fossal surface. In the juvenile and the adult alike, the zygoma stands away from the cranial vault as a postorbital bar. Not to scale. (© J. Schwartz)

of the cranial vault wall such that is has both orbital and temporal fossal surfaces. Although, during growth, the lateralmost extremity of the external (ali)sphenoid may expand vertically (in part, perhaps, because of the relation of musculature to bone [cf. Ross, 1995]) and extend a bit farther away from the brain case to face upon the temporal fossa, the configuration of the frontal process of *Tarsius*'s zygoma (being broad superiorly) and its separation from the lateral wall of the brain case remain constant.

Tarsius's large and unusual interorbital septal region also appears correlated with the enlargement and medial convergence of its orbits (Starck, 1975, 1984). Although the presence of an interorbital septum has been used to unite *Tarsius* with anthropoids (e.g., Cartmill, 1978, 1980), not all anthropoids develop one (e.g., *Aegyptopithecus, Alouatta,* hominoids), and, when they do, it is not the same as in *Tarsius* (Simons and Rasmussen, 1989). In addition, not only do some prosimians (*Loris,* various omomyids, *Galago senegalensis,* and *Microcebus murinus*) develop an interorbital septum, but so, too, do most mammals. In general in mammals, in conjunction with orbital enlargement, reduction of the nasal cavity, and changes in craniofacial orientation, the interorbital septum is retained in the adult (Starck, 1975) rather than incorporated into the ethmoid complex, as in *Macaca* (Zingeser and Lozanoff, 1989. Consequently, absence, not presence, of an interorbital septum in any mammal—primates included—is the phylogenetically significant character state.

The point of greatest interorbital constriction in *Tarsius* and anthropoids is supposed to lie below the olfactory lobes, while in strepsirhines it is above. Hence the development of an "apical" interorbital septum has been cited as a derived feature of Haplorhini (e.g. Cartmill, 1978). However, the interorbital septum of *Galago senegalensis* and *Microcebus murinus* lies below the olfactory lobes, as does the region of greatest constriction in *Hapalemur griseus* (Simons and Rasmussen, 1989).

Starck (1975, 1984) suggested that much of *Tarsius*'s craniofacial morphology—e.g., reduction of the nasal cavity and paranasal sinuses, development of an extensive interorbital septum, encasement of the long olfactory nerves in a bony canal derived from the frontal bones, as well as partial postorbital closure—is related to the extreme hypertrophy of its eyeball and orbit. Thus, seemingly favorable comparisons between *Tarsius* and anthropoids are likely superficial.

The Nasal Region

As originally conceived by Pocock, Haplorhini subsumed primates whose nostrils are continuous or aborally rounded along the lateral margin; similar to carnivores and insectivores, the lateral narial margins of strepsirhine primates are creased or slit. Pocock also noted in haplorhine primates the absence of a naked rhinarium and the presence of a narrow internarial region and a continuous upper lip beset with hairs, but it was Hill (1955) who emphasized these features in the definition of Haplorhini. Subsequently, however, Hofer (1980) demonstrated that *Tarsius* and various New World monkeys are morphologically strepsirhine in having laterally creased (not aborally rounded) nostrils (particularly well developed in *T. dianae* [Niemitz et al., 1991]), and also questioned whether features of *Tarsius*'s nasal region were homologous with those of anthropoid primates.

Hofer (1977) pointed out that many nonprimate mammals are haplorhine in lacking a rhinarium and in having mobile and medially fused upper lips that bear hairs. He also demonstrated that, in contrast to catarrhines, which have a smoothly continuous upper lip, various New World monkeys are similar to lorisids in retaining a midline rhinarial groove coursing from the narial region to the lip margin—which suggests a morphocline from a more deeply fissured upper lip (as in lemurs), to a grooved upper lip (as in lorisids and various New World monkeys), to a smoothly continuous upper lip (as in catarrhine primates). Since Anthropoidea appears to be a monophyletic group (and would be even if *Tarsius* were its sister taxon) (e.g., Schwartz, 1986), one would have to hypothesize that the last common ancestor of extant anthropoids had a medially fissured upper lip. As such, total fusion of the upper lip (i.e., complete coalescence of embryonically separate median nasal prominences) would have been independently derived in most species of *Tarsius* as well as in the last common ancestor of extant catarrhines. In fact, the deeply fissured internarial region of *T. dianae* could reflect different ontogenies of median nasal prominence fusion in the species of this primate genus and in catarrhine primates.

Hofer (1976, 1979, 1980) also noted that, in contrast to New World monkeys, *Tarsius* is restrictively similar only to catarrhine primates in having typical body hairs rather sinus hairs (vibrissae, "whiskers") within the internarial region. The conclusion from these data alone is either that *Tarsius* is most closely related to catarrhines or that each taxon independently evolved whiskerless internarial regions.

As in prosimians and New World monkeys, the nasal capsule of *Tarsius* protrudes anteriorly and the nostrils are laterally directed and separated by a wide internarial region (i.e., the nose is platyrrhine) (Hofer, 1980; Klauer, 1984). As in most mammals, including prosimians, New World monkeys, and even some catarrhine primates, adult tarsiers retain a functional vomeronasal organ. In concert with the absence of a rhinarium, *Tarsius*, anthropoids, and various artiodactyls and perissodactyls also have hair (but different kinds in different taxa) distributed across the upper lip and they also lack a philtrum, which, when present, tethers an otherwise mobile upper lip to the gum (Hofer, 1980; Maier, 1981; Klauer, 1984). Rasmussen's (1986) study of the interincisal diastema of various Eocene adapids (i.e., potential fossil relatives of lemurs) indicates that some of these fossil taxa may have lacked a rhinarium and, thus, had an untethered upper lip. Taking the entire oro-nasal region into consideration, *Tarsius* cannot be regarded either as a morphological intermediate between lower and higher primates or as the sister taxon of anthropoids.

The Auditory Region

The auditory region of *Tarsius* is noteworthy (Schwartz, 1984; MacPhee and Cartmill, 1986; Simons and Rasmussen, 1989). In general, the bulla and the anterior accessory cavity are quite inflated and, especially in the larger species, right and left anterior accessory cavities almost meet each other at the midline of the basicranium (Schwartz, in press). Although while still in its ring stage the ectotympanic is encased by and fuses to the inner wall of the ontogenetically expanding petrosal bulla, it subsequently ossifies into a tubular extension that protrudes laterally beyond the lateral margin of the bulla itself (MacPhee and Cartmill, 1986). A mediolaterally oriented bony septum separates the anterior accessory cavity from the tympanic cavity behind it; the promontory lies medially within the anthropoid tympanic cavity and posterior to the septum.

Anthropoids have a septum, but it extends the anteroposterior length of the bulla and over the promontory; consequently, the tympanic cavity lies lateral and the accessory cavity medial to it. In *Tarsius,* the internal carotid artery penetrates the septum laterally and centrally (Schwartz, 1984) and courses anteriorly through the anterior accessory cavity, pursuing a prepromontorial route (Simons and Rasmussen, 1989). In prosimians and anthropoids, the internal carotid artery pierces the bulla posteriorly and takes a transpromontorial course as it proceeds anteriorly. In anthropoids, the internal carotid artery courses within a bony canal that is incorporated into the septum. Thus, although *Tarsius* and anthropoids can be described as having a perbullar course of the internal carotid artery because the artery is associated with the septum (MacPhee and Cartmill, 1986; MacPhee et al., 1995), the details of bullar penetration and the path of the carotid artery relative to the promontorium and the septum are distinctly different (Simons and Rasmussen, 1989).

In various prosimians, the internal carotid artery bifurcates within the bulla into promontory and stapedial branches. In *Tarsius,* anthropoids, and virtually all lorisiforms (including cheirogaleids), the stapedial artery atrophies prenatally. *Tarsius,* anthropoids, and various lorisiforms also develop an anterior accessory cavity in the middle ear, although, in anthropoids, it is primitively quite trabeculated. The anterior accessory cavity become larger from *T. pumilus, T. spectrum,* and perhaps *T. dianae, T. syrichta,* to *T. bancanus* (cf. Musser and Dagosto, 1987).

The medial and lateral pterygoid plates, although involved with muscles of mastication, can be discussed here because of their peculiar association with the auditory region. In *Tarsius,* the strikingly reduced medial pterygoid plates converge at the midline of the basicranium where they fuse together as well as to the basisphenoid (Schwartz, in press). The medial pterygoid plates thus form a somewhat funnel-shaped structure that appears to extend the nasal cavity posteriorly. *Tarsius*'s lateral pterygoid plates are huge and winglike. They fan out from their articulation with the medial pterygoid plates to embrace and partially fuse with the anterior accessory cavities as far laterally at times as the articular fossae (ibid.). *Tarsius* is thus unique among primates because other prosimians and anthropoids develop individually distinct medial and lateral pterygoid plates that are either subequal in length or of unequal length, with the medial plate being moderately shorter than the lateral plate (ibid.).

Cartmill et al. (1981) and MacPhee and Cartmill (1986) suggested that seven features of the auditory region support the monophyly of extant Haplorhini: (1) absence of an annular bridge; (2) small tympanic cavity; (3) entry of the carotid artery into the bulla posteromedially; (4) presence of an anterior accessory cavity; (5) perbullar path of the carotid artery; (6) promontory artery unreduced in size but stapedial artery vestigial or absent; and (7) primary supply of the middle meningeal artery via the maxillary artery. Of these, MacPhee and Cartmill suggested that only features 1, 2, and 4 are identical in *Tarsius* and anthropoids; accepting the other features as potential synapomorphies requires complicated arguments. Although the claim of tarsier-anthropoid similarity might be correct, features 1 and 2 are also characteristic of lorisids and galagids, and feature 4 is found in all lorisiforms, including cheirogaleids. Interestingly, and to the exclusion of *Tarsius,* lorisiforms and anthropoids are similarly unique in having a complex epitympanic recess and a large mastoid cavity.

From the preceding review, the following emerges as being particularly important. In lorisiforms, the tympanic cavity and accessory chamber lie medial to the septum; in anthropoids, the tympanic cavity and accessory chamber lie lateral to the septum; and, in *Tarsius,* the anterior accessory chamber lies anterior, and the tympanic cavity posterior to the septum (Simons and Rasmussen, 1989). The entry of the carotid artery in *Tarsius* is lateral and central with regard to the entire bullar region (but anterior and central if only the true bulla is considered), and variably posteromedial in anthropoids, as it also is in lorisids and galagids. The perbullar course of the carotid artery is transpromontorial in anthropoids but prepromontorial in

Tarsius (ibid.). Cross-taxic comparison leads to the conclusion that possession of an unreduced promontory artery is primitive for primates, whereas
a vestigial or totally absent stapedial artery represent derived character
states. Since *Tarsius,* anthropoids, and lorisiforms are characterized by the
latter state, there is only one feature of the auditory region that supports
Haplorhini: the maxillary artery is the primary source of the middle meningeal artery. As noted above, other basicranial features suggest that the
relatedness of lorisiforms and anthropoids and others unite *Tarsius,* lorisiforms, and anthropoids.

The Jaws and Teeth

Tarsius's oddly laterally oriented coronoid process, which is most elevated
in *T. syrichta,* barely rises above the level of the mandibular condyle, which
is so mediolaterally compressed from side to side that, even in the large
T. syrichta, it is barely wider mediolaterally than the neck of the ramus
(Schwartz, 1984). In other primates, and mammals in general, the condyle
is mediolaterally broad (Schwartz, 1984). In profile, *Tarsius*'s smooth and
slightly domed condyle lies in the same plane as the ramus and is barely delineated from the bone from which it emerges; in the large tarsier species
its articular surface extends anteriorly for some distance in the direction
of the coronoid process, as well as down the condylar spine of the ramus
(Schwartz, in press).

Tarsius's uniquely configured mandibular condyle articulates in a crisply
excavated articular fossa that is bounded posteriorly by the tubular ectotympanic, medially by the lateral wall of the anterior accessory cavity, and laterally by the root of the zygomatic portion of the temporal bone (Schwartz, in
press). The mandible is thus essentially limited to a hingelike motion during
mastication (also see Jablonski and Crompton, 1994). In all other primates,
the zygomatic root lies superior to the articular fossa, which, in turn, is never
as restricted as in *Tarsius* (Schwartz, in press). Thus, even though lorisids
and catarrhines develop to varying degrees a tubular ectotympanic, this
structure, in conjunction with the postglenoid plate, present to some extent
in all primates, does not impinge significantly on their articular fossae.

The dental formula of adult *Tarsius* is typically given as 2.1.3.3/1.1.3.3.
Morphologically, however, none of *Tarsius*'s anterior teeth—upper or
lower—is "incisiform." Although differing in size, the first six upper teeth
are essentially premolariform in shape, and their bases are ringed with cin-
gulum. They are also pointed and to some extent conical. The first upper
tooth is the tallest and most conical in the upper jaw, and it is also more
trenchant than the tooth identified as the "canine." The second tooth ("second incisor") is the smallest upper tooth; it and the two teeth behind it
(the "canine," which is the second tallest tooth in the upper jaw, and "first

premolar") may bear a tiny posterior heel. The fifth and sixth upper teeth (the "second" and "third premolars") bear an additional cusp lingually, which is somewhat larger in the sixth tooth. Although primitive in overall design and transverseness, M^{1-3} possess distinctively compressed and U-shaped protocristae that emanate from a broad and somewhat compressed protocone; M^{1-2} also bear a tiny prehypocone crista. Among primates, prehypocone cristae are found only in extant lorisiforms and various presumed fossil tarsiiforms (Schwartz, 1984, 1986).

Tarsius's first and third lower teeth (the incisor and first premolar) are the smallest lower teeth; the first tooth is relatively very small in the large species and relatively larger in the smaller species (Schwartz, in press). The distal edge of the first mandibular tooth flares laterally and bears a thin margocristid while the mesial edge is essentially straight. The lingual surface bears a variably distinct longitudinal keel, which is rather mesially positioned and thus subdivides the crown quite asymmetrically. These features are most pronounced on the relatively large lower anterior tooth of *T. pumilus* and less pronounced on the relatively smaller lower anterior tooth of the larger species. Although I (e.g., 1984) previously argued that the homologue of the lower anterior tooth of *Tarsius* does not exist in other primates, this appears not to be the case. The features of *Tarsius*'s lower anterior tooth, which are so clearly illustrated in Musser and Dagosto's (1987) scanning electron micrographs of *T. pumilus*, are undeniably present in the lateral tooth of lemur, loris, and indriid toothcombs as well as in the large lower anterior tooth of various fossil tarsiiforms.

Since the lateral tooth of the prosimian toothcomb is usually identified as a canine (the underlying notion being that the canine and incisors became modified into elongate and narrow toothcomb teeth), logic would demand a similar identification of the lower anterior tooth of *Tarsius* (and of relevant fossil primates, as well). However, of greater importance than giving these teeth a name is recognizing that they are morphologically similar, and that, on this basis, they should be identified as homologues of the same tooth.

The second-fifth lower teeth of *Tarsius* (the canine and premolars) are similar morphologically in being premolariform and bearing a small but distinct heel that is incorporated into the cingulid ringing the base of the crown. The heel and cingulid of the second tooth are least pronounced buccally. The second tooth is the tallest, the next tooth behind is the shortest, and the somewhat bulkier fourth and fifth teeth are intermediate in height. As in lorisiforms and various fossil tarsiiforms, the third-fifth lower teeth (the premolars) are basically ovoid in outline. In all species of *Tarsius*, the lower canine and premolars bear a small but distinct heel incorporated into the cingulid surrounding the base of the crown.

M_{1-3} primitively retain distinct paraconids which remain separated from the metaconid throughout the series. M_3 is also primitive in having a moderately enlarged and somewhat centrally emplaced hypoconulid (posterior lobe). M_{1-3} are distinctive in having a sharply angular hypoconid, a well-defined buccal cingulid, and a tall protoconid and metaconid that are connected by a crest and form a steep wall facing the talonid. The latter cusp and crest configurations are also distinctive of lorisiforms and various fossil tarsiiforms (Schwartz, 1984, 1986).

Given the number of autapomorphic features of *Tarsius*'s jaws and teeth, as well as the obvious primitive aspects of molar morphology, few potential synapomorphies can be delineated. Of particular note are similarities with lorisiforms (especially the above-cited lower molar features, a prehypocone crista on M^{1-2}, and ovoid lower premolariform teeth). More broadly, *Tarsius*'s lower anterior tooth (the incisor) is strikingly similar to the lateral tooth of the toothcomb of toothcombed prosimians—which, if synapomorphic, not only unites these taxa as a clade, but also demonstrates that *Tarsius* has the most derived toothcomb among prosimians.

The Postcranial Skeleton

Postcranially, *Tarsius* possesses enlarged digital pads on hands and feet [being least pronounced in the small *T. pumilus* (Musser and Dagosto, 1987)], grooming claws on the second and third pedal digits, fusion of the tibia and fibula, profound elongation of the calcaneus and navicular bone, and relatively short metatarsals (e.g., Hill, 1955; Jouffroy et al., 1984). Degeneration of the nails on nongrooming clawed digits—often claimed as distinctive of tarsiers—is severe only in *T. bancanus* and *T. syrichta*, whereas *T. spectrum, T. dianae,* and especially *T. pumilus* possess nails that are quite well developed, compressed laterally, and strongly keeled centrally, as in *Daubentonia, Euoticus, Galago, Allocebus,* and various marmosets (ibid.; Soligo and Müller, 1999).

Tibiofibular fusion occurs in the smaller mouse lemurs and bush babies (Howell, 1944), which are remarkably similar to *Tarsius* in their degree of tarsal elongation and relative metatarsal truncation (Jouffroy et al., 1984). Marked elongation of the calcaneus and navicular is noted in the East African Miocene galagids *Progalago* and *Komba* (Walker, 1974) as well as in various fossil tarsiiforms (*Necrolemur, Nannopithex, Hemiacodon, Tetonius, Teilhardina, Arapahovius*) (see review in Schwartz, 1986). Galagids, *Microcebus, Phaner,* and *Tarsius* are collectively distinguished among extant primates in having an extremely elongated calcaneus and a much-reduced cuboid; in *Tarsius,* the orientation of the cuboidal trochlear facet prevents lateral deviation of the foot (Jouffroy et al., 1984). *Tarsius,* galagids, and *Microcebus* are the only prosimians, and *Callithrix* and *Pongo* the only anthropoids, in which the foot is longer than either the femur or tibia. The extremely long hindlimb of *Tarsius* and most galagids is reflected in their having the lowest intermembral indices (femur length/humerus length) among primates (ibid.).

Among extant primates, the femora of *Tarsius* and galagids are distinguished in having a cylindrical head, a more anteriorly expanded greater trochanter that extends onto the shaft, a deeper triangular depression at the intersection of the greater trochanter and the proximoanterior margin of the shaft, bowing of the proximal shaft anteriorly, a relatively short neck and high neck angle, and a relatively pronounced posterior projection of the lesser trochanter (Dagosto and Schmid, 1996). *Tarsius* and galagids also have the highest knee index (>100, reflecting a very deep knee). Many of these femoral features are also seen to some extent in various omomyids and microchoerines (ibid.). *Tarsius*'s talus is unique among primates in being extremely compressed superoinferiorly and in exhibiting little articular morphology superiorly (Schwartz, 1992).

Tarsius is outstanding in the great length of its forelimb (e.g., as measured relative to the length of the precaudal portion of its vertebral column) (Schultz, 1984). It also has the relatively longest hand of any living primate. As in anthropoid hands, *Tarsius*'s third finger is the longest, but, as in prosimian feet, its fourth digit is the longest. Thus *Tarsius* is unique among primates in not having structural similarity between its hand and foot in digital proportions. As in prosimians, *Tarsius* primitively retains a prepollex in the carpal region. *Tarsius* is apparently apomorphically similar to various lorisids in having pisiform-radius contact (Schwartz, 1992), but is primitively similar to most prosimians and anthropoids in having a proximally peaked and triangular os centrale (Schwartz and Yamada, 1998). *Tarsius*'s long scapula, together with its scapulohumeral and scapular indices, suggests an animal that should be terrestrial and quadrupedal. This describes *T. dianae* (which preferentially locomotes quadrupedally along horizontal supports), but not the clinging and leaping *T. bancanus* (thus suggesting caution in interpreting the locomotory behavior of extinct taxa).

A survey of various primates reveals that the nails on *Tarsius*'s nongrooming claw pedal digits I and IV (digit V was not analyzed) are similar to at least *Nycticebus*, *Galago*, and *Microcebus*, among prosimians, and *Cercopithecus aethiops* (but not *C. cephus*), *Papio anubis*, *Homo*, and *Hylobates*, among anthropoids, in having only one layer (Soligo and Müller, 1999). All other primates studied are (primitively) similar in having pedal digit nails composed of two layers (ibid.), which, therefore, makes this character useless in terms of resolving phylogenetic relationships.

The grooming claw of the second and third pedal digits is large in all species of *Tarsius* and similar in shape and structure (primitively retaining two layers) to the single grooming claw present on only the second pedal digit of extant prosimians (Soligo and Müller, 1999; Spearman, 1985). Although Soligo and Müller (1999) claimed that their data support the grouping of *Tarsius* with anthropoid primates and *Microcebus* with *Lemur* rather than with *Galago* and *Nycticebus*, this is clearly not the case. Synapomorphy at some level of common ancestry would be the more reasonable explanation of why *Microcebus, Galago,* and *Nycticebus* have a single-layered nail not only on digits I and IV, but on digit III as well. Assuming that prosimians and anthropoids are, respectively, monophyletic groups (e.g., see review in Schwartz, 1986)], *Tarsius*'s single-layered nail of nongrooming-clawed pedal digits links it not as the sister taxon of either Prosimii or Anthropoidea, but as a member of a subclade either of lorisiform or catarrhine primates.

Regardless of whether *Tarsius* is more closely related (somehow) to either strepsirhines or anthropoids, one must still deal with the fact that this primate possesses two pedal grooming claws. Thus, the following scenarios emerge. (1) The last common ancestor of all extant primates possessed a grooming claw on the second pedal digit, and this condition was retained in strepsirhine primates and in *Tarsius*. *Tarsius* either developed a second grooming claw autapomorphically or retained it from a last common ancestor that it shared with anthropoids; and, thus, the last common ancestor of anthropoids would have "lost" one or two grooming claws (and gained nails). (2) The last common ancestor of all extant primates did not possess a grooming claw. The last common ancestor of strepsirhine primates developed a grooming claw on the second pedal digit independently of either *Tarsius* or the last common ancestor of *Tarsius* and anthropoids. *Tarsius* either developed the second grooming claw on the third pedal digit autapomorphically or retained it from a last common ancestor it shared with anthropoids. Thus, the last common ancestor of anthropoids lost one or even two grooming claws (and "gained" nails). Or (3), as recently proposed by Soligo and Müller (1999), the last common ancestor of all extant primates had a grooming claw on the second pedal digit. This was retained in all strepsirhines and the last common ancestor of *Tarsius* and anthropoids. *Tarsius* developed a grooming claw on the third pedal digit, and the last common ancestor of extant anthropoids lost the grooming claw (and gained a nail).

The simplest suggestion is (4) that *Tarsius* and strepsirhines possess a grooming claw on the second pedal digit because they inherited this unique feature from a common ancestor they shared to the exclusion of anthropoids. As for which state—one or two grooming claws—preceded which, since both of *Tarsius*'s grooming claws and the single grooming claw of strepsirhines are (primitively) composed of two layers, while the nail (whether flat or keeled and compressed) on a strepsirhine's third pedal digit is (derivedly) single layered, it would appear that the last common ancestor of all prosimians would have been *Tarsius*-like in possessing two double-layered grooming claws and *Lemur*- and *Daubentonia*-like in having double-layered structures on digits I and IV. Within this hypothesized clade (prosimians), the situation is not as clear cut. Either (1) the last common ancestor of strepsirhine prosimians was characterized by the loss of the grooming claw on the third pedal digit and *Tarsius* and lorisiforms independently developed single-layered nails on digits I and IV, or (2) *Tarsius* and lorisiforms are united by their common development of single-layered nails on digits I and IV, with reduction to a single grooming claw occurring independently in lemuriforms and lorisiforms. Truly, beyond the fact that *Tarsius* and strepsirhine prosimians appear synapomorphic in having a grooming claw on the second pedal digit, the distribution of different nail and claw morphology and histological detail does not lend itself to an easy resolution of other potential phylogenetic relationships.

Placentation

Luckett (e.g., 1974, 1976), essentially following Hubrecht (1898), argued that *Tarsius* and anthropoids are closely related because the end product of their placental development is a hemochorial type, in contrast to the versions of epitheliochorial placentation seen in strepsirhine primates. When faced with such notable differences between *Tarsius* and anthropoids as in blastocyst attachment and amnion development, Luckett suggested that they were due to *Tarsius* retaining the primitively mammalian bicornuate uterine configuration and anthropoids developing a simplex uterus. If *Tarsius* were to have a simplex uterus, these substantial differences would vanish.

From Luckett's studies, and those he cites, the attainment of hemochorial placentation appears correlated with the following events: the establishment of a chorioallantoic placenta, which bypasses a transitory choriovitelline stage, in conjunction with the rudimentary development of an allantoic diverticulum and the precocious differentiation of a mesodermal body stalk. Differences between the diversity of mammals that achieve hemorchorial placentation lie in details of blastocyst implantation. Interestingly, not only do various nonprimate mammals (e.g., tenrecs, hedgehogs, elephant shrews, flying lemurs, and some bats) mirror anthropoids more precisely than does *Tarsius* in their course of hemochorial placental development, these mammals more faithfully reproduce the course of hemochorial placental development seen in anthropoids—and they do so in the environment of a bicornuate uterus.

Also in anthropoids and these nonprimate mammals, the blastocyst attaches by the embryonic trophoblast to the orthomesometrial pole of the uterine endometrium. *Tarsius* is distinguished from anthropoids and many other mammals in that its blastocyst attaches by the paraembryonic trophoblast to the mesometrial wall of the uterine endometrium. Although Luckett (1993) continues to dismiss alternative interpretations of the placental data and to maintain that *Tarsius* would develop hemochorial placentation in the same manner as anthropoids if it had a simplex uterus, the demonstration of homology between *Tarsius* and anthropoids in development of hemochorial placentation is certainly not self-evident (also see Cartmill, 1994). If, however, as Martin (e.g., 1968, 1990) has maintained, hemochorial placentation is primitive for primates, the issue is moot—but one must still be specific with regard to developmental details.

The Brain

The brain of *Tarsius* presents a series of contradictory comparisons (Stephan, 1984). It is relatively smooth and ungrooved, as is typical of small, nocturnal prosimians (e.g., *Microcebus murinus, Cheirogaleus medius,* and *Galago demidovii*), and the commissural system and simple cerebellum are (primitively?) reminiscent of insectivores. Oddly, the temporal and occipital lobes are larger than the frontal lobe. As otherwise seen in birds, *Tarsius*'s brain is excavated rostrally and its olfactory nerves, as measured between the olfactory mucosa and the secondary olfactory cortices, are long and thin. *Tarsius*'s olfactory peduncles, which course between the olfactory bulbs and the cerebral hemispheres, are short and broad and, as in lemurs, the ventricles are obliterated. In anthropoids, the olfactory nerves are short and the olfactory peduncles long. As in some mouse lemurs, *Tarsius*'s olfactory bulbs project beyond the frontal poles of the cerebrum. In *Tarsius* and all other prosimians, the inner face of the occipital lobe bears a triradiate arrangement of deep calcarine sulci that emanate from its center. Among primates, the cerebral hemispheres of only *Tarsius,* other prosimians, and marmosets develop a marked occipital extension. In general, it appears that the brain of *Tarsius* is most like that of other prosimians, but whether this reflects phylogenetic affinity or primitive retention from the common primate ancestor is unclear. Stephan (1984) commented that, for the most part, whereas *Tarsius* differs from prosimians in brain morphology, it also deviates from the configuration seen in anthropoids.

Miscellany

In a study on vitamin C biosynthesis, *Tarsius* and the anthropoids in the sample emerged as being nonsynthesizers, whereas the lemurs and lorises analyzed were synthesizers (Pollock and Mullin, 1987). Synthesizing vitamin C is apparently the primitive condition among mammals. Pollock and Mullin point out, however, not only that more data should be collected before taking these results as unequivocal, but that, among New World hystricomorph rodents—which systematists accept as constituting a monophyletic group—the guinea pig stands out autapomorphically as the only nonsynthesizer of vitamin C.

Fossil *Tarsius* and *Tarsius*-like Relatives

Tarsius eocaenus (middle Eocene, PRC) is represented by isolated teeth (two M_3 s, two $M_{1\ or\ 2}$ s, and a P^3) (Beard et al., 1994) and diagnosed in comparison with omomyids (broadly construed) and *Afrotarsius*. Living Tarsius and the fossil taxon are similar in having paraconids on M_{1-3} and an extended but narrow hypoconulid on the M_3 (pers. obs.). In stark contrast with *Tarsius*, however, *T. eocaenus* lacks distinct M_{1-2} hypoconulids, its M_{1-2} cristids obliquae course to the metaconid, the upper premolar is dominated by the paracone, and all teeth bear cingula/-ids (pers. obs.). Further, in *Tarsius*, the M_{1-3} are somewhat compressed and melded at their bases, forming a sheer wall to the talonid. *Tarsius*'s M_3 trigonid and talonid are almost at the same level, and not very disparate in height on M_{1-2}. In *T. eocaenus*, the metaconid and protoconid are bulbous and melded only on M_3, and the trigonids tower over the talonids on M_{1-3}. In *Tarsius*, the paraconid is bulbous, vertically oriented, and more centrally than lingually situated; it maintains the same relative distance from the metaconid on all molars and is connected to the protoconid by a complete and arcuate paracristid. In *T. eocaenus* the paraconid is more compressed, not connected to the protoconid, lingually placed, and markedly inclined forward; it is also closer to the metaconid on M_3 than on M_{1-2}. In *Tarsius*, the protoconid and metaconid are subequal in height. In the fossil M_3, at least, the protoconid is the taller of the two cusps. In *Tarsius*, the hypoconid is angular and lies slightly mesial to the entoconid; both cusps are well separated from the trigonid. In *T. eocaenus*, the hypoconid isrounded and lies quite distal to the very mesially placed entoconid, which extends essentially to the base of the metaconid. Lower molar buccal cingulids are thick in *Tarsius*, but only moderately developed in *T. eocaenus*. In *Tarsius*, the upper penultimate premolar paracone is centrally placed and compressed and it bears pre- and postparacristae. In the fossil P^3, the paracone is lingually placed and not compressed; the tooth is distended distally by a distinct metacone swelling. Since the differences between these fossil teeth far exceed the differences between extant species of *Tarsius*, it seems premature to assign the former to a species of this genus. *T. eocaenus* differs from all traditionally recognized fossil tarsiiforms, but is similar to *Afrotarsius*, in having well-separated paraconids and metaconids on all lower molars.

   An isolated right lower molar from the early Miocene of Thailand referred to the species *Tarsius thailandica* (Ginsburg and Mein, 1987) appears to represent an extinct species of this genus. It differs from the extant species in that its paraconid is lingually placed and its hypoconid lies opposite the entoconid, which, in turn, lies close to the metaconid.

Xanthorhysis tabruni (middle Eocene, PRC) (Beard, 1998) is reminiscent of *Tarsius* in dental morphology. The type and only specimen consists of a partial left mandibular corpus lacking the ramus but including much of the symphysis. It retains two posterior premolars, M_{1-3}, and two anterior single alveoli, with the mesial one being large and the other moderate in size. Discrepancies between the published description and the specimen require a review of the relevant morphologies.

Xanthorhysis is similar to *Tarsius* in having a slender corpus. Both have (1) a distinct, somewhat bulbous and vertically oriented paraconid on M_{1-3} that lies more centrally than lingually, (2) relatively tall, pointed protoconids on the distal two premolars, (3) a strong, mesially directed crest running down the cusp on the ultimate premolar whose relatively short talonid basin is enclosed, (4) cingulids ringing the premolars and coursing the length of the buccal sides of the molars, (5) entoconids on all molars that lie well behind the metaconid, (6) molar protoconids and metaconids that are melded at their bases and form a steep wall upon the talonid, (7) relatively broad M_{1-2} talonids enclosed by cresting systems, and (8) cristids obliquae that course to the metaconids on all molars.

Xanthorhysis differs from *Tarsius* in that the hypoconid on all molars is not as angular and it lies just distal to the entoconid. The paracristid diminishes in prominence in the series M_{1-3}, and the penultimate premolar is relatively larger with a more prominent heel. The last premolar is much taller and somewhat compressed, and M_{1-2} possess a low and flat hypoconulid. Also, the M_3 trigonid stands well above the level of the talonid, and, despite the fact that the M_3 is distended into a centrally placed heel, this tooth is much smaller, particularly in overall length and talonid width. Among the published Chinese specimens, *Xanthorhysis* is certainly the most similar to *Tarsius* with the best clues to possible relatedness seen in the molar trigonids (especially the nature of the paraconid) and the last lower premolar.

Afrotarsius chatrathi (early Oligocene, the Fayum, Egypt) has been interpreted as being perhaps closely related to *Tarsius* (Simons and Bown, 1985). The fossil resembles *Tarsius* mostly in primitive features (e.g., distinct paraconids on all molars), although its lower molar metaconids and protoconids are tall and melded at their bases (forming a steep wall facing the talonid basin) and its paraconids are low set. *Afrotarsius* differs from *Tarsius* in (1) having trigonids that are very tiny relative to talonid breadth and width, (2) talonid basins that are broad and deeply but smoothly excavated, (3) deep and relatively broad notches between protoconid and hypoconid and metaconid and entoconid, and (4) an M_3 that is smaller than M_1 and

M_2. Kay and Williams (1994) suggested that possession of a buccally oriented cristid obliqua is synapomorphic of *Afrotarsius, Tarsius,* and anthropoid primates, but the widespread distribution of this feature among fossil and living primates (cf. Schwartz, 1986; Schwartz and Tattersall, 1985) indicates that it is symplesiomorphic.

Fossils traditionally identified as tarsiiforms were grouped with *Tarsius* not only because of general similarities in molar morphology, but also because the skulls of three fossils in particular—*Pseudoloris, Necrolemur, Tetonius*—were reminiscent of *Tarsius*'s in general dental arcade and cranial shape, as was the auditory region of the skull of *Necrolemur* (Gregory, 1922; Simpson, 1940; see review by Schwartz, 1984). Through increasingly wider spheres of comparison between these fossils and others, *Tarsius* accrued numerous extinct relatives whose dental primitiveness was adapted to theories of anthropoid origins. Generally overlooked, however, is the fact that those who supported a tarsiiform assemblage often also pointed to the many details of the dentition as well as of the orbital, auditory, and mastoid regions in which the relevant fossils differed significantly from their presumed, extant relative. In addition, various paleontologists (e.g., Gregory, 1922) noted that the fossils could be compared equally well (e.g., in rostral elongation) or even more accurately (e.g., petromastoid inflation) with *Galago*.

According to Beard and MacPhee (1994) and Beard et al. (1992), cranial morphology indicates that *Shoshonius* is more closely related to *Tarsius* than other traditionally recognized tarsiiform or omomyid taxa. Although Ross (1994) argued that this conclusion is based on questionable homology and determination of character polarity, as well as a posteriori weighting of characters, there are at least some features that need to be reconciled. In light of errors in Beard et al.'s (1991) identifications and descriptions of various cranial landmarks, the discussion below relies on personal observations that both provide new information and corroborate the corrections provided by Beard and MacPhee (1994).

Beard and MacPhee (1994) united *Shoshonius* with *Tarsius* on the basis of (1) development of a basioccipital flange that overlaps the medial aspect of the auditory bulla, (2) an extremely narrow central stem (defined as the width of the basisphenoid and basioccipital bones between bullae), (3) an alisphenoid flange with bullar overlap (defined as the extension of the lateral pterygoid plate onto the bulla), (4) narrow and peaked choanae, and (5) reduction of the snout.

Beard and MacPhee (1994, p. 82) defined snout length as snout extension "beyond the alveolar border of the anteriormost teeth" and judged it to be apomorphically short in *Shoshonius, Tarsius,* and anthropoids because they thought it was plesiomorphically long in *Galago*. A broad comparative survey of primates reveals, however, that extension of the snout beyond the alveolar region of the anteriormost teeth is infrequently encountered, characterizing only *Arctocebus, Loris, Allocebus* (slight), *Phaner* and, to varying degrees, galagids, but not most prosimians or anthropoids (see illustrations in Schwartz and Tattersall, 1985). Thus, as defined by Beard and MacPhee, *Tarsius's* reduced snout is plesiomorphic.

With regard to lateral pterygoid plate-bullar contact (the alisphenoid flange), the common condition in mammals is lack of contact. Among extant primates, only prosimians develop a contact between the lateral pterygoid plate and the anterior or anterolateral portion of the bulla (see, for example, illustrations in Gregory, 1922; Saban, 1963; Schwartz and Tattersall, 1985). In indriids, the contact is broad and complete, with the added apomorphy of appression both superiorly and inferiorly of the short medial pterygoid plate to the lateral plate such that a narrowly ovoid funnel-shaped structure is formed. A shallow and much more compressed fossa between the pterygoid plates is found in *Daubentonia, Hapalemur, Lepilemur, Varecia, Lemur,* and *Eulemur,* and contact between the lateral pterygoid plate and the bulla may be long anteroposteriorly and complete vertically, or interrupted by a foramen of sometimes large size that is functionally a continuation of the foramen ovale (through which the mandibular branch of the trigeminal nerve courses).

In galagids and more so in lorisids, the contact between the lateral plate and the bulla can be relatively extensive. However, because the foramen ovale in these taxa is situated in line with the posterior end of the lateral pterygoid plate, and thus more medially than in most primates, lateral pterygoid plate-bullar contact may be tenuous in some specimens (e.g., as Beard and MacPhee described for *Galago*). Since the articular fossa is more medially situated in *Loris* and *Arctocebus* than it generally is in primates, it may intervene between the lateral pterygoid plate and the bulla and further disrupt the bridge between these two structures.

In subfossil Malagasy prosimians, substantial contact between the lateral pterygoid plate and the bulla is preserved in *Pachylemur, Mesopropithecus,* and *Hadropithecus,* as well as in the significantly airorhynchous *Megaladapis.* In palaeopropithecines, some contact exists between the lateral pterygoid plate and the flattened bulla. Contact in *Archaeolemur* is disrupted by the linear arrangement of the foramen ovale and anterior carotid foramen between the lateral plate and the inflated bulla. Of note in archaeolemurines and *Mesopropithecus* is that the medial and lateral pterygoid plates form a distinct funnel-shaped structure, as seen in extant indriids. *Adapis, Notharctus, Necrolemur, Tetonius,* and *Rooneyia* display substantial contact between the lateral pterygoid plate and the bulla (pers. obs.; see also Gregory, 1922; Saban, 1963), although the contact may not be as extensive as in *Shoshonius* (pers. obs.; cf. Beard and MacPhee, 1994).

Since lack of lateral pterygoid plate-bullar contact characterizes most mammals, contact between these structures emerges among primates as synapomorphic of *Tarsius* and other prosimians. Beard and MacPhee (1994) suggested that the configuration of incomplete contact seen in galagids is primitive for primates, with *Tarsius* and *Shoshonius* being similarly derived in their degree of contact. However, the reason galagids (and lorisids) have incomplete contact is because they possess the derived condition of a medially situated foramen ovale lying between the lateral pterygoid plate and the bulla. Consequently, within prosimians, broad contact (as seen in indriids and others) emerges as the primitive character state, while the configuration seen in galagids and lorisids is synapomorphic for them.

A final note, though. Since the broad contact seen in *Tarsius* is actually between the lateral pterygoid plate and the vastly enlarged anterior accessory chamber (rather than the bulla itself), it might not be proper to suggest equivalence of this contact with contact in other prosimians between the lateral pterygoid plate and the bulla.

Beard and MacPhee (1994) considered the development of a basioccipital flange overlapping the medial aspect of the auditory bulla a synapomorphy of *Tarsius, Shoshonius,* and *Tetonius.* This configuration is, however, present in various galagids and cheirogaleids, and is occasionally seen in lorisids. Consequently, if development of a basioccipital flange is apomorphic (rather than a feature related to small size), it points to a clade within Prosimii. Depending on how one assesses the character "narrow and peaked choanae," one can include various lorisiforms in the comparison with *Tarsius* and *Shoshonius.* Thus, none of the features Beard and MacPhee (1994) proposed as uniting either *Tarsius* and *Shoshonius* as sister taxa or, alternatively, *Tarsius, Shoshonius,* various omomyids and anthropoid primates as a group hold up under scrutiny. There is, however, a potential synapomorphy of *Tarsius* and *Shoshonius* that has not been previously pointed out.

Tarsius is unique among extant primates in having short medial pterygoid plates that coalesce to form a funnel-shaped structure (see above). In galagids and lorisids, there is some approximation of the medial pterygoid plates toward the midline of the basisphenoid, but the prevalent condition among primates is well-separated medial pterygoid plates. Thus, there is among extant primates an apparent morphocline of narrowing of the basisphenoid/basioccipital region (creating Beard and MacPhee's "central stem"). A central stem is seen in the early Paleogene *Ignacius* as well as in *Shoshonius,* which, like *Tarsius,* also fuses the medial pterygoid plates into a funnel-shaped structure (pers. obs.; Beard, pers. com.).

With regard to tympanic morphology, *Tarsius* is unique among primates in that, after the tympanic ring becomes ontogenetically internalized and fused to the inner wall of the auditory bulla, it elongates laterally beyond the bulla as a tubular extension (see above). *Shoshonius* is generally similar to other Paleogene taxa (e.g., *Necrolemur, Rooneyia, Plesiadapis, Adapis*) as well as extant lorisiforms and tupaiids in having an internalized ring that is connected to the inner bullar wall by an annular bridge (MacPhee and Cartmill, 1986). In *Shoshonius,* the bridge is much narrower than it is in *Rooneyia* and *Plesiadapis* (Beard and MacPhee, 1994; pers. obs.). In tupaiids and extant primates with an annular bridge, a gap (recessus dehiscence) separates it from the tympanic ring, whereas in *Necrolemur, Rooneyia, Adapis,* and *Plesiadapis,* the bridge between the ring and bullar wall is complete. *Ignacius* is *Tarsius*-like in fusing the ring to the inner wall of the bulla (cf. Kay et al., 1992). It cannot be determined if *Shoshonius* had a recessus dehiscence. Beard and MacPhee (1994) concluded that the configuration seen in *Plesiadapis* and other primates is primitive for primates. The broader comparison indicates, however, that either the trait "lack of an annular ring" or the trait "an annular ring with recessus dehiscence" is the primitive condition *within* a clade that is united first and foremost by a change in the configuration of the auditory bulla from its primitive position medial to the tympanic ring to its development laterally around the ring (cf. Cartmill, 1975; Schwartz, 1986). If internalization of the tympanic ring is apomorphic for extant Prosimii, then the possession of this configuration by various Eocene forms (e.g., *Adapis, Notharctus, Necrolemur, Rooneyia, Shoshonius*) unites them with this clade.

Fossil tarsiiforms compare well with both *Tarsius* and *Galago* in details of the proximal femur (Dagosto and Schmid, 1996) as well as in calcaneal and navicular elongation and distal tibiofibular fusion (see above). *Tarsius* differs from these fossils in lacking the cuboidocalcaneal facets and a "well-developed socket for the pivot" of the cuboidocalcaneal articulation that are characteristic not only of other extant primates, but also of those extinct tarsiiforms for which these postcranial elements are known (the omomyid *Hemiacodon* and the anaptomorphids *Teilhardina* and perhaps *Tetonius*) (Szalay, 1976 401; Szalay and Decker, 1974).

In its talus, *Tarsius* bears a (primitively) short, broad facet, whereas in fossil tarsiiforms, *Galago,* and other lorisiforms, the trochlear facet is long and narrow (S. Ford, pers. com.). *Tarsius*'s talus is also compressed proximodistally, as is also seen in lorisids (Schwartz, 1992). Postcranials attributed to *Shoshonius* are not as apomorphically *Tarsius*-like as they are *Galago*-like (Dagosto et al., 1999). In general, there appears to be even less postcranial synapomorphy uniting *Tarsius* alone with anthropoids or *Tarsius* and omomyids with anthropoids (e.g., Dagosto and Schmid, 1996; Covert and Williams, 1994) than there is potential cranial synapomorphy, which is quite slim (see above, and Cartmill and Kay, 1978; MacPhee and Cartmill, 1986; Ross, 1994).

Inasmuch as studies on fossil tarsiiforms have tended not to test the hypothesis of monophyly for a group consisting of omomyids, anaptomorphids, and microchoerines, or the monophyly of each of these three groups, it is appropriate to raise these questions here.

A potential synapomorphy of some fossil tarsiiforms is the development in the lower jaw of a large, semiprocumbent anterior tooth which, in bearing a longitudinal keel and a flared lateral edge with margocristid, is virtually morphologically identical to the lateral tooth of the strepsirhine/prosimian toothcomb (Schmid, 1983; Schwartz, 1980, 1984). If the large anterior lower tooth of microchoerines and various omomyids and anaptomorphids is synapomorphic for them, then, like *Tarsius*, they, too, possessed an extremely derived toothcomb, in which only the homologue of the lateral tooth of the prosimian toothcomb remained (cf. Schmid, 1983; Schwartz, 1984). Since these fossils also share other potential dental apomorphies seen in *Tarsius* and lorisiforms (Schwartz, 1984, 1986; see above), there is reason for recognizing a prosimian clade and a *Tarsius*-lorisiform-fossil tarsiiform clade within it.

Tarsiiforms that do not have this enlarged lower anterior tooth—the omomyids *Loveina*, *Washakius*, *Chumashius*, and *Asiomomys* and the anaptomorphids *Chlororhysis* [type] and *Anaptomorphus*—are similar, for example, to *Pelycodus*, *Smilodectes*, *Notharctus*, *Adapis*, *Europolemur*, *Protoadapis*, and *Pronycticebus* in retaining the (primitive) configuration of two small, spatulate, incisiform teeth with a larger, trenchant, caniniform tooth behind (or specimens at least preserve the alveoli for two small anterior teeth and a larger tooth behind) (cf. Beard and Banyue, 1991; Covert and Williams, 1991, 1994; Rasmussen et al., 1995; Schwartz, 1984, 1986; Szalay, 1976). Consequently, the phylogenetic relationships of these taxa may not lie with large anterior-toothed tarsiiforms, but, rather, with various adapids, with which they share derived upper and lower molar morphologies, such as distinct protocone folds and lingually directed cristids obliquae (Schwartz, 1984, 1986).

In this light, it is noteworthy that the first-known lower jaw of *Caenopithecus* (traditionally thought of as an adapid) bears a pair of anterior alveoli that would have housed enlarged teeth (e.g., see illustrations in Gregory, 1922, and Schwartz and Tattersall, 1985), which were later found partially preserved in another specimen (Franzen, 1994). Since, as Robinson (1968) remarked, identifying a specimen as a tarsiiform, adapid, or notharctid should not be based on size or paleobiogeography, it seems productive to reconsider the relationships of *Caenopithecus* (Schwartz, 1986). For example, in having prehypocone cristae on M^{1-2}, U-shaped upper molar protocristae, distinct upper molar lingual cingula, *Caenopithecus* is broadly similar to *Omomys*-like omomyids (cf. Schwartz, 1984, 1986; see below). More specifically, *Caenopithecus* and *Macrotarsius* are similarly more derived in having (particularly on M^{1-2}) extraordinarily broad protocristae, strong meso-styles, a series of thin cristae in the prehypocone region, and pre- and postcingula that thicken at their lingual extremities and "square up" the crown. Since *Tarsius* and extant lorisiforms also develop upper molar pre-hypocone cristae, these features may be broadly synapomorphic of a clade that includes these extant prosimians, at least some fossil tarsiiforms, and *Caenopithecus.*

The relationships of the omomyid *Dyseolemur* are difficult to assess because, uniquely among primates, all of its antemolar teeth are premolar-iform, increasing in size and some morphological complexity distally (Ras-mussen et al., 1995). Somewhat similar to *Tarsius, Dyseolemur's* anterior teeth lie one behind the other, not across the front of the jaw as in toothcombed prosimians and anthropoids (callithricids included). The anteriormost lower teeth of *Shoshonius* are not known, but the single alveoli are arranged one behind the other. If the small anterior alveoli housed teeth similar to those of *Tarsius,* this would lend some support to Beard and MacPhee's (1994) claim of a special relationship between these taxa. *Dyseolemur's* uniquely configured anteriormost lower tooth does, however, empha-size that anterior tooth morphology contradicts traditional groupings of tarsiiforms.

The removal of nontoothcombed taxa from Omomyidae and Anap-tomorphidae allows for discussion of monophyly and the potential re-lationships of these groups on the basis of consistent dental morphology throughout the jaws (Schwartz, 1984, 1986). An omomyid clade (includ-ing *Omomys, Ourayia, Macrotarsius, Mytonius, Stockia, Dyseolemur, Lushius,* and *Caenopithecus*) can be delineated on the basis, for example, of broad and parabolic upper molar protocristae, M^{1-2} with lingual cingula swollen below the protocone as well as distolingually, anterobuccally displaced prehypo-cone cristae, and anteroposteriorly compressed M^3s (Schwartz, 1986). Ex-cluded from this clade is *Macrotarsius macrorhysis* (middle Eocene, China), which Beard et al. (1994) identified on the basis of an isolated left M_1 and a right lower posterior premolar. The M_1 has an arcing paracristid, a lingually open and expansive trigonid with well-separated cusps, a rela-tively vertical buccal side, a hypoconid and entoconid that rise above the cristids, and a centrally placed distal flexure of the talonid, rather a distinct hypoconulid.

The M_1 of the type specimen of *Macrotarsius* (*M. montanus,* Carnegie Mu-seum of Natural History #9592), however, has a stout and straight para-cristid, a tightly compact and relatively small trigonid with closely appressed cusps, a cristid that closes off the trigonid lingually, a generally bulbous crown base with notable buccal selling, talonid cusps that are incorporated into the cristids, and a distinct hypoconulid that is situated toward the hypoconid. Similarities between the two *"Macrotarsius"* essentially lie in their being larger than most omomyids and in their M_1s having a cristid obliqua that courses to the base of the protoconid and a buccal cingulid, notably around the trigonid. The Chinese lower premolar is dominated by the protoconid, on the sides of which lie a moderately low metaconid and a very low paraconid; these three cusps subtend a steep and lingually oriented trigonid basin. In *M. montanus,* the trigonid cusps of the last lower premolar are subequal in size and height and subtend a tall and level trigonid basin. In having somewhat bulbous cusps, a not very towering trigonid, and a talonid that is longer and wider than the trigonid, the Chinese M_1 is morphologically primate (see review in Schwartz, 1986). It is premature to speculate which and how many taxa these isolated teeth represent.

An anaptomorphid clade (including *Tetonius, Absarokius, Pseudotetonius, Utahia, Anemorhysis, Arapahovius, Trogolemur, Aycrossia, Gazinius, Strigorhysis, Uintanius, Steinius,* and *Altanius*) could be distinguished by the possession of, for example, an M_1 cristid obliqua that arcs toward the metaconid and a definitive M^1 protocone fold (Schwartz, 1984), but these features also characterize the microchoerines *Nannopithex, Necrolemur,* and *Microchoerus* and the presumed omomyid, *Hemiacodon,* thereby suggesting that all of these taxa form a clade (Schwartz, 1984, 1986). The latter three taxa are further derived in having some enamel crenulation and M_2 cristid obliqua that arcs to the metaconid, and *Necrolemur* and *Microchoerus* are set apart by their marked enamel crenulation and quadrate M^{1-2}. In its lack of all lower molar paraconids, *Pseudoloris* emerges as the most derived large anterior-toothed fossil tarsiiform.

To summarize, if the large lower anterior tooth of microchoerines and various omomyids and anaptomorphids is synapomorphic for them, then we can describe these taxa as having an extremely derived toothcomb (cf. Schmid, 1983; Schwartz, 1984). As in *Tarsius,* this toothcomb would consist of only one pair of teeth, which presents itself as the morphological homologue of the pair of lateral teeth of the prosimian toothcomb. If *Shoshonius* had the derived prosimian auditory configuration of an internalized tympanic ring, it was perhaps derived within that clade in lacking anterior, toothcomb teeth altogether, as is the case in *Dyseolemur.*

Attention to tooth morphology in systematic endeavors cannot be overemphasized (e.g., Schwartz, 1980). Although there has been a long tradition in mammalian paleontology of identifying teeth by their positions and occlusal relationships, with crown shape having limitless malleability (e.g., Butler, 1974), studies on the regulation of tooth formation suggest otherwise (e.g., see Vaahtokari et al 1996; Thomas et al., 1997; Ferguson et al., 1998; Jernvall et al., 1998; Mitsiadis et al., 1998; Tucker et al., 1998; Pispa et al., 1999; Jernvall, 2000; Jernvall and Thesleff, 2000). A case in point involves the *Barx-1* gene, which is normally active posteriorly in the jaws as part of a cascade of molecular communication that produces molariform teeth. When *Barx-1* is experimentally activated in the presumptive incisor region of mice, true molar teeth develop. These teeth are not incisors that have been transmuted to look like molars. Rather, they are molars because they derive from a specific sequence of regulatory molecule communication that produces molariform teeth. Thus, wherever in the jaw a molarlike tooth is found, it is a molar, both genetically and developmentally. By extension, teeth that look like incisors, canines, and premolars have the morphologies they do because of regulatory constraints and should be identified as the teeth they look like.

In addition to identifying teeth by their morphology, we should rid ourselves of the notion that the only teeth a mammal can have are incisor, canine, and molar-class teeth (the latter including molars and their successors, premolars). Clearly, if morphology is our guide to homology, then oddly shaped teeth that do not conform to these long-standing expectations (such as the so-called lower incisors of *Eosimias* [see below]) should be regarded as representing different tooth classes. Consequently, we should acknowledge that comparability in lower anterior tooth morphology between *Tarsius* and various tarsiiforms, and the extension of this comparability to the lateral toothcomb tooth of all non-*Daubentonia* prosimians, is of definite phylogenetic potential in revealing not only homology, but also synapomorphy.

Fossils, *Tarsius*, and Anthropoid Origins

Since a consideration of *Tarsius*'s phylogenetic relationships cannot be divorced from the question, "What unites Anthropoidea?" it is necessary to discuss briefly specimens that have been offered up as potential anthropoid ancestors, such as *Eosimias* (Beard et al., 1994; MacPhee et al., 1995; Beard et al., 1996; Gebo et al., 2000), *Siamopithecus* (Chaimanee et al., 1997; Ducrocq, 1998), *Wailekia* (Ducrocq et al., 1995), and *Bahinia*, which Jaeger et al. (1999) consider an eosimiid. Embedded in the presentations of these specimens as "basal" or ancestral anthropoids is, however, the assumption that they are primates.

Beginning with Simpson's (1940) classic studies on early fossil primates, these mammals have been identified by their possession of cheek teeth with relatively low and bulbous cusps, upper molars (especially M^{1-2}) with somewhat straightened mesial and distal sides, lower molar trigonids that do not tower markedly above the talonids, lower molar paraconids that more closely approximate the metaconid from M_{1-3}, and lower molar talonids that are at least as long mesiodistally and wide buccolingually as the trigonids (see Schwartz, 1986, for review); (extant as well as fossil prosimians and anthropoids have different derived representations of these character states).

Although the type specimens *Eosimias sinensis* and *E. centennicus* do not represent the same genus morphologically, neither species is dentally primate. Both have tall and dominant trigonids on all preserved lower molars (M_{1-2} for *E. sinensis,* and M_{1-3} for *E. centennicus*). The nonprimate nature of the lower molars is even more pronounced in *E. centennicus* in which, throughout the series M_{1-3}, the trigonid remains large with fully expressed cusps (including a well-developed paraconid that stands apart from the metaconid on all three molars) while the talonid becomes buccolingually markedly narrower than the trigonid. Isolated upper teeth referred to *E.* cf. *centennicus* (Tong, 1997; pers. obs.) are not blatantly primatelike. The premolars and molar bear very tall, pointed cusps, the buccal cusps of the molar are oriented distally away from the protocone, and the distal side of the molar is deeply waisted. Thus, in light of the suggestion that there is a *Tarsius* and omomyid clade that is related to an *Eosimias* and anthropoid clade (Beard et al., 1994, 1996; also Beard and MacPhee, 1994, and MacPhee et al., 1995), the upper and lower molars of *Eosimias* are certainly more primitive than in *Tarsius* or any omomyid, much less any anthropoid primate.

The configuration of the two anteriormost teeth of *E. centennicus* are of particular note. The first tooth is noticeably smaller than the second, but the two teeth form a size-shape gradient in that their relatively small crowns bear pointed tips that recurve posteriorly and somewhat concave and vertical lingual surfaces that are completely ringed by a margocristid (cf. Beard et al., 1996). With the essentially completely preserved mandibular symphysis of *E. centennicus* properly oriented anteroposteriorly (not obliquely as in the published illustration), the lower jaw assumes a V-shaped configuration, and the linear arrangement of the two anterior teeth, one in front of the other, is immediately appreciated. No known primate has lower anterior teeth that are even vaguely morphologically similar to, or in the same relative positions as those of *E. centennicus.* Thus, outgroup comparison indicates that, even if *Eosimias* were a primate, it would be more derived than any anthropoid in lower anterior tooth design and thus excluded from ancestry of this group.

Inasmuch as *Eosimias*'s upper and lower molars do not conform to the basic criteria of being primate, to what mammalian group might this taxon be related? Given the fact that paleontology in China has only scraped the surface of potential mammalian diversity, this taxon or taxa could represent endemic clades that left no living descendants. The curious lower anterior dentition aside, the general configurations of the lower molars of *Eosimias* (as well as of the only known lower molar of *Bahinia,* an M_1) are very similar to tupaiids and talpids, both of which are found in Asia.

Bahinia differs from fossil and living primates in its preserved upper dentition in having expanded buccal and lingual moieties (with a waisted midsection in between), with tall, crest-connected, buccolingually compressed paracones and metacones that tilt in toward the well-excavated trigon basins, very mesially situated protocones that are compressed and incorporated into the arced protocristae that course to the buccal cusps, and massively ledge like lingual cingula. In addition, the apparent upper canine is disproportionately massive at its base and would probably have been quite long and trenchant. Although the upper molars of *Bahinia* are not as fully dilambdomorphic as they are in some tupaiids and talpids, the skewing of the protocone and basic configuration of the buccal cusps are consistent with the essential elements of dilambdomorphy. Importantly, the upper molars of *Bahinia* (as is the upper molar referred to *Eosimias*) are not euthemorphic, which excludes them from being considered primate (cf. Hershkovitz, 1977). As for lower teeth, even though only M_1 is known and the hypoconid region is somewhat damaged, in contrast to primate teeth, the talonid is rather narrow relative to the trigonid, and the trigonid dominates the crown.

Whatever similarities *Eosimias* and *Bahinia* share with anthropoids, they are not synapomorphies. In addition, the supposedly anthropoidlike individual features cited for *Eosimias* are themselves ambiguous. For instance, although *Eosimias* has been described as having an "anteroposteriorly abbreviated, dorsoventrally deep symphysis" (cf. Beard et al., 1996: 84), its symphysis is angled forward and not fully vertical, and, by this criterion, those New World monkeys with more primitively configured symphyseal regions would be excluded from Anthropoidea while some prosimians would be more anthropoid than *Eosimias*. Having "large and projecting [lower] canines" (Beard et al., 1996) (if this is what these teeth really are in *Eosimias*) also describes Eocene prosimians, whose lower canines are often more projecting than most anthropoids'. *Eosimias*'s vertically implanted lower incisors (Beard et al., 1996) (if this is what they are) is also not indicative of a basal anthropoid. The lower anterior teeth of most anthropoids are procumbent to some degree while the those of other mammals (carnivores, for example) are much more orthally oriented. It is rare indeed to find an anthropoid with lower anterior teeth as vertically implanted as in *Eosimias*.

Eosimias's single-rooted first lower premolar may have been the smallest of the three teeth identified as premolars (Beard et al., 1994), but single rootedness does not universally describe the lower anterior premolar of New World monkeys, in whom this tooth is typically the largest of the three premolars (Hershkovitz, 1977). *Neosaimiri*'s last and penultimate lower premolars may be similar to *Eosimias'* in being "slightly exodaenodont and obliquely oriented in the tooth row" (Beard et al., 1996: 84), but in most anthropoids (New World monkeys included) the ultimate and sometimes also penultimate premolars are oriented with one root buccally and the other lingually. Clearly, singling out and representing *Neosaimiri* as a typical anthropoid is inappropriate.

It is also incorrect to claim that *Eosimias* is a basal anthropoid because it and *Neosaimiri* have large lower molar protoconids and mesially placed entoconids (cf. Beard et al., 1994). This description does not even apply to all lower molars attributed to *Eosimias*. Nevertheless, even if this description were correct, these features do not characterize the molars of anthropoids as a group.

They do, however, describe the molars of tupaiids and various prosimians (including *Macrotarsius montanus,* but not *"M." macrorhysis*), as well as the isolated molars attributed to *"Tarsius" eocaenus*. The development of a premetacristid, as seen in *Eosimias* (Beard et al., 1994), is commonplace among fossil and extant primates as well as various insectivores and scandentians.

With regard to the remaining features that Beard et al. (1994, 1996) offer as synapomorphies of *Eosimias* and anthropoids, the observation that the "M_3 trigonid . . . [is] . . . appreciably wider than the talonid" (Beard et al., 1996: 84) cannot be taken out of context. This configuration characterizes all lower molars of *Eosimias* and thus excludes this taxon from the primate clade (see above). As for "hypoconulid lobe on M_3 being reduced both mesiodistally and buccolingually" (ibid.), this is a restatement of the preceding feature. No anthropoid or other primate has such a truncated M_3 talonid, although various tupaiids do. Finally, *Eosimias* may be similar to anthropoids in having "a rounded, nonprojecting angular region providing expanded area for insertion of pterygoid muscles" (ibid.), but so are fossil and extant indriids, *Adapis* and related species, lorisids, *Hapalemur, Daubentonia,* the plesiadapiforms *Platychoerops* and *Chiromyoides,* and numerous perissodactyls and artiodactyls.

Given the dubious association of eosimiids with primates in general (much less with a specific primate subclade), it is appropriate to reconsider claims that a petrosal fragment (MacPhee et al., 1995) and isolated tarsal bones (Gebo et al., 2000) discovered in the same fissure fill as *Eosimias* belong to that taxon and, therefore, that these specimens reflect the primitive anthropoid configurations of the bullar and ankle regions. With regard to the petrosal, MacPhee et al. (1995) argued that it must have come from an eosimiid because it could not be anything else. It is much smaller than expected of the omomyid from that locality (the supposed *Macrotarsius*), and it is not very *Tarsius*-like. This petrosal is primatelike in having a "long, plate-like continuation of the otic capsule in the plane of the tympanic cavity" (ibid.: 508). It differs from fossil tarsiiforms (or lorisiforms) in having a relatively large stapedial artery (as is found, for example, in lemuriforms, including *Adapis* and kin), but the course of the promontory artery is most reminiscent of *Necrolemur* and *Rooneyia* (ibid.).

Although the authors state in their article's abstract that "the element does present arterial features consistent with its being haplorhine," in their text, the most favorable comparisons are with omomyids. It is only because it was assumed that this petrosal fragment came from an eosimiid, which, in turn, was assumed to be a basal anthropoid, that the otherwise ambiguous morphologies of this bony element were then interpreted as being primitively anthropoid. That is, because these general features are found in a presumed anthropoid ancestor, they must represent the configuration from which the apomorphically anthropoid condition evolved. Clearly, this is tautological.

A similar assumption underlies the interpretation of the talus and calcaneus assigned to *Eosimias* (Gebo et al., 2000). First, these postcranial elements were assumed to be eosimiid (and thus a basal anthropoid) because of their minuscule size, and then their morphologies were explained. Given that Gebo et al. (2000) offered only one feature of the talus as being potentially diagnostic of anthropoids—a reduced medial facet on its body for the tibial malleolus—it might be that the tiny talus is actually from an anthropoid primate, but one that would be more derived than *Saimiri*, at least, because, as illustrated, the facet in the eosimiid is relatively much smaller than in this New World monkey.

Since, however, Gebo et al. (2000) basically compared their specimen only to one New World monkey, *Saimiri,* and to a tarsally very derived fossil tarsiiform, *Hemiacodon,* the single shared talar feature is overshadowed by the greater number of similarities seen in the tali of their eosimiid and *Hemiacodon:* e.g., a moderate talar neck angle, moderately high talar body, shallow trochlea, small posterior trochlear shelf, and relatively narrow talar body (p. 277). Gebo et al. (2000) interpreted these features as apomorphies of their group adapiforms, but, if they are derived compared to other primates, the comparison also subsumes the tali of their eosimiid and *Hemiacodon.* Consequently, there is again cause to recognize a prosimian clade. It is unfortunate that Gebo et al. (2000) did not expand their relatively small comparative sample to include extant lemurs and lorises because this would have clarified the extent to which these apomorphies are shared more broadly among prosimians.

With regard to the eosimiid calcaneus, the limited published comparisons make the delineation of synapomorphy difficult. In relative length of the distal segment, the eosimiid calcaneus is relatively longer than in *Adapis,* shorter than omomyids, and more like *Saimiri,* while, in transverse width, it is narrower than in New World monkeys and wider than in omomyids. The calcaneocuboid joint of eosimiids is flat, as in *Notharctus* and omomyids, but it bears a nonarticular region in its medioplantar section, as in anthropoids. Although Gebo et al. (2000) state that the articular surface in their eosimiid calcaneus is similar to that of anthropoids (being represented only by *Saimiri*), their illustration suggests the better comparison is between *Saimiri* and *Hemiacodon.*

Review of the features that Gebo et al. (2000) list as synapomorphic of the haplorhine clade they reconstructed using PAUP reveals contradictions with their earlier published interpretation of character polarity. There, Beard (1988) argued that all nonstrepsirhine primates retain the primitive eutherian talar configuration of having both a vertical talofibular facet from which a small plantar process protrudes and a groove for the m. flexor hallucis longus tendon that lies plantad and central to the tibiotalar joint. In Gebo et al. (2000), these primitive retentions are used to support the grouping Haplorhini. Regardless of the interpretation of character polarity, only the talus attributed to *Hemiacodon* can be described by this list of features. The tali of *Tarsius* and those assigned to the anaptomorphids *Tetonius* and *Teilhardina* cannot (Schwartz, 1992).

In addition, the platyrrhine *Aotus* displays the presumed strepsirhine condition of a slightly laterally oriented and broad m. flexor hallucis tendon groove, but various prosimians (e.g., *Eulemur, Propithecus*) do not (cf. Beard, 1988; Schwartz, 1992). With regard to other aspects of the PAUP analysis (in which scandentian, dermopteran, and plesiadapiform tarsal morphology is taken collectively as reflecting the primitive outgroup state), adapiforms and haplorhines are presented as sister groups. If this relationship is indeed supported by synapomorphy, then the eosimiid and the omomyid (*Hemiacodon*) are subsumed in it.

What, then, of *Eosimias* and a potential eosimiid clade? Although the teeth attributed to *Eosimias* and *Bahinia* are not apomorphically primate (and thus cannot be basal anthropoids), the eosimiid petrosal fragment and postcranials appear to be. Could these elements be from the same, tiny taxon? Since MacPhee et al. (1995) and Gebo et al. (2000) point specifically to various morphological similarities between these specimens and omomyids (i.e. tarsiiforms as used here), these specimens may represent a taxon or taxa whose relationships lie close to or within this larger group. In this regard, although the few known teeth attributed to *Tarsius eocaenus* may not represent a species of the extant taxon, they are of the right size (cf. MacPhee et al., 1995; Gebo et al., 2000) to be associated with the petrosal and postcranials. More complete specimens are needed to assess these phylogenetic possibilities. With regard to the family Eosimiidae (Beard et al., 1994), since *"Eosimias" centennicus* and *Bahinia* are generally similar in dental morphology to the type specimen of the genus and species *E. sinensis,* these taxa provide a small window onto a long-lost group of extinct, nonprimate Asian mammals.

Wailekia and *Siamopithecus* (both middle-late Eocene, Thailand) are clearly dentally primate. Although Ducrocq et al. (1995) suggested that *Wailekia* (known from a partial lower jaw with the last two molars and an isolated lower molar) was an early anthropoid, there are clear dental similarities between it and the Asian sivaladapid-like *Hoanghonius* (Ducrocq, 1998; C. Beard, pers. com.; pers. obs.), including the development on the lower molars of a stout, anteriorly arcing paracristid and a distolingual entoconid-hypoconulid notch—both compelling synapomorphies. In terms of preserved mandibular morphology (e.g., relatively shallow and long corpus, thin and high-rising coronoid process, relatively lower condyle, and forwardly inclined symphysis), *Wailekia* retains the primitive mammalian configuration.

Siamopithecus is regarded as a basal anthropoid with possible links to another supposed early anthropoid, the Burmese *Pondaungia* (Chaimanee et al., 1997; Ducrocq, 1998). Presumed anthropoid features of *Siamopithecus* include "large body size; upper cheek teeth that are not medially waisted and lack conules, but have an external expansion of the paracone; very bunodont lower molars that lack a paraconid with a mesially directed cristid obliqua; and a very deep mandible" (Ducrocq, 1998: 99). Nevertheless, "some of these features are found in some prosimians, but this suite of features is more likely that of an anthropoid" (ibid.). With an estimated body mass of 6500–7000 g, *Siamopithecus* was large, but, then, so were some species of *Pelycodus, Notharctus,* and *Leptadapis* and even more so (often tenfold more) all subfossil Malagasy prosimians. Even the extant *Indri* falls into this size range. Since morphological comparisons between *Siamopithecus* and prosimians were limited to Eocene adapiforms, rather than to a greater taxonomic representation of primates, the anthropoid features of the former deserve comment.

For instance, "deep mandible" is not in and of itself diagnostic, as many fossil prosimians (including *Adapis* and *Leptadapis*) and extant prosimians also have deep mandibles. The lack of a paraconid and the possession of a mesially directed cristid obliqua may typify anthropoids, but many extant and fossil prosimians can be similarly described (Schwartz, 1984, 1986; Schwartz and Tattersall, 1985). Since only the last two molars of *Siamopithecus* are intact, and the M_1 lacks all but the distal portion of the talonid, one cannot know if the cristid obliqua on the latter tooth did not course lingually and even contact the metaconid, as is the case in *Pelycodus* and most anaptomorphids, which also have mesially directed cristids obliquae on M_{2-3}. In their general bunodonty and distinct, paraconid-less paracristids that enclose mesiodistally short but relatively deep trigonid basins, the preserved M_{2-3} of *Siamopithecus* are comparable to these teeth in *Necrolemur* and *Microchoerus*. A buccally expanded upper molar paracone is not a characteristic of extant anthropoids, but it is a feature of the upper molars of the Eocene *Moeripithecus* and *Algeripithecus,* which are presumed early anthropoids. Anaptomorphids had expanded upper molar buccal surfaces, and many of these taxa also had other characteristics noted in the upper dentition of *Siamopithecus,* such as upper molars with long lingual slopes (particularly on M^2, in which the protocone is quite centrally placed) and constricted trigon basins (also on the last upper premolar), distinct protocone folds that became confluent with the postcingulum, buccolingually exaggerated M^2s, reduced M^3s, and enlarged last premolars. In short, the preserved dental features of *Siamopithecus* display more potential synapomorphy with members of the anaptomorphid and microchoerine clade hypothesized here than wit anthropoids. Among *Siamopithecus*'s autapomorphies are large body mass and a deep mandible (which might be a consequence of large size). Since it appears that some kind or kinds of omomyids or omomyidlike primates were present in Asia during the middle-late Eocene (such as "*Macrotarsius*" *macrorhysis* and "*Tarsius*" *eocaenus,* as well as *Asiomomys* [Beard and Banyue, 1991]), it is not surprising that an anaptomorphid or two would also be sympatrically present, as is clearly the case among the North American Eocene faunas.

Conclusions

Without a doubt, *Tarsius* is an extraordinarily unique mammal. Yet, while its uniquenesses make it an intriguing primate, scientists have obscured its potential phylogenetic relationships, which are dictated by a presumed scala naturae envisioning transformation series from lower to higher primates that, somehow, include a tarsierlike phase. Taken on their own, however, *Tarsius*'s morphologies data do not easily lend themselves to any version of this scenario.

Tarsius shares with extant prosimians a grooming claw on the second pedal digit and a toothcomb (autapomorphically reduced in the former), as well as ontogenetic internalization of the ectotympanic. Among extant prosimians, *Tarsius* resembles lorisiforms in having (1) a prehypocone crista at least on M^1, (2) on M_1, a tall protoconid and metaconid that are melded at their bases, (3) an angular and buccally distended hypoconid on the lower molars, (4) some rostral elongation, (5) a downward distension of the premaxillary alveolar region, and (6) marked orbital enlargement with medial constriction (cf. Schwartz, 1984; Schwartz and Tattersall, 1985). *Tarsius* also shares with galagids and cheirogaleids some degree of tarsal elongation and at least tibiofibular syndesmosis if not total fusion.

If lorisids and galagids are sister taxa (e.g., Schwartz, 1986), then a consequence of the latter presumed synapomorphy is secondary tarsal shortening in lorisids. If *Tarsius* is related specifically to a lorisid and galagid clade to the exclusion of cheirogalids, then its lack of an ascending pharyngeal artery (cf. Cartmill, 1975) demands explanation. *Tarsius* and galagids alone share a number of femoral uniquenesses (Dagosto and Schmid, 1996), but this is contradicted by *Tarsius*'s lack of the lorisiforms' long, narrow trochlear talar facet. *Tarsius* shares with all lorisiforms reduction/loss of the stapedial artery in the adult and with lorisids and galagids development of a small tympanic cavity and lack of an annular bridge. However, these features also occur in anthropoids (MacPhee and Cartmill, 1986).

The most clear-cut similarities between *Tarsius* and anthropoids include (1) primary supply of the middle meningeal artery via the maxillary artery, (2) lack of a tapetum, (3) lack of a moist, hairless rhinarium, and (4) the inability to synthesize vitamin C. Accepting features of the auditory region, the eye, placentation, the oro-nasal region, and the skull in general as synapomorphic of *Tarsius* and anthropoids requires a posteriori explanations.

As the preceding review has hopefully illuminated, there is still much to be sorted out about the phylogenetic relationships of *Tarsius* to both extinct and extant taxa.

Acknowledgments

I thank Sharon Gursky, Elwyn Simons, and Pat Wright for inviting me to contribute to their timely effort; Colin Groves and Susan Ford for sharing unpublished data; and Chris Beard for discussing *Shoshonius* and the Chinese specimens with me and giving me full access to originals and casts. I also thank the curators and their assistants at the American Museum of Natural History (Departments of Anthropology and Mammals), New York City, National Museum of Natural History (Smithsonian Institution, Department of Mammals), Washington, D. C., and the Rijksmuseum van Natuurlijke Historie, Leiden (Department of Mammals) for allowing me to study their collections.

References

Ankel-Simons F. 2000. Primate anatomy: an introduction. San Diego: Academic Press.

Beard KC. 1988. The phylogenetic significance of strepsirrhinism in Paleogene primates. In J Primatol 9: 83–96.

Beard KC. 1998. A new genus of Tarsiidae (Mammalia: Primates) from the Middle Eocene of Shanxi Province, China, with notes on the historical biogeography of tarsiers. Bull Carnegie Mus Nat Hist 34: 260–277.

Beard KC, Banyue W. 1991. Phylogenetic and biogeographic significance of the tarsiiform primate *Asiomomys changbaicus* from the Eocene of Jilin Province, People's Republic of China. Am J Phys Anthrop 85: 159–166.

Beard KC, MacPhee RDE. 1994. Cranial anatomy of *Shoshonius* and the antiquity of Anthropoidea. In Fleagle JG, Kay RF, editors, Anthropoid origins, 55–97. New York: Plenum.

Beard KC, Krishtalka L, Stucky RK. 1992. First skulls of the early Eocene primate *Shoshonius cooperi* and the anthropoid-tarsier dichotomy. Nature 349: 64–67.

Beard CK, Tao Q, Dawson MR, Banyue W, Chuankuei L. 1994. A diverse new primate fauna from middle Eocene fissure-fillings in southeastern China. Nature 368: 604–609.

Beard CK, Yonsheng T, Dawson MR, Jingwen W, Xueshi H. 1996. Earliest complete dentition of an anthropoid primate from the late middle Eocene of Shanxi Province, China. Science 272: 82–85.

Blainville HM, Ducrotay de. 1839. Ostéographie au description iconographique comparé e du squelette et du système dentaire de cinq classes d'animaux vertèbrès rècents et fossiles pour servir de base à la zoologie et à la géologie. Mammifères. I. Primates. Paris: Arthus Bertran.

Brehm AE. 1868. La vie des animaux illustrée. Paris: Baillere.

Buffon GL, Comte de. 1765. Histoire naturelle, générale et particulière. Paris: L'Imprimerie du Roi.

Butler PM. 1974. Molar cusp nomenclature and homology. In Butler PM, Joysey KA, editors, Development, function and evolution of teeth, 439–453. New York: Academic Press.

Camel GJ. 1706–8. De quadrupedibus phillipensibus. Phil Trans Lond 25: 2197.

Cartmill MC. 1975. Strepsirhine basicranial structures and the affinities of the Cheirogaleidae. In Luckett WP, Szalay FS, editors, Phylogeny of the primates: a multidisciplinary approach, 313–354. New York: Plenum Press.

Cartmill M. 1980. Morphology, function and evolution of the anthropoid postorbital septum. In Ciochon R, Chiarelli A, editors. Evolutionary biology of the New World monkeys and continental drift, 243–274. New York: Academic Press.

Cartmill M. 1978. The orbital mosaic in prosimians and the use of variable traits in systematics. Folia Primatol 30: 89–114.

Cartmill M. 1994. Anatomy, antinomies, and the problem of anthropoid origins. In Fleagle JG, Kay RF, editors, Anthropoid origins, 549–566. New York: Plenum Press.

Cartmill M, Kay RF. 1978. Craniodental morphology, tarsier affinities, and primate suborders. In Chivers DJ, Joysey KA, editors, Recent advances in primatology, vol. 3, 204–214. London: Academic Press.

Cartmill M, MacPhee RDE, Simons EL. 1981. Anatomy of the temporal bone in early anthropoids, with remarks on the problem of anthropoid origins. Am J Phys Anthro 56: 3–22.

Castenholz A. 1984. The eye of *Tarsius*. In Niemitz C, editor, Biology of tarsiers, 303–317. Stuttgart: Gustav-Fischer-Verlag.

Chaimanee Y, Suteethorn V, Jaeger J-J, Ducrocq S. 1997. A new late Eocene anthropoid primate from Thailand. Nature 385: 429–431.

Covert HH, Williams BA. 1991. The anterior lower dentition of *Washakius insignis* and adapid-anthropoid affinities. J Hum Evol 21: 463–467.

Covert HH, Williams BA. 1994. Recently recovered specimens of North American Eocene omomyids and adapids and their bearing on debates about anthropoid origins. In Fleagle JG, Kay RF, editors, Anthropoid origins, 29–54. New York: Plenum Press.

Dagosto M, Schmid P. 1996. Proximal femoral anatomy of omomyiform primates. J Hum Evol 30: 29–56.

Dagosto M, Gebo DL, Beard KC. 1999. Revision of the Wind River faunas, early Eocene of central Wyoming. Part 14. Postcranium of *Shoshonius cooperi* (Mammalia: Primates). Ann Carnegie Mus 68: 175–211.

Delantey M, Ross C. 2000. The phylogenetic position of *Tarsius:* a total evidence approach (abstract). J Vert Paleonto 20 (suppl. to no. 3): 38A.

Ducrocq S. 1998. Eocene primates from Thailand: are Asian anthropoideans related to African ones? Evol Anthrop 7: 97–104.

Ducrocq S, Jaeger J-J, Chaimanee Y, Suteethorn V. 1995. New primate from the Palaeogene of Thailand, and the biogeographical origin of anthropoids. J Hum Evol 28: 477–485.

Erxleben JCP. 1777. Systema regni animalis. Leipzig: Impenesis Weygandianis.

Ferguson CA, Tucker AS, Christensen L, Lau AL, Matzuk MM, Sharpe PT. 1998. Activin is an essential early mesenchymal signal in tooth development that is required for patterning of the murine dentition. Genes Devel 12: 2636–2649.

Fitzinger LJ. 1861. Wissenschaftlich-populäre Naturgeschichte der Säugethiere. Wien: IK und K Hof- und Staatsdruckerei.

Fleagle JG, Kay RF, editors. 1994. Anthropoid origins. New York: Plenum Press.

Flower WH. 1883. On the arrangement of the orders and families of existing Mammalia. Proc Zool Soc Lond 1883: 178–186.

Franzen JL. 1994. The Messel primates and anthropoid origins. In Fleagle JG, Kay RF, editors, Anthropoid origins, 99–122. New York: Plenum Press.

Gebo DL, Dagosto M, Beard KC, Tao Q, Jingwen W. 2000. The oldest known anthropoid postcranial fossils and the early evolution of higher primates. Nature 404: 276–278.

Geoffroy-Saint Hilaire E. 1812. Suite au tableau des quadrumanes. Ann Mus Hist Natl Paris 19: 156–170.

Geoffroy-Saint Hilaire E, Cuvier G. 1795. Memoire sur les rapports naturels du Tarsier (*Didelphis macrotarsus* Gm.), lu à la Société d'Histoire Naturelle, le 21 Messidor an III.

Ginsburg L, Mein P. 1987. *Tarsius thailandica* nov. sp., Tarsiidae (Primates, Mammalia) fossile d'Asie. C R Acad Sci (Paris) 304: 1213–1215.

Gray JE. 1870. Catalogue of monkeys, lemurs, and fruit-eating bats in the collection of the British Museum. Brit Mus (Nat Hist) Lond 1870: 1–137.

Gregory WK. 1922. The origin and evolution of the human dentition. Baltimore: Williams and Wilkins.

Hershkovitz, P. 1977. Living New World monkeys (Platyrrhini), with an introduction to the Primates, vol. 1. Chicago: University of Chicago Press.

Hill WCO. 1953. Primates, comparative anatomy and taxonomy, vol. I: Strepsirrhini. Edinburgh, UK: Edinburgh University Press.

Hill WCO. 1955. Primates, comparative anatomy and taxonomy, vol. II: Haplorhini: Tarsioidea. Edinburgh, UK: Edinburgh University Press.

Hofer HO. 1976. Preliminary study of the comparative anatomy of the external nose of South American monkeys. Folia Primatol 25: 193–214.

Hofer HO. 1977. The anatomical relations of the ductus vomeronasalis and the occurrence of taste buds in the papilla palatina of *Nycticebus coucang* (Primates, Prosimiae) with remarks on strepsirhinism. Geg Morph Jahrb Leipzig 123: 836–856.

Hofer HO. 1979. The external nose of *Tarsius bancanus borneanus* Horsfield 1821 (Primates, Tarsiiformes). Folia Primatol 33: 180–192.

Hofer HO. 1980. The external anatomy of the oro-nasal region of primates. Z Morph Anthro 71: 233–249.

Horsfield T. 1821. Zoological researches in Java. London: Black, Kingsbury, Parbury, Allen.

Howell AB. 1944. Speed in animals (1965 facsimile of 1944 edition). New York: Hafner.

Hubrecht AAW. 1898. Über die Entwicklung der Placenta von *Tarsius* und *Tupaia,* nebst Bemerkungen über deren Bedeutung als haematopoietische Organe. 4th Int Congr Zool Cambridge 1898: 343–411.

Illiger C. 1811. Prodromus systematis mammalium et avium additis terminis zoographicis utriudque classis. Berlin: C. Salfeld.

Izard MK, Wright PC, Simons EL. 1985. Gestation length in *Tarsius bancanus.* Am J Primatol 9: 327–331.

Jablonski NG, Crompton RH. 1994. Feeding behavior, mastication, and tooth wear in the western tarsier (*Tarsius bancanus*). Int J Primatol 15: 29–59.

Jaeger J-J, Thein T, Benammi M, Chaimanee Y, Naing Soe A, Lwin T, Tun T, Wai S, Ducrocq S. 1999. A new primate from the Middle Eocene of Myanmar and the Asian early origin of anthropoids. Science 286: 528–530.

Jernvall J. 2000. Linking development with generation of novelty in mammalian teeth. Proc Natl Acad Sci USA 97: 2641–2645.

Jernvall J, Thesleff I. 2000. Reiterative signaling and patterning during mammalian tooth morphogenesis. Mech Devel 92: 19–29.

Jernvall J, Aberg T, Kettunen P, Keranen S, Thesleff I. 1998. The life history of an embryonic signaling center: BMP-4 induces *p21* and is associated with apoptosis in the mouse tooth enamel knot. Development 125: 161–169.

Jones F Wood. 1920. Discussion on the zoological position and affinities of *Tarsius.* Proc Zool Soc Lond 1920: 491–949.

Jouffroy F-K, Berge C, Niemitz C. 1984. Comparative study of the lower extremity in the genus *Tarsius*. In Niemitz C, editor, Biology of tarsiers, 167–189. Stuttgart: Gustav-Fischer-Verlag.

Kay RF, Williams BA. 1994. Dental evidence for anthropoid origins. In Fleagle JG, Kay RF, editors, Anthropoid origins, 361–445. New York: Plenum Press.

Kay, RF, Thewissen JGM, Yoder, AD. 1992. Cranial anatomy of Ignacius-Graybullianus and the affinities of the plesiadapi forms. Am J Phys Anthropol 89: 477–498.

Klauer G. 1984. The macroscopial and microscopial anatomy of the external nose in *Tarsius bancanus*. In Niemitz C, editor, Biology of tarsiers, 291–301. Stuttgart: Gustav-Fischer-Verlag.

Kolmer W. 1930. Zur Kenntnis des Auges der Primaten. Anat Entwickl 93: 679–722.

Linnaeus C. 1758. Systema naturae per regna tria naturae, secundum classes, ordines, genera, species cum characteribus, differentiis, synonymis, locis. Editio decima, reformata. Stockholm: Laurentii Salvii.

Luckett WP. 1974. Comparative development and evolution of the placenta in primates. Contr Primatol 3: 142–234.

Luckett WP. 1976. Cladistic relationships among primate higher categories: evidence of the fetal membranes and placenta. Folia Primatol 25: 245–276.

Luckett WP. 1993. Developmental evidence from the fetal membranes for assessing archontan relationships. In MacPhee RDE, editor, Primates and their relatives in phylogenetic perspective, 149–186. New York: Plenum Press.

MacPhee RDE, Cartmill MC. 1986. Basicranial structures and primate systematics. In Swindler DR, Erwin J, editors, Comparative primate biology, vol. 1: Systematics, evolution, and anatomy, 219–275. New York: Alan R. Liss.

MacPhee RDE, Beard KC, Tao Q. 1995. Significance of primate petrosal from Middle Eocene fissure-fillings at Shanghuang, Jiangsu Province, People's Republic of China. J Hum Evol 29: 501–514.

Maier W. 1981. Nasal structures in Old and New World primates. In Ciochon RL, Chiarelli AB, editors, Evolutionary biology of the New World monkeys and continental drift, 219–241. New York: Plenum Press.

Martin RD. 1968. Towards a new definition of primates. Man 3: 377–401.

Martin RD. 1990. Primate origins and evolution: a phylogenetic reconstruction. London: Chapman and Hall.

McNab BK, Wright PC. 1987. Temperature regulation and oxygen consumption in the Philippine tarsier, *Tarsius syrichta*. Physiol Zool 60: 596–600.

Miller GS Jr., Hollister N. 1921. Twenty new mammals collected by H. C. Raven in Celebes. Proc Biol Soc Wash 34: 67–76.

Mitsiadis TA, Mucchielli M-L, Raffo S, Proust J-P, Koopman P, Goridis C. 1998. Expression of the transcription factors *Otlx2*, *Barx1* and *Sox9* during mouse odontogenesis. Eur J Oral Sci 106 (suppl. 1): 112–116.

Mivart St. G. 1864. Notes on the crania and dentition of the Lemuridae. Proc Zool Soc Lond 1864: 611–648.

Musser GG, Dagosto M. 1987. The identity of *Tarsius pumilus*, a pygmy species endemic to the montane mossy forests of central Sulawesi. Am Mus Nov 2867: 1–53.

Niemitz C. 1979. Outline of the behavior of *Tarsius bancanus*. In Doyle CA, Martin RD, editors, The study of prosimian behavior, 631–660. New York: Academic Press.

Niemitz C. 1984a. Synecological relationships and feeding behaviour of the genus *Tarsius*. In Niemitz C, editor, Biology of tarsiers, 59–75. Stuttgart: Gustav-Fischer-Verlag.

Niemitz C. 1984b. Locomotion and posture of *Tarsius bancanus*. In Niemitz C, editor, Biology of tarsiers, 191–225. Stuttgart: Gustav-Fischer-Verlag.

Niemitz C, Klauer G, Eins S. 1984. The interscapular brown fat in *Tarsius bancanus*, with comparisons to *Tupaia* and man. In Niemitz C, editor, Biology of tarsiers, 257–273. Stuttgart: Gustav-Fischer-Verlag.

Niemitz C, Nietsch A, Warter S, Rumpler Y. 1991. *Tarsius dianae:* a new primate species from Central Sulawesi (Indonesia). Folia Primatol 56: 105–116.

Pallas PS. 1778. Novae species quad e glirium ordine cum illustrationibus variis complurium ex hoc ordine animalium. Erlangen: W Walther.

Peters A. Preuschoft H. 1984. External biomechanics of leaping in *Tarsius* and its morphological and kinematic consequences. In Niemitz C, editor, Biology of tarsiers, 225–255. Stuttgart: Gustav-Fischer-Verlag.

Pispa J, Jung H-S, Jernvall J, Kettunen P, Mustonen T, Tabata MJ, Kere J. Thesleff I. 1999. Cusp patterning defect in *Tabby* mouse teeth and its partial rescue by FGF. Devel Biol 216: 521–534.

Pocock RI. 1918. On the external characters of the lemurs and *Tarsius*. Proc Zool Soc Lond 1918: 19–53.

Pollock JI, Mullin RJ. 1987. Vitamin C biosynthesis in prosimians: evidence for the anthropoid affinity of *Tarsius*. Amer J Phys Anthro 73: 65–70.

Rasmussen DR. 1986. Anthropoid origins: a possible solution to the Adapidae-Omomyidae paradox. J Hum Evol 15: 1–12.

Rasmussen DT, Shekelle M, Walsh SL, Riney BO. 1995. The dentition of *Dyseolemur*, and comments on the use of the anterior teeth in primate systematics. J Hum Evol 29: 301–320.

Roberts M. 1994. Growth, development, and parental care in the western tarsier (*Tarsius bancanus*) in captivity: evidence for a "slow" life-history and nonmonogamous mating system. Int J Primatol 15: 1–28.

Robinson P. 1968. The paleontology and geology of the Badwater Creek area, central Wyoming. Part. 4. Late Eocene primates from Badwater, Wyoming, with a discussion of material from Utah. Ann Carnegie Mus 39: 307–326.

Rosenberger A, Szalay FS. 1980. On the tarsiiform origins of the Anthropoidea. In Ciochon RC, Chiarelli AB, editors, Evolutionary biology of the New World monkeys and continental drift, 139–157. New York: Plenum Press.

Ross C. 1994. The craniofacial evidence for anthropoid and tarsier relationships. In Fleagle JG, Kay RF, editors. Anthropoid origins, 469–547. New York: Plenum Press.

Ross C. 1995. Allometric and functional influences on primate orbit orientation and the origins of the Anthropoidea. J Hum Evol 29: 201–227.

Saban, R. 1963. Contribution à llétude de llos temporal des Primates. Mem Mus Natl Hist Nat, ser. A 29: 1–378.

Schmid P. 1983. Front dentition of the Omomyiformes (Primates). Folia Primatol 40: 1–10.

Schultz M. 1984. Osteology and myology of the upper extremity of *Tarsius*. In Niemitz C, editor, 143–165. Biology of tarsiers. Stuttgart: Gustav-Fischer-Verlag.

Schwartz JH. 1980. A discussion of dental homology with reference to primates. Am J Phys Anthro 52: 463–480.

Schwartz JH. 1984. What is a tarsier? In Eldredge N, Stanley SM, editors, Living fossils, 38–49. New York: Springer-Verlag.

Schwartz JH. 1986. Primate systematics and a classification of the order. In Swindler DR, Erwin J, editors, Comparative primate biology, vol. 1: Systematics, evolution, and anatomy, 1–41. New York: Alan R Liss.

Schwartz JH. 1992. Issues in prosimian phylogeny and systematics. In Matano S, Tuttle RH, Ishida H, Goodman M, editors, Topics in primatology, vol. 3: Evolutionary biology, reproductive endocrinology, and virology, 23–36. Tokyo: University of Tokyo Press.

Schwartz JH. 1996. *Pseudopotto martini:* a new genus and species of extant lorisiform primate. Anthro Pap Am Mus Nat Hist 78: 1–14.

Schwartz JH. In press. *Tarsius:* behavior, morphology, systematics, paleontology, and evolution. In Coppens Y, Senut B, Thomas H, editors, Primates, excluding Hominoidea. Milan: Jaca Books.

Schwartz JH, Tattersall I. 1985. Evolutionary relationships of living lemurs and lorises (Mammalia, Primates) and their potential affinities with European Eocene Adapidae. Anthro Pap Am Mus Nat Hist 60: 1–100.

Schwartz JH, Yamada TK. 1998. Carpal anatomy and primate relations of tarsiers. In Niemitz, C, editor, Biology of tarsiers, 319–343. Stuttgart: Gustav-Fischer-Verlag.

Simons EL, Bown TM. 1985. *Afrotarsius chatrathi,* first tarsiiform primate (?Tarsiidae) from Africa. Nature 313: 475–477.

Simons EL, Rasmussen DT. 1989. Cranial morphology of *Aegyptopithecus* and *Tarsius* and the question of the tarsier-anthropoidean clade. Am J Phys Anthro 79: 1–23.

Simpson GG. 1940. Studies on the earliest primates. Bull Am Mus Nat Hist 85: 185–212.

Soligo C, Müller AE. 1999. Nails and claws in primate evolution. J Hum Evol 36: 97–114.

Spearman RIC. 1985. Phylogeny of the nail. J Hum Evol 14: 57–61.

Starck D. 1975. The development of the chondrocranium in primates. In Luckett WP, Szalay FS, editors, Phylogeny of the Primates: a multidisciplinary approach, 127–155. New York: Academic Press.

Starck D. 1984. The nasal cavity and nasal skeleton of *Tarsius*. In Niemitz C, editor, Biology of tarsiers, 275–289. Stuttgart: Gustav-Fischer-Verlag.

Stephan H. 1984. Morphology of the brain in *Tarsius*. In Niemitz, C, editor, Biology of tarsiers, 319–343. Stuttgart: Gustav-Fischer-Verlag.

Storr GC. 1780. Prodromus methodi mammalium. Tübingen.

Szalay FS. 1976. Systematics of the Omomyidae (Tarsiiformes, Primates): taxonomy, phylogeny, and adaptations. Bull Am Mus Nat Hist 156: 157–450.

Szalay FS, Decker RL. 1974. Origins, evolution, and function of the pes in the Eocene Adapidae (Lemuriformes, Primates). In Jenkins FA, Jr., editor, Primate locomotion, 239–259. New York: Academic Press.

Thomas BL, Tucker AS, Qiu B, Ferguson CA, Hardcastle Z, Rubenstein JL, Sharpe PT. 1997. Role of Dlx-1 and Dlx-2 genes in patterning of the murine dentition. Development 124: 4811–4818.

Tong Y. 1997. Middle Eocene small mammals from Liguanqiao basis of Henan Province and Uanqu basin of Shanxi Province, Central China. Pal Sinica 18 (n.s. C): 42–49, 199–201.

Tucker AS, Mathews KL, Sharpe PT. 1998. Transformation of tooth type induced by inhibition of BMP signaling. Science 82: 1136–1138.

Vaarhtokari A, Aberg T, Jernvall J, Keranen S, Thesleff I. 1996. The enamel knot as a signaling center in the developing mouse tooth. Mech Devel 54: 39–43.

Walker AC. 1974. A review of the Miocene Lorisidae of East Africa. In Doyle GA, Martin RD, Walker AC, editors, Prosimian biology, 435–447. London: Duckworth.

Wolin LR. 1974. What can the eye tell us about behaviour and evolution? or: the aye-ayes have it, but what is it? In Martin DR, Doyle GA, Walker AC, editors, Prosimian biology, 489–497. London: Duckworth.

Woollard HH. 1926. Notes on the retina and lateral geniculate body in *Tupaia, Tarsius, Nycticebus* and *Hapale*. Brain 49: 77–104.

Zingeser MR, Lozanoff S. 1989. Growth of the interorbital septum in fetal rhesus macaques (abstract). Am J Phys Anthro 78: 329.

CHAPTER 4

Morphometrics, Functional Anatomy, and the Biomechanics of Locomotion among Tarsiers

Robert L. Anemone and Brett A. Nachman

The modern tarsiers of Southeast Asia have long played an important role in discussions of the phylogeny, classification, and functional morphology of primates. Their peculiar combination of primitive (e.g., sublingua, grooming claws, tribosphenic molars) and derived traits (e.g., partial post-orbital closure, loss of rhinarium, retina with macula) have resulted in much difference of opinion as to their systematic position within the order primates (Gregory, 1910; Simpson, 1945; Hill, 1955; Clark, 1971; Szalay, 1976; Cartmill, 1982; Wible and Covert, 1987; Fleagle and Kay, 1994). Whether they are classified as prosimians or haplorhines, however, all students agree that their locomotor behavior and associated postcranial musculo-skeletal anatomy are some of the most fascinating aspects of the biology of the genus *Tarsius*.

While much of the modern interest in the functional morphology of tarsiers dates to Napier and Walker's (1967) seminal paper on vertical clingers and leapers, a series of earlier studies laid the groundwork for comparative and functional analyses of the musculo-skeletal system of the tarsiers. Much of this work is in German, including Mollison's (1911) important monograph on limb proportions among primates, and Haffer's (1929) discussion of the elongated tarsus seen in galagos and tarsiers and its biomechanical significance in leaping. The earliest comprehensive discussion in English of the morphology of a modern tarsier is Woollard's (1925) monograph on *Tarsius spectrum*. Hill (1955) and later Clark (1971) provided "state of the art" summaries of the comparative anatomy of tarsiers (and other primates) that have only recently been supplanted by Fleagle (1999) and Martin (1990). The edited volume by Niemitz (1984a) pulled together the results of morphological and behavioral analyses in order to address phylogenetic, classificatory, and functional issues raised by our growing knowledge of modern and fossil tarsiiforms.

The Vertical Clinging and Leaping (VCL) hypothesis of Napier and Walker (1967) stated that bush babies (Galagidae), tarsiers (Tarsiidae), lepilemurids (Lepilemuridae), and indriids (Indriidae) all shared a similar locomotor adaptation involving clinging to vertical supports and hindlimb-propelled jumps between these supports. Napier and Walker (1967) also suggested that the morphology associated with VCL was already present among some Eocene primates (both Adapidae and Omomyidae) and that VCL may in fact have been the primitive pattern of locomotion for all Euprimates. In the rapid response to the VCL hypothesis, a number of important facts about both extant and extinct saltatory primates were established (Cartmill, 1972, 1974; Martin, 1972; Szalay, 1972; Stern and Oxnard, 1973; Jouffroy, 1975; Jouffroy and Lessertisseur, 1978, 1979): (1) the four groups of modern VCL primates use subtly different postures and locomotor behaviors; (2) the four groups of modern VCL primates exhibit at least two and possibly three different hindlimb morphologies; and (3) the postcranial skeletons of Eocene primates are poorly known and their testimony as to the primitive nature of the VCL adaptation is equivocal at best.

Over the next two decades, a number of investigators revisited the VCL hypothesis and further clarified aspects of the functional morphology of saltatory locomotion among modern prosimians and tarsiers (Jungers et al., 1980; Szalay and Dagosto, 1980, 1988; McArdle, 1981; Gebo, 1987a, 1987b, 1988, 1993a, 1993b; Anemone, 1988, 1990, 1993; Fleagle and Anapol, 1992; Dagosto, 1993; Demes et al., 1998). In this chapter, we analyze the musculoskeletal anatomy of modern tarsiers in an attempt to better understand the functional adaptations related to their peculiar locomotor habits. Our approach involves a review of the substantial literature on the functional anatomy of leaping primates, with an emphasis on those papers that include consideration of tarsiers, and a reanalysis of a large (N = 277) comparative dataset (Anemone, 1988) on the morphometrics of limb proportions and the hindlimb of prosimians, tarsiers, and tupaiids.

Materials and Methods

The analyses presented here are based on measurements taken by the senior author of 277 postcranial skeletons of prosimian primates and tree shrews from the U.S. National Museum of Natural History, the American Museum of Natural History, and the Harvard Museum of Comparative Zoology (Anemone, 1988, 1990, 1993). The skeletons were mostly (75%) fully adult, wild-shot, nonpathological specimens (including all eight tarsiers): roughly one quarter were of adult, nonpathological, zoo or lab-reared animals. The sexes were pooled in all analyses since prosimians do not typically exhibit significant postcranial dimorphism (Smith and Jungers, 1997), and because many museum specimens are of unknown sex. Morphometric data collected included maximum lengths for all forelimb and hindlimb long bones, skeletal trunk length (Biegert and Maurer, 1972), tarsal bone lengths, and a series of biomechanically oriented measurements of joint surfaces, bone segments, and muscle attachment sites in the hindlimb (Anemone, 1988). This morphometric data set was analyzed in a number of ways: (1) through the calculation of indices and ratios reflecting biomechanical aspects of skeletal design, and (2) with the linear regression techniques of allometric analysis. Throughout the analysis, our goals have been to determine functionally and biomechanically significant aspects of skeletal design and muscular function related to the positional behavior of tarsiers, and to locate these morphological features within the comparative context of prosimian anatomy and behavior.

Results: The Functional Morphology of Leaping among Tarsiers

Forelimb

The osteology and myology of the tarsier forelimb has been described by Schultz (1984), while details concerning the relative length of the forelimb and of its segments are also available (Jouffroy and Lessertisseur, 1979; Niemitz, 1984d). While Schultz (1984) provides an excellent description of the gross anatomy of the tarsier forelimb, he includes little or no discussion of functional aspects of tarsier morphology in relation to their specialized locomotor and postural behaviors. Niemitz (1984c) describes some aspects of the use of the hand during feeding and during locomotor and postural activities that (Niemitz, 1984b) based on his observations of both captive and wild western tarsiers (*Tarsius bancanus*).

The main joint complex of the forelimb associated with positional behavior of tarsiers and other vertical clingers and leapers is the elbow joint, and its comparative anatomy has been nicely summarized by Rose (1993). Building upon earlier work by Szalay and Dagosto (1980) as well as his own work (Rose, 1988), Rose summarizes the characters thought to be related to stability in the flexed elbow joint associated with the vertical clinging postures of tarsiers, galagos, and indriids. Like other vertical clingers and leapers, tarsiers have an inflated capitulum, a marked medial lip on the radial head, a cone-shaped trochlea that is high anteriorly with a very narrow zona conoidea separating the trochlea from the capitulum anteriorly (Rose, 1993). Szalay and Dagosto (1980) stress the great anterior height of the trochlea and its shallowness posteriorly with a shallow facet for the olecranon process, the deep groove separating trochlea from capitulum, and the diagonal ridge running across the distal surface of the trochlea as traits that facilitate stability in the elbow during flexed-arm grasping of vertical supports.

The intrinsic musculature of the hands of tarsiers bears an interesting specialization that is associated with their specialized ability for manual gripping (Day and Iliffe, 1975). The contrahentes muscles among primates are innervated by a deep branch of the ulnar nerve, typically vary in number between zero and four, and insert onto the proximal phalanges where they facilitate convergence of the digits (Day and Napier, 1963; Swindler and Wood, 1973). Day and Illife (1975) noted significant differences from the typical primate configuration of the contrahentes group in the five tarsier hands they dissected. The modal pattern among these tarsiers included five contrahentes muscles with short and long tendinous slips inserting into the dorsal surface of the proximal and distal phalanges, respectively. In addition, the contrahens muscles of digits 2 and 3 also send tendinous slips to attach to the medial aspect of the proximal phalanx of the adjacent digit.

Day and Iliffe (1975) suggest that the contrahentes muscles facilitate the grip of tarsiers by increasing the pressure brought to bear between the extensive digital pads and the surface being gripped. This is accomplished through the simultaneous flexion of the metacarpophalangeal joint and the proximal interphalangeal joint (by both short and long slips), and the extension of the distal interphalangeal joint brought about by the long slip. Day and Iliffe (1975:248) suggest that "this mechanism, together with prehensility of the hand and the use of the tail against the support, would be of great advantage to *Tarsius* in movement through its arboreal habitat where the smooth, moist boles of trees would demand an effective mechanism for vertical clinging." It might be added here that the manual grip of tarsiers is further strengthened by the specialized condition of the large terminal digital pads with their extensive and fine papillary ridges, backed by small, triangular nails (Clark, 1971) and by their intrinsically long hands (Jouffroy and Lessertisseur, 1979).

Axial Skeleton

The axial skeleton of tarsiers has not been well studied, in spite of the fact that behavioral (Niemitz, 1977, 1979, 1984b) and biomechanical (Preuschoft et al., 1979) investigations have suggested the importance of the vertically hanging tail in maintaining vertical clinging postures and in body mechanics during leaping and landing (see Shapiro, 1993, for a review of the biomechanics of primate vertebral columns). The naked, papillary ridge covered region found on the ventro-proximal region of the tarsier tail (Sprankel, 1965) would appear to be a clinging adaptation. Contrary to the suggestion by Napier and Walker (1967) that it acts as a passive counterweight during leaping, the tail plays an active role as a rudder and steering mechanism during leaping locomotion among tarsiers as well as in midflight body rotations that prepare the animal for landing (Niemitz, 1984b). Preuschoft et al. (1979) explore the biomechanical ramifications of vertical and horizontal postures for leaping primates and develop a model specifically for tarsiers. They do not, however, present or analyze any morphological data on the axial skeleton of tarsiers beyond suggesting that both the cross-sectional area of the dorsal trunk musculature and the diameters of the vertebral bodies increase caudally among primate leapers (Schultz, 1961; Ankel, 1972; Preuschoft et al., 1979). Ankel (1972:228) suggests that the cervical region of tarsiers is specialized to allow for a greater degree of rotation of the head upon the vertebral column by virtue of the orientation of the zygapophyses and the shape of the vertebral centra. This feature may be important since, when tarsiers leap from vertical supports, their direction of travel is essentially backwards (Niemitz, 1984b). Enhanced mobility in the cervical region allows them the ability to turn around and look in the direction of the jump prior to and during takeoff.

Limb Proportions

Prosimian limb proportions have been well studied by a variety of different investigators (e.g., Stern and Oxnard, 1973; Jouffroy and Lessertisseur, 1979; McArdle, 1981; Anemone, 1988). Jouffroy and Lessertisseur (1979) compared limb proportions from 127 prosimian skeletons in two different ways: they calculated the lengths of limbs and limb segments relative to precaudal vertebral length, and they calculated intralimb relative proportions. Tarsiers are outstanding among primates in this dataset by virtue of their possession of both the longest forelimbs (135%) and the longest hindlimbs (231%) among primates, relative to the length of the precaudal vertebra. The extreme length of the tarsier forelimb is sometimes underappreciated or obscured by a focus on the intermembral index, which measures the length of the forelimb (humerus plus radius length) relative to that of the hindlimb (femur plus tibia length). Tarsiers and the other vertical clinging and leaping prosimians have the longest hindlimbs and, therefore, the lowest intermembral indices among primates (Napier and Walker, 1967; Jouffroy and Lessertisseur, 1979; McArdle, 1981; Anemone, 1988). But only in *Tarsius* are both fore- and hindlimbs extremely elongated, while the typical pattern seen in other leapers like galagos, lepilemurids, and cheirogaleids involves long hindlimbs and short forelimbs (Jouffroy and Lessertisseur, 1979).

Within the forelimb, tarsiers have very short humeri (27% of the length of the forelimb), moderately short radii (35%), and very long hands (38%). They are closest to indriids and *Daubentonia* in these relative proportions. Within the hindlimb, tarsiers have short femora (36%) and tibiae (29%), and very long feet (35%) and resemble other leapers like cheirogaleids and, especially, galagos. The great length of tarsier and galago feet is, of course, mainly the result of tarsal elongation, and will be further discussed below.

We calculated a series of standard limb proportion ratios on our

Table 4.1. Comparative Limb Proportions

Taxon	Inter-membral index	SD	Humero-femoral index	SD	Crural index	SD	Brachial index	SD	N
VCL[a]	60.20	5.56	55.44	6.10	93.73	6.01	110.55	6.94	123
Quadrupeds[b]	70.77	2.07	71.11	5.05	101.48	8.11	102.18	8.14	106
Slow Climbers[c]	88.37	4.33	85.00	4.22	94.89	2.57	102.90	6.76	55
Tupaiidae	72.65	1.90	78.45	2.56	108.03	3.06	93.72	5.51	20
Lemuridae	68.02	3.89	64.84	5.04	94.74	2.71	105.22	7.33	58
Lepilemuridae	59.00	1.58	55.14	1.31	94.65	2.31	108.29	3.41	21
Cheirogaleidae	70.58	1.45	70.81	2.17	108.24	6.15	107.46	6.64	28
Lorisidae	88.37	4.33	85.00	4.22	94.89	2.57	102.90	6.76	55
Galagidae	60.63	7.62	57.15	7.84	95.22	6.19	107.31	5.38	67
Indriidae	59.36	3.45	51.68	3.46	86.36	1.62	114.28	6.95	27
Tarsiidae	58.12	1.55	51.62	1.06	99.24	4.79	124.50	3.66	8

[a] Includes Tarsiidae, Galagidae, Lepilemuridae, and Indriidae.
[b] Includes Tupaiidae, Lemuridae, and Cheirogaleidae.
[c] Includes Lorisidae.

comparative sample of 284 prosimian (including tupaid) skeletons in order to investigate the length of limbs and limb segments of tarsiers relative to other families of prosimian primates: the results are presented in table 4.1. The intermembral index measures the length of the long bones of the forelimb (humerus and radius) relative to those of the hindlimb (femur and tibia), and clearly separates the four leaping families (Lepilemuridae, Indriidae, Galagidae, and Tarsiidae) from both the three active quadrupedal families (Tupaiidae, Lemuridae, and Cheirogaleidae) and the slow-climbing Lorisidae.

While all prosimian families have longer hindlimbs than forelimbs reflected by intermembral indices of less than 100, vertical clinging and leaping taxa have forelimbs that are roughly 60% of the length of their hindlimbs, while active quadrupeds have forelimbs that are roughly 70% as long as their hindlimbs. Tarsiers have a significantly lower intermembral index (reflecting relatively longer hindlimbs) than each of the quadrupedal families (P < 0.001, ANOVA), but are not significantly different from any of the other VCL families. Slow-climbing lorisids have the highest intermembral indices among prosimians, with forelimbs that are 88% as long as their hindlimbs.

The humero-femoral index presents essentially the same picture of quadrupeds and slow climbers having higher values than leapers, with tarsiers again having a significantly smaller index than all nonleaping families (P < 0.001, ANOVA). Both the intermembral and humero-femoral indices reflect the biomechanical importance of a relatively long hindlimb among leaping prosimians, and allow good discrimination between quadrupedal and leaping prosimians. Long hindlimbs are common among leapers since they allow a longer time or greater distance of force generation against the substrate, providing greater acceleration in leaping.

The crural and brachial indices document the length of the distal element of the hind and forelimb respectively, relative to that of each limb's proximal bony element (table 4.1). The crural index is a poor discriminator of locomotion among prosimian primates and reflects a large amount of variability among prosimian families in the length of the tibia relative to that of the femur. While tarsiers typically possess a tibia that is nearly equal in length to that of the femur (crural index of 99.2), the leaping indriids are significantly different from all prosimian families including Tarsiidae ($P < 0.001$, ANOVA) by the possession of a femur approximately 15% longer than the tibia. Galagidae, Lepilemuridae, Lemuridae, and even the slow-climbing Lorisidae all have femora roughly 5% longer than tibiae, and none are significantly different from tarsiers. Tarsiers have a significantly higher crural index than Indriidae, and a significantly lower index than Cheirogaleidae and Tupaiidae, both of whom have a tibia that is 8% longer than the femur ($P < 0.001$, ANOVA).

While brachial index is another relatively poor discriminator of quadrupeds and vertical clingers and leapers, tarsiers have a significantly higher brachial index than all other prosimian families ($P < 0.001$, ANOVA), reflecting the relatively short humerus mentioned by Jouffroy and Lessertisseur (1979). Most prosimian families possess a radius that is between 5% and 15% longer than the humerus. Tupaiids are the only family with a brachial index of less than 100, indicating a longer humerus than radius, and they are significantly different from all prosimian families on this index ($P < 0.001$, ANOVA). Slow climbers and quadrupeds are essentially indistinguishable on brachial index.

While these limb proportion ratios can reveal adaptive design features in mammalian skeletons, by their very nature they have the potential to obscure other important information concerning the absolute length of limbs and limb segments. While biomechanical analysis strongly suggests that a long hindlimb lever is of critical importance in enabling a hindlimb-propelled leaper to attain takeoff velocity (Hall-Craggs, 1964, 1965a,b; McArdle, 1981; Anemone, 1993), these indices tell us nothing about either the absolute or size-corrected length of the segments or whether the value of a particular ratio is due to change in the numerator or denominator. By themselves, the intermembral and humero-femoral indices do not reveal that tarsiers not only possess the longest hindlimb but they also have the longest forelimb (relative to vertebral length) among the Prosimii (Jouffroy

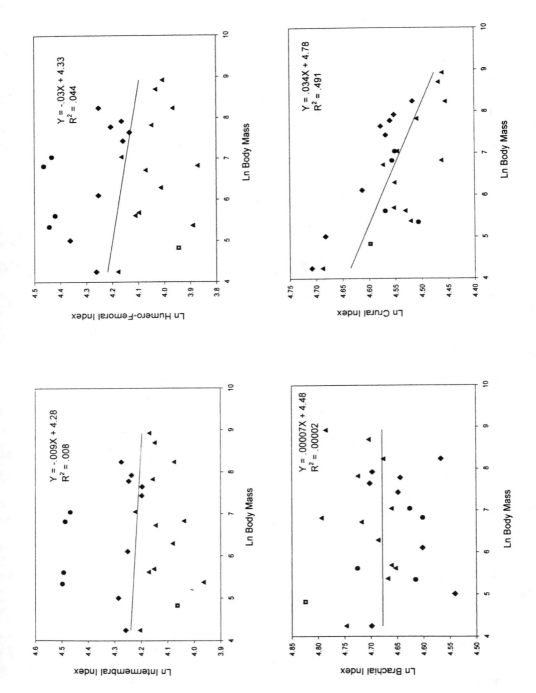

and Lessertisseur, 1979; Anemone, 1988). Similarly, the brachial index does not tell us if tarsiers have a long radius or a short humerus. Limb proportion ratios are also unable to tell us anything about size-related trends in relative limb proportions among animals spanning a wide range of body sizes.

Linear regression of limb lengths (dependent or Y variable) against body mass (independent or X variable) provides a method to investigate the length of limbs or limb segments among prosimians in relation to body size. Least squares regressions of limb proportion indices against species' mean mass clearly documents the lack of a relationship between body size and intermembral, humero-femoral, and brachial index, and a slightly negative relationship ($R^2 = 0.491$) between crural index and body size (fig. 4.1). These regressions suggest that while tarsiers possess distinctive limb proportion indices, they are basically similar to other leapers regardless of issues of body size and allometry.

Hindlimb

The tarsier pelvis is most similar among prosimians to the other small-bodied leapers of the family Galagidae (Anemone, 1988). Like the bush babies, tarsiers have a long and narrow ileum with a narrow iliac crest, and a slightly concave gluteal surface that faces mainly laterally. The sacro-iliac joint is positioned cranially, with roughened areas for ligamentous attachments extending to the iliac crest. The anterior inferior iliac spine is a small bump on the lateral surface of the ilium, while the anterior superior iliac spine is a small ridge just caudal to the ventral edge of the iliac crest. In prosimians, only among galagos and tarsiers is the anterior superior iliac spine separated from the iliac crest. The long and thin pubic rami enclose an elongate, oval obturator foramen, and the pubic symphysis is moderately long and slopes caudally. The ischium is short and stout and projects mostly dorsally as in other vertical clingers and leapers (Fleagle and Anapol, 1992), with a well-developed ischial spine and tuberosity.

Morphometric analysis of the pelvis of tarsiers and other prosimians strongly supports the suggestion that the relative lengths of the osteological elements that comprise the pelvis are shaped by locomotor behavior. Vertical clingers and leapers possess significantly longer ilia and shorter ischia relative to total pelvic length than do active quadrupeds ($P < 0.001$, ANOVA),

Figure 4.1. (facing page) Regressions of limb proportion indices against each species' average body mass (all values are natural logarithms). No linear relationship with body mass exists for intermembral, brachial, and humero-femoral index (slopes not significantly different than zero and very low R^2 values). A weak negative relationship with body mass exists only for the crural index. Symbols: circles = slow climbers, triangles = leapers, diamonds = quadrupeds, square = tarsiers.

Table 4.2. Pelvic Ratios

Taxon	Relative ilium length*	SD	Relative ischium length	SD	Relative pubis length	SD	N
VCL	72.97	7.29	28.22	1.90	31.30	4.52	113
Quadrupeds	68.59	3.94	32.47	5.45	31.35	3.27	100
Slow climbers	74.23	11.75	26.16	4.89	45.14	8.05	52
Tupaiidae	62.08	1.03	38.34	8.09	30.37	2.14	20
Lemuridae	71.54	1.34	29.61	1.31	30.05	1.68	56
Lepilemuridae	72.52	3.27	30.05	1.54	25.69	1.47	20
Cheirogaleidae	70.13	1.96	32.56	1.77	35.34	2.71	24
Lorisidae	75.23	11.75	26.16	4.89	45.13	8.05	52
Galagidae	73.01	8.58	27.55	1.45	32.76	4.31	65
Indriidae	71.57	9.45	28.03	1.35	29.24	2.09	20
Tarsiidae	77.52	1.27	25.60	1.18	37.96	3.00	8

*Each of these ratios uses total pelvic length as the denominator.

and tarsiers are distinguished by having the longest ilium and the shortest ischium among all prosimian families (table 4.2).

The long ilium among leapers is thought to be related to two different but complementary biomechanical aspects of hip flexion and extension (Anemone, 1988, 1993). First, it increases the moment arm of *M. tensor fasciae femoris,* an important flexor of the hindlimb that helps retract the hindlimb in preparation for landing after a jump. Second, it may allow *M. gluteus medius* an increased average muscle fiber length, thus increasing the physiological speed of contraction of this important hip extensor. Both functions would seem to be of great value to small-bodied vertical clingers and leapers like tarsiers. The short ischium among leapers clearly affects the length of the hamstrings lever in hindlimb extension. A short ischium would provide a short moment arm, specialized for maximizing speed of contraction to aid in hindlimb extension from the extremely flexed posture that precedes the leap among VCL primates. The dorsal projection of the ischium would allow effective leverage through a wide range of hindlimb extension, and would also be of great benefit to prosimian leapers (Fleagle and Anapol, 1992). Interestingly, slow climbers have slightly longer ilia and shorter ischia than leapers.

Although the relative length of the pubis does not clearly differentiate between leapers and quadrupeds, tarsiers have a significantly longer pubis than all other prosimian families ($P < 0.001$, ANOVA) except Lorisidae and Cheirogaleidae. The length of the pubis closely correlates with the moment arm length of the adductor musculature, resulting in long adductor muscular lever arms in slow-climbing lorises, quadrupedal cheirogaleids, and leaping tarsiers. This functional arrangement may be advantageous to tarsiers as a postural adaptation involved in maintaining balance and posture while clinging to vertical supports.

The dichotomy that characterizes morphological adaptations among VCL prosimians is also in evidence in the myology and kinematics of the hip joint. Demes et al. (1998) showed that, with regard to the extensor musculature of the hip joint, tarsiers and galagos have relatively small hip extensors (16% and 17% of total hindlimb muscle mass, respectively) compared to indriids (between 20% and 23% of hindlimb muscle mass).

The opposite pattern is seen in the plantarflexors. In both tarsiers and galagos, the plantarflexors account for 7% of total hindlimb muscle mass, while in the indriids only 4% of hindlimb musculature is devoted to plantarflexion. Furthermore, with the hamstrings excluded from the comparison, the absolute mass of the hip extensors in tarsiers is less than the absolute mass of the plantarflexors; once again, the reverse is true in indriids. Similarly, the kinematic data suggest that during a leap, tarsiers go through a smaller range of motion at the hip joint than do indriids (Demes et al., 1996, 1998).

The functional morphology of the femur has long been at the center of arguments concerning the adaptations involved in vertical clinging and leaping among tarsiers and other prosimian primates. In their initial discussion, Napier and Walker (1967) suggested that VCL primates could be identified by their possession of the following traits: (1) a long and straight femoral shaft with a thick neck set perpendicular to the shaft, (2) posterior expansion of the articular surface of the head onto the neck, (3) greater trochanter overhanging the ventral aspect of the shaft, (4) narrow patellar groove with a prominent lateral margin, and (5) ventro-dorsally deep femoral condyles. While these traits are all found among galagos and tarsiers, other leapers (e.g., indriids and lepilemurids) do not possess all of these traits, suggesting a dichotomy of leaping adaptations among Malagasy and non-Malagasy leaping prosimians (Jouffroy, 1962; Cartmill, 1972; Martin, 1972; Stern and Oxnard, 1973; McArdle, 1981; Godfrey, 1988; Anemone, 1988, 1990, 1993).

Tarsiers most closely resemble galagos in possessing a long and straight femur that is well adapted to dealing with the large sagittal plane compressive stresses involved in saltatory, bipedal locomotion (Burr et al., 1982). The femoral head in tarsiers is cylindrical, sits on a thick and short neck set perpendicular to the shaft, and has a substantial degree of expansion of its articular surface onto the posterior aspect of the neck. This morphology is also found among galagos and has long been associated with vertical clinging and leaping in these taxa (Napier and Walker, 1967; McArdle, 1981; Anemone, 1988, 1990, 1993), although its functional significance continues to be debated.

Demes et al. (1996) have recently suggested that this suite of morphological features confers stability on the hip joint in response to the stresses associated with leaping among small-bodied prosimians, while Anemone (1988, 1990, 1993) suggested that this morphology is a postural adaptation associated with the flexed, abducted, and laterally rotated position of the femur in vertical clinging. Malagasy prosimians, whether leapers or quadrupeds, all have spherical femoral heads with little or no expansion of articular surface onto the neck, and are easily distinguished from galagos and tarsiers in proximal femoral morphology. The spherical femoral head of indriids may allow even greater amounts of lateral rotation and abduction than is allowed by the cylindrical femoral head found in tarsiers and galagos (Demes et al., 1996). The cantilevered beam model of femoral neck mechanics recently developed by Rafferty (1998) among larger prosimians (including some Malagasy leapers) holds promise for exploring the mechanical significance of the tarsier (and galago) femoral neck.

Another functionally important aspect of the proximal femoral morphology of tarsiers is the presence of a greater trochanter that overhangs the ventral aspect of the femoral shaft. This trait is shared by the other leaping prosimians as well as by the Malagasy quadrupedal taxa (i.e., all hindlimb-dominant prosimians, sensu Martin, 1972). The significance of this trait is related to the direction of pull and leverage of the *M. vastus lateralis*, which arises along the ventral aspect of the greater trochanter. The *M. vastus lateralis* is hypertrophied in many leapers, including all lemuriformes (Jouffroy, 1962, 1975; Jungers et al., 1980), galagos (McArdle, 1981; Anemone, 1988), and tarsiers (Woollard, 1925), as well as in leaping rodents (Alezais, 1900; Howell, 1932) and macropodids (Hopwood and Butterfield, 1976). Both *M. vastus lateralis* and *M. vastus medialis* have a proximally restricted origin from the femoral shaft among tarsiers (Woollard, 1925), as well as among all other primate leapers (Jouffroy, 1962, 1975; Jungers et al., 1980; McArdle, 1981; Anemone, 1988).

This trait serves to maximize the length of the strongly parallel-running fibers in these muscles. Stern (1971, 1974) has shown that the combination of long fiber length and large cross-sectional area allows a muscle to maximize force generation at high contraction velocity, an obvious advantage for leapers who must generate enormous muscular forces to attain takeoff velocity (Hall-Craggs, 1964, 1965a,b, 1974; Jouffroy et al., 1974). The importance of *M. vastus lateralis* in leaping by *Eulemur fulvus* was confirmed by Jungers and colleagues through the use of telemetered electromyography (Jungers et al., 1980).

Table 4.3. Femoral Ratios

Taxon	Third trochanter[a]	SD	Lesser Trochanter	SD	Greater Trochanter	SD	Condylar index[b]	SD	N
VCL	1.438	.204	1.575	.078	1.668	.189	109.62	7.12	121
Quadrupeds	1.535	.472	1.485	.215	1.564	.235	98.55	7.06	106
Slow climbers	1.284	.010	1.515	.167	1.327	.078	84.11	5.06	55
Tupaiidae	2.482	NA	1.362	NA	1.373	NA	90.94	3.08	20
Lemuridae	1.547	.182	1.570	.229	1.711	.150	101.19	6.27	58
Lepilemuridae	1.483	.142	1.515	.042	1.707	.063	114.54	2.65	21
Cheirogaleidae	1.031	.128	1.334	.130	1.292	.064	102.06	5.66	28
Lorisidae	1.284	.010	1.515	.167	1.327	.078	84.11	5.06	55
Galagidae	1.295	.204	1.592	.070	1.479	.050	107.77	5.78	67
Indriidae	1.623	.069	1.601	.110	1.889	.099	111.46	9.87	27
Tarsiidae	1.266	NA	1.566	NA	1.614	NA	117.02	5.57	8

[a] Cube root of mean species body mass was used as the denominator in each of the trochanteric ratios. Body mass data from Dagosto and Terranova, 1992.

[b] Anterior-posterior width divided by medio-lateral breadth of condyles times 100.

The proximal femur includes several morphological features that have been thought to be related to leaping locomotion among prosimian primates. Earlier studies by Anemone (1988, 1990, 1993) indicated that the femoral trochanters were positioned more proximally among leapers than among quadrupedal prosimians. It was suggested that this might be an adaptation for decreasing the mechanical advantage of the hip muscles inserting into the trochanters and for reducing the moment of inertia of the hindlimb, in both cases providing extra speed in flexion and extension of the thigh on the hip as an aid to leapers in attaining takeoff velocity. The earlier results were based on ratios of trochanteric position relative to femoral length and were complicated by the fact that leapers have the longest femora among prosimians. As a result, it was difficult to know if the low trochanteric ratios among leapers were the result of proximally placed trochanters (i.e., small numerators) or long femora (i.e., large denominators) (McArdle, 1981; Anemone, 1993).

These data were reanalyzed (table 4.3) through the calculation of a different set of ratios in which femoral trochanteric length was standardized by dividing it by the cube root of mass, rather than by femur length. The results of these analyses suggest that, overall, leapers and quadrupeds are not significantly different ($P < 0.01$, ANOVA) in the distance from the hip joint to any of the femoral trochanters. When the comparisons are made at the family level, with a focus on the relative position of tarsiers (table 4.3), the picture is essentially the same. The only consistently important difference that arises from this analysis suggests that tree shrews have a significant distally positioned third trochanter compared to all prosimian families (P < 0.001, ANOVA), while neither tarsiers nor other leaping families possess significantly more proximally positioned trochanters than do quadrupedal prosimian taxa.

The distal femur in tarsiers is also characterized by a derived morphology that is clearly associated with the biomechanical demands of hindlimb-dominated leaping behavior. The condylar index is a measure of the antero-posterior diameter of the femoral condyles relative to their medio-lateral width. Table 4.3 indicates that leapers have significantly higher condylar indices than quadrupeds, and that tarsiers have the highest condylar index among the prosimians, significantly higher (P < 0.001, ANOVA) than that of all other prosimian families except Indriidae and Lepilemuridae. In addition, tarsiers possess a patellar articular surface that is raised ventrally from the shaft of the femur, a specialized trait that is only shared by the most specialized leaping prosimians (e.g., Galagidae, Lepilemuridae, and Indriidae), and that further enhances the mechanical advantage of extension at the knee by the *quadriceps* group (Anemone, 1988, 1990, 1993).

Tarsiers also share with both leapers and quadrupedal prosimians a narrow and deep patellar groove with a raised lateral ridge. These traits reflect an emphasis on powerful flexion and extension at the knee (Tardieu, 1981), with a particularly powerful component of force derived from the contraction of *M. vastus lateralis* (Napier and Walker, 1967; Jouffroy, 1975; Jouffroy and Lessertisseur, 1979; Jouffroy et al., 1984; Jungers et al., 1980; McArdle, 1981; Anemone, 1988, 1990, 1993). All of these traits of the distal femur are involved in maximizing the moment arm of the *Mm. quadriceps femoris* in extension of the flexed knee, a critical component of locomotion in any leaping primate (Hall-Craggs, 1964, 1965a,b; Jouffroy et al., 1973; Jouffroy and Gasc, 1974; Jungers et al., 1980; Niemitz, 1985). These skeletal adaptations combine with the hypertrophy among tarsiers and other leapers of the *Mm. quadriceps,* and particularly of the *M. vastus lateralis,* to create a musculo-skeletal system capable of generating the enormous forces required to attain takeoff velocity (Hall-Craggs, 1964, 1965a; Preuschoft et al., 1979; Jungers et al., 1980; Niemitz, 1985).

One of the more striking features of the tarsier hindlimb is the fusion of the tibia and fibula, a trait unique among extant primates. Among tarsiers, fusion occurs over the distal 60% of the tibial shaft, and it is generally thought to be advantageous in leaping, limiting movement at the tibio-talar joint to the antero-posterior plane (i.e., flexion and extension). Until recently, true tibio-fibular fusion was known in only one other primate, the Eocene microchoerine *Necrolemur* (extending 37% of tibial length according to Dagosto, 1985). However, Rasmussen et al. (1998) have recently described tarsierlike fusion in the Fayum primate *Afrotarsius* (extending 60%

of tibial length). Fusion of the tibia and fibula in tarsiers is found in conjunction with a suite of characters that is also associated with frequent leaping. These include a mediolaterally compressed shaft, a retroflexed tibial plateau, and proximal insertion of the *Mm. gracilis* and *sartorius*. This latter feature shortens the moment arms and effectively increases the speed of these knee flexors. Interestingly, this suite of features is also found in *Afrotarsius*, demonstrating that the leaping specializations of the Tarsiidae date back at least to the Oligocene (Rasmussen et al., 1998).

The uniqueness of the tarsier foot has been known for well over a century. In fact, it was the unusually elongated anklebones that first led Buffon, in 1765, to informally label this primate "tarsier" (cited in Niemitz, 1985). Since that time, and increasingly in the latter part of the twentieth century, the anatomy of the tarsier foot as well as that of *Galago* has been intensely studied. These studies have typically focused on the leaping specializations of the foot (Hall-Craggs, 1965b, 1966; Napier and Walker, 1967; Jouffroy et al., 1984). More recently, however, Gebo (1987a) showed that, while tarsiers and galagos share certain leaping adaptations, tarsiers have a "fundamentally different method of bone rotation" for foot inversion, as well as a unique pattern of musculature associated with vertical clinging.

Using dissected cadavers from the Duke Primate Center to determine relative muscle weights, Gebo (1987a) demonstrated that tarsier foot musculature shows a different arrangement than that of *Galago*. With regard to the extrinsic musculature of the foot, tarsiers have nearly twice the superficial flexors versus deep flexor muscles, while galagos are almost equal in these two muscle groups. As these superficial flexors (*Mm. gastrocnemius, soleus,* and *plantaris*) are associated with powerful leaping, this arrangement illustrates the primary nature of leaping in the tarsier locomotor repertoire. The equal weights in galago musculature reflect a more varied locomotor repertoire with deep flexors aiding in grasping and climbing activities.

As the label "vertical clinger and leaper" implies, tarsiers are not only capable of powerful leaps, but are also quite capable of grasping behaviors. To this end, tarsiers have adapted by increasing their intrinsic musculature for grasping (Gebo, 1987a). Tarsiers display the highest values among prosimians for the intrinsic/extrinsic muscle ratio. Gebo (1987a:15) points out that this increase serves to resist "the greater forces produced from the interaction of body mass and gravity" when vertically clinging. It should be noted that the next two highest values for the intrinsic/extrinsic ratio are *Galago senegalensis* and *Propithecus*—two other prosimians with a very high frequency of vertical clinging. In addition, tarsiers have modified the I-V grip (Gebo, 1985) with a specialized insertion pattern of the contrahentes muscles discussed above in the forelimb section (Day and Iliffe, 1975).

This pattern, present in the feet as well as the hands, allows the digits to be arched against a vertical support—effectively placing more pressure on the oversized digital pads and thus providing more friction for vertical clinging to a slippery substrate (e.g., bamboo). Tarsiers have also modified the extensor retinacular bands of the foot, combining the A and B bands into one broad band (Gebo, 1987a). Gebo suggests that this modified retinaculum efficiently keeps the extensor tendons in place during the extreme dorsiflexion of the tarsier ankle that occurs while clinging to a vertical support.

The highly specialized nature of the tarsier foot is most readily apparent in its bony anatomy and associated joint structures. Some of these specializations are associated with frequent leaping and are shared with galagos; others are unique to tarsiers and are related to joint movements involved in inversion. Perhaps one of the most unique aspects of tarsier anatomy is the way in which the navicular rotates during inversion (Gebo, 1987a, 1993a). Tarsiers posses a suite of anatomical features that allow for a method of foot inversion unique among primates. Most prominent among these, however, is the morphology of the navicular.

The typical method of foot inversion performed by all prosimians centers on the Transverse Tarsal Joint (TTJ), which is comprised of the calcaneo-cuboid joint and the talo-navicular joint. Hall-Craggs (1966) has demonstrated that in galagos, inversion occurs with the navicular, cuboid, and more distal parts of the *pes* rotating around a stable calcaneal axis through the TTJ. Most rotation of the navicular occurs at its proximal end, as the distal navicular (with its three-pronged shape articulating securely with the cuneiforms) allows for very little mobility. During inversion, the distal navicular actually locks up, allowing rotation to continue at the mobile TTJ (especially at the calcaneo-cuboid "pivot" joint). The midfoot is strengthened in this process by the presence of a synovial joint between the calcaneus and navicular.

Tarsiers, on the other hand, have developed features that limit mobility at the transverse tarsal joint and increase mobility at the distal navicular. TTJ mobility is limited in tarsiers by the fact that they possess neither the cuboid process nor the calcaneal pit that makes this such a mobile joint in other prosimians. The calcaneo-cuboid joint of tarsiers is flat, allowing limited rotational mobility. Moreover, the talar head is not rounded in tarsiers as it is in other prosimians, and the anterior plantar talar facet is not smooth and continuous, but rather separated by a ridge. All of these traits function to limit mobility at the TTJ. In tarsiers, the proximal and distal navicular, rather than the calcaneo-cuboid joint, are responsible for the majority of foot rotation. During inversion, tarsiers are able to rotate both the proximal and the distal navicular, with the calcaneus rotating very little.

Tarsiers have evolved a completely remodeled distal navicular to accomplish this. It has a rounded, ball-like entocuneiform facet, a modified dished ectocuneiform facet, and a mesocuneiform facet that only articulates with the mesocuneiform in extreme dorsiflexion. All of these traits provide for greater mobility at this joint. The proximal navicular is mobile as well, and there exists a space between the calcaneus and navicular that allows for the increased rotation of the latter. Gebo (1987a) points out that the navicular rotation of tarsiers is just one adaptation to a VCL locomotor repertoire that is accompanied by extreme tarsal elongation. To mitigate the mechanical stresses placed on the inverted foot while clinging, galagos developed an enhanced calcaneal-cuboid pivot combined with a synovial joint for added strength. Tarsiers responded by remodeling the joints used to rotate the foot—developing an anatomy that "evolved only once in the order and is a truly unusual primate phenomenon" (Gebo, 1987a:28).

With regard to leaping, tarsiers possess the longest calcaneus and navicular of any primate (Jouffroy, 1975; Jouffroy and Lessertisseur, 1979; Jouffroy et al., 1984; McArdle, 1981; Gebo, 1987a, 1993a; Anemone, 1988), followed closely by galagos. Hall-Craggs (1965b) showed how the elongated calcaneus of *Galago* (and by extension *Tarsius*) acts to increase the functional length of the hindlimb. A longer hindlimb increases the distance or time over which the force of a leap is applied to the ground, resulting in higher accelerations, greater takeoff velocity, and longer leaps (Hall-Craggs, 1964, 1965a,b; McArdle, 1981). Furthermore, the tarsus has elongated distally without a resultant lengthening of the proximal segment. This distal elongation effectively lengthens the load arm (or resistance arm) of plantarflexion, while the proximal segment, representing the lever arm of the plantarflexors, remains short. The biomechanical resultant is a significant decrease in mechanical advantage, sacrificing power in favor of speed (Smith and Savage, 1956), and is best interpreted as a leaping adaptation (Hall-Craggs, 1965a,b; McArdle, 1981; Anemone, 1988).

The anatomical (as well as kinematic) data regarding tarsier and galago leaping have led researchers to recognize a dichotomy among VCLs—small-bodied leapers versus large-bodied leapers. This dichotomy is characterized by the "foot-powered" leaping of the small variants (galagos and tarsiers) and the "thigh-powered" leaping of the larger animals (indriids) (Cartmill, 1972; Stern and Oxnard, 1973; McArdle, 1981; Anemone, 1988; Gebo and Dagosto, 1988). Recent studies have underscored the importance of the foot in tarsier leaping. Unlike the large-bodied leapers of Madagascar, which lack the tarsal elongation of tarsiers and galagos, tarsiers use their feet for propulsion, go through a greater range of motion at the ankle joint (Demes et al., 1996), and possess larger ankle plantarflexors versus hip extensors (Demes et al., 1998).

Aside from tarsal elongation, tarsiers and galagos share several other traits associated with proficient leaping. The plantar calcaneal surface is straight rather than curved in the middle (Gebo, 1987a). Also, the talus has a straight neck (as opposed to medially deviated) and a high talar body (dorso-plantar). The high talar body is thought to be a leaping adaptation—allowing for quick pivots around the talocrural joint (Hall-Craggs, 1965b).

Two tarsal indices have been calculated in order to elucidate the morphometric relationships among tarsiers and other prosimians (table 4.4). The foot lever index is a true measure of the mechanical advantage of plantarflexion at the ankle joint and is calculated as the ratio of proximal calcaneal length (i.e., moment arm of the *Mm. triceps surae*) to the combined distal length of the calcaneus and the length of the cuboid (i.e., resistance arm of plantarflexion). The results clearly indicate the uniqueness of tarsiers among prosimian primates. Tarsiers have the smallest mechanical advantage of plantarflexion and, as a result, the most speed adapted ankle joint in the suborder. Tarsiers have a significantly smaller foot-lever index than tree shrews and all other prosimian families except Galagidae (P < 0.001, ANOVA). Malagasy leapers are clearly distinguished from the leapers of Africa (i.e., bush babies) and Asia (i.e., tarsiers) by their lack of tarsal elongation, perhaps making up for this deficit with extreme elongation of the femur (Jouffroy, 1975). Cheirogaleids are intermediate between the extremely long tarsals of tarsiids and galagids and the short tarsals of lepilemurids and indriids.

Examination of the calcaneal index reveals essentially the same story of distal elongation of the calcaneus in small-bodied, non-Malagasy vertical clingers and leapers. This index is particularly useful for comparisons of fossil taxa, where individual calcanea are commonly found without associated cuboids. It is calculated as the ratio of the distal calcaneus to the total calcaneus length and directly reflects the degree of distal elongation of this bone. Tarsiers are again distinguished by virtue of their possession of the highest values on this index, followed closely by galagids and then by cheirogaleids. These differences are again highly significant (P < 0.001, ANOVA) when comparing tarsiers to all other families except Galagidae.

Conclusions

Tarsiers are among the most specialized of living primates in terms of the musculo-skeletal anatomy associated with locomotor behavior, and as such they provide a model system in which to explore questions related to the biomechanics and evolution of locomotion in the order. In particular, they play a central role in considerations of Napier and Walker's (1967) VCL hypothesis, especially as this hypothesis pertains to the origins of primate locomotion in the early Cenozoic. Specifically, for Napier and Walker's hypothesis to be supported, there would have to be clear evidence that all

Table 4.4. Comparative Foot Metrics

Taxon	Foot lever index[a]	SD	Calcaneal Index[b]	SD	N
VCL	49.70	18.72	56.71	15.06	123
Quadrupeds	62.55	13.98	47.23	8.36	106
Tupaiidae	76.38	6.62	38.70	1.65	20
Indriidae	74.42	3.65	37.65	8.27	27
Lemuridae	66.44	6.50	43.99	2.61	58
Lepilemuridae	64.67	3.44	46.18	1.41	21
Cheirogaleidae	46.10	8.92	57.65	5.11	28
Galagidae	32.91	5.24	69.53	5.02	67
Tarsiidae	27.20	0.95	75.83	1.00	8

[a] Proximal calcaneus length divided by combined distal calcaneus and cuboid length times 100.
[b] Distal calcaneus length divided by total calcaneal length times 100.

modern VCL primates share an essentially similar morphology related to locomotor behavior, and that morphology would have to be evident to some extent among the earliest euprimates, the Adapidae and Omomyidae (Anemone, 1990). What this review of tarsier morphology clearly documents is that modern VCLs do not share the same set of musculo-skeletal adaptations related to locomotion, and that although they are closest to galagos in most respects, tarsiers have independently evolved a unique set of musculo-skeletal adaptations related to vertical clinging and leaping. A similar range of morphological adaptations exists among Eocene omomyids, ranging from the highly derived, tarsier-like European microchoerine *Necrolemur* (Gebo, 1988; Dagosto and Schmid, 1996; Dagosto and Terranova, 1992), to North American omomyines like *Omomys* and *Hemiacodon* which are most similar to modern cheirogaleids (Anemone and Covert, 2000).

Rather than evolving once, vertical clinging and leaping most likely evolved independently in the ancestors of modern tarsiers, galagos, and Malagasy leapers (i.e., indriids and lepilemurids), resulting in three different, convergent solutions to the biomechanical challenges facing saltatory mammals. While the most significant differences between these approaches to saltation involve the presence (in bush babies and tarsiers) or absence (in indriids) of tarsal elongation, in most other respects tarsiers have clearly evolved a higher degree of anatomical specialization to this demanding form of locomotion.

Tarsiers are extremely small-bodied primates who have adapted to vertical clinging and leaping in the small branch milieu (Cartmill, 1972) through a combination of traits involving adaptive modification of limb proportions and the muscles and bones of the forelimb and especially of the hindlimb. Forelimb traits that are associated with vertical clinging and leaping in tarsiers include (1) the unique arrangements of the contrahentes muscles and large digital pads in the hands, (2) the long forelimb skeleton, and (3) the specialized features of the elbow joint.

The hindlimb is the main anatomical locus of specialized features associated with VCL behaviors. These features include (1) a long and narrow ilium and short and dorsally projecting ischium, (2) a long hindlimb with a long and straight femur, (3) a cylindrical femoral head with posterior expansion of its articular surface onto the neck, (4) deep femoral condyles and raised patellar articular surface, (5) a fused tibia-fibula with retroflexed tibial plateau, and (6) tarsal elongation and specialized rotational mechanisms in the midfoot.

While some of these traits are also found in indriids, lepilemurids, and especially bush babies, tarsiers are unique in possessing this particular suite of morphological features, all of which are clearly related to the demands of a small-bodied vertical clinging and leaping primate. The presence of alternative morphological adaptations to vertical clinging and leaping among living primates suggests that homoplasy has been the rule in the evolution of locomotion among leaping prosimians.

References

Alezais H. 1900. Le quadriceps fémoral des sauteurs. C R Acad Sci Paris 52: 510–511.

Anemone RL. 1988. The functional morphology of the prosimian hindlimb: some correlates between anatomy and positional behavior. Ph.D. dissertation, University of Washington.

Anemone RL. 1990. The VCL hypothesis revisited: patterns of femoral morphology among quadrupedal and saltatory prosimian primates. Am J Phys Anthro 83: 373–393.

Anemone RL. 1993. The functional anatomy of the hip and thigh in primates. In Gebo DL, editor, Postcranial adaptation in nonhuman primates, 150–174. DeKalb: Northern Illinois University Press.

Anemone RL, Covert HH. 2000. New skeletal remains of *Omomys* (Primates, Omomyidae): functional morphology of the hindlimb and locomotor behavior of a Middle Eocene primate. J Hum Evol 38: 607–633.

Ankel F. 1972. Vertebral morphology of fossil and extant primates. In Tuttle R, editor, The functional and evolutionary biology of primates, 233–240. Chicago: Aldine-Atherton.

Biegert J, Maurer R. 1972. Rumpfskelettlange, Allometrien und Körperproportionen bei catarrhinen Primaten. Folia Primatol 17: 142–156.

Burr DB, Piotrowski G, Martin RB, Cook PN. 1982. Femoral mechanics in the lesser bushbaby (*Galago senegalensis*): structural adaptations to leaping in primates. Anat Rec 202: 419–429.

Cartmill M. 1972. Arboreal adaptations and the origin of the Order Primates. In

Tuttle R, editor, The functional and evolutionary biology of primates, 97–122. Chicago: Aldine-Atherton.

Cartmill M. 1974. Pads and claws in arboreal locomotion. In Jenkins FA, editor, Primate locomotion, 45–83. New York: Academic Press.

Cartmill M. 1982. Assessing tarsier affinities—Is anatomical description phylogenetically neutral? Geobios Mem Spec 6: 279–287.

Clark WEL. 1971. The antecedents of man. Chicago: Quadrangle Books.

Dagosto M. 1985. The distal tibia of primates with special reference to the Omomyidae. Int J Primatol 6: 45–75.

Dagosto M. 1993. Postcranial anatomy and locomotor behavior in Eocene primates. In Gebo DL, editor, Postcranial adaptation in nonhuman primates, 199–219. DeKalb: Northern Illinois University Press.

Dagosto M, Schmid P. 1996. Proximal femoral anatomy of omomyiform primates. J Hum Evol 30: 29–56.

Dagosto M, Terranova CJ. 1992. Estimating the body size of Eocene primates: a comparison of results from dental and postcranial variables. Int J Primatol 13: 307–344.

Day MH, Iliffe SR. 1975. The contrahens muscle layer in *Tarsius*. Folia Primatol 24: 241–249.

Day MH, Napier JR. 1963. The functional significance of the deep head of the flexor pollicis brevis in primates. Folia Primatol 1: 122–134.

Demes B, Fleagle JG, Lemelin P. 1998. Myological correlates of prosimian leaping. J Hum Evol 34: 385–399.

Demes B, Jungers WL, Fleagle JG, Wunderlich RE, Richmond BG, Lemelin P. 1996. Body size and leaping kinematics in Malagasy vertical clingers and leapers. J Hum Evol 31: 367–388.

Fleagle JG. 1999. Primate adaptation and evolution. New York: Academic Press.

Fleagle JG, Anapol F. 1992. The indriid ischium and the hominid hip. J Hum Evol 22: 285–306.

Fleagle JG, Kay RF. 1994. Anthropoid origins. New York: Plenum Press.

Gebo DL. 1985. The nature of the primate grasping foot. Am J Phys Anthro 67: 269–277.

Gebo DL. 1987a. Functional anatomy of the tarsier foot. Am J Phys Anthro 73: 9–31.

Gebo DL. 1987b. Locomotor diversity in prosimian primates. Am J Primatol 13: 271–281.

Gebo DL. 1988. Foot morphology and locomotor adaptation in Eocene primates. Folia Primatol 50: 3–41.

Gebo DL. 1993a. Functional morphology of the foot in primates. In Gebo DL, editor, Postcranial adaptation in nonhuman primates, 175–196. DeKalb: Northern Illinois University Press.

Gebo DL. 1993b. Postcranial adaptation in nonhuman primates. DeKalb: Northern Illinois University Press.

Gebo DL, Dagosto M. 1988. Foot anatomy, climbing, and the origin of the Indriidae. J Hum Evol 17: 135–154.

Godfrey LR. 1988. Adaptive diversification of Malagasy strepsirhines. J Hum Evol 17: 93–134.

Gregory WK. 1910. The orders of mammals. Bull Am Mus Nat Hist 27: 1–524.

Haffer A. 1929. Bau und Funktion des Affenfusses. Ein Beitrag zur gelenk- und muskelmechanik. 1. Die Prosimier. Z Anat Entw Gesch 90: 46–51.

Hall-Craggs ECB. 1964. The jump of the bush baby. Med Biol Illus 14: 170–174.

Hall-Craggs ECB. 1965a. An analysis of the jump of the lesser galago (*Galago senegalensis*). J Zool 147: 20–29.

Hall-Craggs ECB. 1965b. An osteometric study of the hind-limb of the jump of the Galagidae. J Anat 99: 119–126.

Hall-Craggs ECB. 1966. Muscle tension relationships in *Galago senegalensis*. J Anat 100: 699–700.

Hall-Craggs ECB. 1974. Physiological and histochemical parameters in comparative locomotor studies. In Martin RD, Doyle GA, Walker AC, editors, Prosimian biology, 829–845. London: Duckworth.

Hill WCO. 1955. Primates: comparative anatomy and taxonomy, 2. Haplorhini: Tarsioidea. Edinburgh, UK: Edinburgh University Press.

Hopwood PR, Butterfield RM. 1976. The musculature of the proximal pelvic limb of the eastern grey kangaroo *Macropus major* (Shaw) and *Macropus giganteus* (Zimm). J Anat 121: 259–272.

Howell AB. 1932. The saltatorial rodent *Dipodomys*: the functional and comparative anatomy of its muscular and osseous systems. Proc Am Acad Arts Sci 67: 377–536.

Jouffroy FK. 1962. La musculature des membres chez les lémuriens de Madagascar. Etude descriptive et comparative. Mammalia 26: 1–326.

Jouffroy FK. 1975. Osteology and myology of the lemuriform postcranial skeleton. In Tattersall I, Sussman RW, editors, Lemur biology, 149–192. New York: Plenum Press.

Jouffroy FK, Gasc JP. 1974. A cineradiographical analysis of leaping in an African prosimian (*Galago alleni*). In Jenkins FA, editor, Primate locomotion, 117–142. New York: Academic Press.

Jouffroy FK, Lessertisseur J. 1978. Etude ecomorpologique des proportions des membres des primates et specialement des prosimiens. Anal Sci Nat Zool, Paris 20: 99–128.

Jouffroy FK, Lessertisseur J. 1979. Relationships between limb morphology and locomotor adaptations among prosimians: An osteometric study. In Morbeck ME, Preuschoft H, Gomberg N, editors, Environment, behavior, and morphology: dynamic interactions in primates, 143–181. New York: Gustav-Fischer-Verlag.

Jouffroy FK, Berge C, Niemitz C. 1984. Comparative study of the lower extemity in the genus *Tarsius*. In Niemitz C, editor, Biology of tarsiers, 167–190. Stuttgart: Gustav-Fischer-Verlag.

Jouffroy FK, Gasc JP, Decombas M, Oblin S. 1974. Biomechanics of vertical leaping from the ground in *Galago alleni*: a cinereadiographic analysis. In Martin RD, Doyle GA, Walker AC, editors, Prosimian biology, 817–827. London: Duckworth.

Jouffroy FK, Gasc JP, Oblin S. 1973. An application of cineradiography to the study of locomotion in Primates. Am J Phys Anthro 38: 527–530.

Jungers WL, Jouffroy FK, Stern JT. 1980. Gross structure and function of the quadriceps femoris in *Lemur fulvus:* an analysis based on telemetered electromyography. J Morphol 164: 287–299.

Martin RD. 1972. Adaptive radiation and behaviour of the Malagasy lemurs. Phil Trans R Soc Lond 264: 295–352.

Martin RD. 1990. Primate origins and evolution: a phylogenetic reconstruction. Princeton, NJ: Princeton University Press.

McArdle JE. 1981. Functional morphology of the hip and thigh of the Lorisiformes. Contrib Primatol 17: 1–132.

Mollison, T. 1911. Die Körperproportionen der Primaten. Morphol Jahrb 42: 79–300.

Napier JR, Walker AC. 1967. Vertical clinging and leaping—A newly recognized category of locomotor behavior of primates. Folia Primatol 6: 204–219.

Niemitz C. 1977. Zur Funktionsmorphologie und Biometrie der Gattung *Tarsius* Storr, 1780. Cour Forsch Senkenberg 25: 1–161.

Niemitz C. 1979. Outline of the behavior of *Tarsius bancanus*. In Doyle GA, Martin RD, editors, The study of prosimian behavior, 621–660. New York: Academic Press.

Niemitz C. 1984a. Biology of tarsiers. Stuttgart: Gustav-Fischer-Verlag.

Niemitz C. 1984b. Locomotion and posture of *Tarsius bancanus*. In Niemitz C, editor, Biology of tarsiers, 191–225. Stuttgart: Gustav-Fischer-Verlag.

Niemitz C. 1984c. Synecological relationships and feeding behavior of the genus *Tarsius*. In Niemitz C, editor, Biology of tarsiers, 59–75. Stuttgart: Gustav-Fischer-Verlag.

Niemitz C. 1984d. Taxonomy and distribution of the genus *Tarsius* Storr, 1780. In Niemitz C, editor, Biology of tarsiers, 1–16. Stuttgart: Gustav-Fischer-Verlag.

Niemitz C. 1985. Leaping locomotion and the anatomy of the tarsier. In Kondo S, editor, Primate morphophysiology, locomotor analyses and human bipedalism, 235–250. Tokyo: University of Tokyo Press.

Preuschoft H, Fritz M, Niemitz C. 1979. Biomechanics of the trunk in primates and problems of leaping in *Tarsius*. In Morbeck ME, Preuschoft H, Gomberg N, editors, Environment, behavior, and morphology: Dynamic interactions in primates, 327–345. New York: Gustav-Fischer-Verlag.

Rafferty KL. 1998. Structural design of the femoral neck in primates. J Hum Evol 34: 361–383.

Rasmussen DT, Conroy GC, Simons EL. 1998. Tarsier-like locomotor specializations in the Oligocene primate *Afrotarsius*. Proc Natl Acad Sci USA 95: 14848–14850.

Rose MD. 1988. Another look at the anthropoid elbow. J Hum Evol 17: 193–224.

Rose MD. 1993. Functional anatomy of the elbow and forearm in primates. In Gebo DL, editor, Postcranial adaptation in nonhuman primates, 70–95. DeKalb: Northern Illinois University Press.

Schultz AH. 1961. Vertebral column and thorax. Primatologia 4: 1–66.

Schultz M. 1984. Osteology and myology of the upper extremity of *Tarsius*. In Niemitz C, editor, Biology of tarsiers, 143–165. Stuttgart: Gustav-Fischer-Verlag.

Shapiro L. 1993. Functional morphology of the vertebral column in primates. In Gebo DL, editor, Postcranial adaptation in nonhuman primates, 121–149. DeKalb: Northern Illinois University Press.

Simpson GG. 1945. The principles of classification and a classification of mammals. Bull Am Mus Nat Hist 85: 1–350.

Smith JM, Savage RJR. 1956. Some locomotory adaptations in mammals. J Linn Soc 42: 603–622.

Smith RJ, Jungers WL. 1997. Body mass in comparative primatology. J Hum Evol 32: 523–559.

Sprankel H. 1965. Untersuchungen an *Tarsius*. 1. Morphologie des Schwanzes nebst ethnologischen Bemerkungen. Folia Primatol 3: 153–188.

Stern JT. 1971. Functional myology of the hip and thigh of cebid monkeys and its implications for the evolution of erect posture. Biblio Primatol 14: 1–319.

Stern JT. 1974. Computer modeling of gross muscle dynamics. J Biomech 7: 411–428.

Stern JT, Oxnard CE. 1973. Primate locomotion: some links with evolution and morphology. Primatologia 4: 1–93.

Swindler DR, Wood CB. 1973. An atlas of primate gross anatomy: baboon, chimpanzee, and man. Seattle: University of Washington Press.

Szalay FS. 1972. Paleobiology of the earliest primates. In Tuttle R, editor, The functional and evolutionary biology of primates, 3–35. Chicago: Aldine-Atherton.

Szalay FS. 1976. Systematics of the Omomyidae (Tarsiiformes, Primates): taxonomy, phylogeny, and adaptations. Bull Am Mus Nat Hist 156: 157–450.

Szalay FS, Dagosto M. 1980. Locomotor adaptations as reflected on the humerus of Paleogene primates. Folia Primatol 34: 1–45.

Szalay FS, Dagosto M. 1988. Evolution of hallucial grasping in the primates. J Hum Evol 17: 1–33.

Tardieu C. 1981. Morpho-functional analysis of the articular surfaces of the knee-joint in primates. In Chiarelli AB, Corruccini RS, editors, Primate evolutionary biology, 68–80. Berlin: Springer-Verlag.

Wible JR, Covert HH. 1987. Primates: cladistic diagnosis and relationships. J Hum Evol 16: 1–22.

Woollard JJ. 1925. The anatomy of *Tarsius spectrum*. Proc Zool Soc Lond 70: 1071–1184.

CHAPTER 5

The Axial Skeleton of Primates: How Does Genus *Tarsius* Fit?

Friderun A. Ankel-Simons and Cornelia Simons

In this chapter, we describe the morphology of the tarsier axial skeleton and discuss its functional implications. Our description provides a qualitative description rather than a quantitative analysis of vertebral morphology. To describe and discuss this integral part of the *Tarsius* postcranium, we establish a comparative baseline and outline the baseline in a discussion of the basic "Bauplan" of the primate axial skeleton.

Our precise definition of basic, unspecialized primate morphology will provide a valid description of what constitutes a specialization in primate spinal morphology in general, and tarsier spinal morphology in particular. To describe specializations—a characteristic that deviates from the morphological "norm" of a particular group in some manner—a "norm" must be established.

The important issue of "functional potential" (Ankel, 1967) will be considered. Functional potential constitutes the capacity of any morphological feature to undertake activities that lie outside the normal habitual behavioral repertoire of the animal. For example, a tail reduced to a mere stump and only a few vestigial vertebral bodies, such as that found in humans, does not have any functional potential. Similarly, the short tails of primates like *Macaca mulatta* do not have a great deal of functional potential. Long tails, however, have quite a high degree of functional potential since they can be used in various ways, such as fifth extremities or prehensile tails (in Alouattinae and Atelinae), as a third strut in upright positions (as in *Tarsius*), or as a balancing and steering appendage in many of the long-tailed, jumping primates (such as galagos or colobines).

The functional potential of any morphological feature, particularly in the postcranium of primates, usually exceeds the habitual functional repertoire. For example, humans have several distinctive and unique spinal morphological features that define and restrict the spine's functional potential in these habitually bipedally walking, standing, and running primates. These features are autapomorphies, specializations, or adaptations to a mode of locomotion unique among living primates. However, the axial skeleton of the only habitually bipedal primate is also quite capable of accommodating very different functional activities (e.g., swimming, cycling, turning somersaults, climbing trees and steep mountains). This means that even a rather derived, or specialized, axial skeleton still has enormous functional potential that obviously exceeds the morphological characteristics of the highly specialized bipedal walker.

This fact holds for less derived primates, such as capuchin monkeys or cercopithecines. In such morphologically derived vertebrates as, for example antelopes or whales, the functional potential is comparatively low, as they are highly adapted to specific environments; they are unable to climb trees or brachiate. For most primates, the axial skeleton does not have any morphological specializations that could effectively restrict functional performance; their morphological potential is very high because they are comparatively nonderived.

As Ankel (1967) suggested, distinct morphological specializations within the primate axial skeleton are few. Cartmill's (1985: 87) statement, "All nonleaping arboreal mammals have markedly specialized vertebral columns," is an extreme assessment. The spines of arboreal nonleaping howling monkeys, capuchin monkeys, or the large colobine monkey *Nasalis larvatus*, for example, are not "markedly specialized." The vertebrate axial skeleton is a functionally stiff, s-shaped, springlike pole, tightly bound on the outside by tendons and ligaments, with only slight and directionally restricted ability to bend and flex and only in rare exceptions to rotate. The intervertebral disks are the points of elastic cushioning between vertebral bodies that provide some padding and shock absorption qualities, while also fastening the vertebral bodies together.

Primate spines are made of (1) vertebral centra or bodies, (2) intervertebral disks, (3) the ossified neural canal that surrounds and protects the spinal cord on the dorsal aspect of vertebrae and multiple bony projections (paired processus articularis also called pre- and post zygapophyses, occur, on the cranial and caudal end of the neural canal, and (4) single processus spinosus projecting from the center of the dorsal roof of the neural canal, also called *spinous processes,* and a pair of lateral processus transversalis, also called *transverse processes* or *diapophyses* (on each side of the vertebral bodies). All of these structures guide and constrain the mobility of the entire spine. These bony processes of the axial skeleton serve as areas of attachment for muscles and their tendons.

To visualize the restricted but incredibly versatile flexibility of axial morphology in primates, we show in this chapter how the shape of the vertebral bodies (or centra), their bony attachments, and the thickness and flexibility of the intervertebral disks fundamentally regulate the ability of the entire spine and its parts to provide the body with both stability and flexibility.

When discussing the spine of primates we encounter a major problem.

Not only the thickness and structure of the integral intervertebral disks, but also their functions are virtually unknown in nonhuman primates and can be properly evaluated only in living, active animals. This, in turn, means that only multidimensional X-ray movies of all the regions of a spine in motion could provide us with true images of the role that intervertebral disks have during locomotion. However, these fundamental details remain unidentified. We therefore must be content to evaluate the mobility of the axial skeleton as a unit by inference.

Size and shape of vertebral bodies and their intervertebral disks crucially influence flexibility and weight-bearing properties of the spine. Simply said, both craniocaudal length and dorsoventral shape (such as circular, dorsoventrally compressed, laterally compressed, and so on) of vertebral bodies together with the thickness and elasticity of the intervertebral disks dictate mobility between one body and the adjacent ones. Generally speaking, long vertebral bodies allow less of an arch, per unit length, when bending in any direction than do short ones. Theoretically, if we look at vertebral bodies that are only connected to each other by intervertebral disks (as in the distal portion of tails) it is obvious that, when flexed as much as possible, long vertebral bodies create a wider arch, with a larger radius, than short vertebral bodies of equal number (fig. 5.1).

Also, vertebral bodies that are round, when viewed from a cranial or caudal perspective, allow less of an angle between two adjacent vertebrae than do dorsoventrally compressed ones. A striking example of the latter situation is the terminal part of the caudal region of prehensile-tailed New World monkeys. Here the spine is essentially reduced to vertebral bodies and their intervertebral disks and the tendons of the extensor and flexor caudalis muscles. In the prehensile tails of Atelines (*Ateles, Brachyteles, Lagothrix*), these vertebral bodies are strongly dorsoventrally flat and comparatively short (Ankel, 1962), thus accommodating predominantly dorsoventral movements between adjacent vertebrae and enabling the animal to curl the tail end into a tight circle in a ventral direction. This would be impossible with long vertebral bodies that are round.

In primates, all bony processes on the vertebral body and the neural arch so severely limit possible movements between adjacent vertebrae that flexibility is often restricted. This is also mandated by the simple fact that the axial skeleton envelops and protects the spinal cord, nerves, and blood vessels, which could easily be damaged by unlimited intervertebral mobility.

Establishing the Baseline

The spine is commonly described according to regions. Beginning at the base of the skull and proceeding toward the tail (or caudally), we distinguish

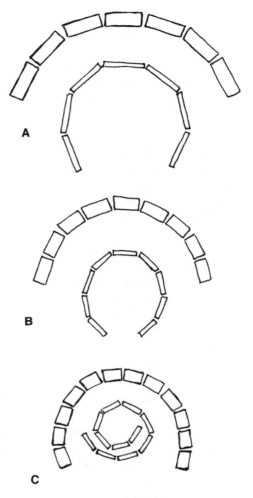

Figure 5.1. This figure documents the influence of the two variables, length and diameter of vertebral bodies, on the ability of regions of the spine to flex in a dorsoventral direction, considering only the vertebral bodies and intervertabral disks. All six units are of the same length. (A) Long vertebral bodies with large and small diameters. (B) Medium length vertebral bodies with large and small diameter compared with (C) short vertebral bodies with large and small diameter.

between the cervical region, thoracic region, lumbar region, sacral region, and caudal region. These regions vary in the presence, absence, and shape of the neural canal, the position and length of attached processes, and in the number of vertebrae. Thus, the axial skeleton acts as a whole, and any perceived movement is in reality the sum of many very small changes of position between adjacent vertebrae that together add up to either bending dorsoventrally, flexing laterally, or even twisting of the entire structure. Movements between adjacent vertebrae in living animals are obscured by the overlaying musculature and epidermis covered with fur, and it is problematical to pinpoint where the movement actually takes place.

The weight-bearing capability of the spine is restricted by the capacity of the intervertebral disks to withstand stresses. These involve compressive forces as well as stretching, twisting, and pulling. However, as muscles, ligaments, and tendons tightly envelop the entire spine, these forces are usually absorbed and diverted. The first part of the spine to incur damage if overly stressed is usually the intervertebral disk.

The articular processes (zygapophyses), which are part of the neural arch, are directed cranially and caudally and extend beyond the vertebral body to connect vertebrae. The prezygopophyses have articular facets that are directed dorsally or laterally while the postzygopophyses are directed ventrally or medially as they meet the articular processes of the adjacent vertebrae. The position of these zygapophyses differs characteristically between regions (e.g., Ankel, 1967; Ankel-Simons, 2000). There is an intermediary vertebra at the end of the cervical and thoracic region that has prezygapophyses with the position characteristic for the region cranial to it and with postzygapophyses in the different and characteristic position of the caudally adjacent axial region.

The change of position of the articular processes between the pre- and postzygapophyses in this transitional vertebra always occurs abruptly within one single vertebra. It would compromise the stability of the axial skeleton if this change of articulations between regions should happen gradually. There would be no functional stability within and between regions, which essentially establishes the crucial functional integrity of the entire spine. Snakes are a good example of animals without transitional vertebrae, with this drastic change of zygophyseal position between the cranial and caudal zygapophyses. Their long axial skeletons are rather uniform as far as zygapophyseal position is concerned, and consequently no regional differences exist in mobility or rigid stability along the entire length of the spine. Without this functional integrity, the primate axial skeleton would be randomly flexible and thus instable and functionally useless for animals with two limb girdles.

Basics of Primate Axial Morphology

Within the cervical region of primates, the two most cranial vertebrae are always characteristic in their morphology as they accommodate the positioning of the skull on the spine. The fundamental morphology of these two extraordinarily well-adapted vertebrae rarely varies in primates.

The first ring-shaped cervical vertebra lacks the vertebral body and has two bowl-shaped articulations on its ventral, anterior aspect that accommodate the occipital condyles on the anterior aspect of the foramen magnum of the skull. The intervertebral disk is not involved in this articulation. This first cervical vertebra is named "atlas," as it carries the head.

This connection constitutes a true joint, which allows forward and rearward movements of the skull against the atlas (such as nodding in agreement in human behavior). Medially on the dorsal aspect of the ventral part of the atlas we find a single articulation that receives the cranially directed dens of the second cervical vertebra. This dens inserts above the ventral arch of the atlas and establishes the only intervertebral articulation that allows free rotary movement between two vertebrae (or shaking of the head as in showing negative reactions in human behavior). This is the second true joint of the spine that also lacks the intervertebral disk so characteristic for all other regions of the axial skeleton except within the sacral region, where the vertebral bodies are fused with each other and intervertebral disks are obliterated. The dens (which ontogenetically is the missing vertebral body of the atlas that is cranially fused onto the vertebral body of the axis) is tied down tightly onto the atlas and thus separated from the lumen of the neural canal by a strong ligament (transverse ligament).

Laterally on the atlas are posteriorly directed articular facets that meet the corresponding articulations on the roots of the neural arch of the axis. Interestingly, these paired lateral articulations between the atlas and axis and the position of the single articulation between the dens of the axis with the inner surface of the ventral arch of the atlas show distinctive differences between upright humans and quadrupedal primates. These differences are directly correlated with the carriage of the head on the axial skeleton in these two disparate locomotor positions. The angle between the dens and the vertebral body of the axis changes in primates with different locomotor habits. The dens is directed straight upward in bipedal humans and angled dorsally in all primates that do not habitually move bipedally (Ankel, 1967; Ankel-Simons, 2000). The number of vertebrae in each region of the axial skeleton varies in a characteristic manner between primate groups. Only the number of cervical vertebrae is the same (seven) in all primates and in most mammals (Schultz, 1961).

Comparative Morphology of the Axial Skeleton of *Tarsius*

Unfortunately, in comparative and descriptive vertebrate morphology, the odds for possible errors seem to increase dramatically with decreasing object size (Miller, 1996). This is a serious problem for the evaluation of small to tiny bones and fossil elements because important structural details become obscured or simply do not develop in very small skeletal elements.

This is, to a great extent, due to the plasticity of living material such as bone. As tarsiers are among the smallest of living primates, our evaluation of the axial skeleton and its parts will be done with appropriate caution.

Hill (1955) described a bony lamina of the atlas that stretches dorsally across the axial dens, separating it from the lumen of the atlas's neural canal and reputedly facilitating the unusual ability of tarsiers to turn their heads 180 degrees backwards. However, no such bony lamina has been found in any tarsier atlas that we have been able to examine.

In 1967, Ankel illustrated the unusual position of the intervertebral zyga-pophyses between the cervical vertebrae 2 to 7, which are very likely crucially implicated in the tarsiers' ability to turn their heads straight backwards, not unlike owls. In the same article, Ankel discusses the extraordinary length of the thoracic transverse processes of the first ten or so vertebrae of tarsiers that have continuous articular grooves (no separation between the capitu-lum and tuberculum costae) ventrally for the associated long and continual articular facets on the first nine ribs that articulate with these exceedingly long transverse processes (see also Ankel-Simons, 2000).

Sprankel (1965) and Grand and Lorenz (1968) discussed the use of the proximal tail as a tripodial strut (the feet plus tail) by *Tarsius bancanus* and *T. syrichta* when they are clinging to upright supports, stating that in *T. banca-nus* it is "the seventh caudal vertebra (which) is the most proximal point of tail support," while "in *T. syrichta*, the use of the tail appears to depend upon the degree of inclination of the branch" (Grand and Lorenz 1968: 165).

The paper by Sprankel (1965) reports, in great detail, behavioral and morphological characteristics of *T. bancanus* and *T. syrichta*. The discussions are mainly based on observations of captive tarsiers but are occasionally supported by information about behavior in their natural habitat. Sprankel (1965) indicates that three species, namely *T. bancanus, T. syrichta,* and *T. spectrum* all use the proximo-ventral aspect of their tails as support while resting in an upright position on vertical branches. According to Sprankel (1965), the two former species have friction skin that covers an area of the proximal underside of their tails, and *T. bancanus* has papillary ridges on this caudal friction pad.

The number of vertebrae in the different spinal regions of tarsiers has been established by Schultz (1961): (1) seven cervical vertebrae; (2) thirteen (sometimes twelve) thoracic vertebrae (counted according to the presence of ribs) as compared to nine (sometimes eight) if counted "functionally" ac-cording to articular position (Washburn and Buettner-Janusch, 1952; Erik-son, 1963; Shapiro, 1993); (3) six lumbar vertebrae counted according to the absence of ribs and ten (or eleven) if counted "functionally" according to articular position; (4) three sacral vertebrae (rarely two); and (4) an av-erage of twenty six tail vertebrae.

Our description is based on the study of three *Tarsius syrichta* and one *T. bancanus* specimens that are in the Osteology Collection of the Duke Primate Center Division of Paleontology. Additionally, many previously noted observations through the years (by FAS) on various specimens of these two species of tarsiers are also incorporated here.

The Cervical Region of Primates

Both of the first two cervical vertebrae (the atlas and axis) have foramina (transverse foramina for the transit of the vertebral arteries and veins) in the base of the transverse processes. These foramina are characteristic for all primate cervical vertebrae. The neural arches are usually cranio-caudally short or clasplike. Length and shape of spinal processes varies widely in this region among primates. The ends of the short transverse processes are often split into two tubercles.

The articular facets are usually shingle-like with the cranial ones directed cranially, dorsally, and slightly medially and the caudally adjacent facets covering them in a shingle-like manner. Seen from cranially or caudally, these articulations appear situated on the circumference of a large circle with its center way above the vertebra (see also Ankel-Simons, 2000: 252). Generally, the cranial and caudal surfaces of cervical vertebral bodies 3 to 7 are not flat. The cranial surfaces have cranially directed lateral extensions that embrace the somewhat indented sides of the cranially preceding vertebral body laterally. Also, the ventrocaudal edge of the bodies is somewhat extended caudally and reaches under the cranioventral edge of the following body, which is accordingly concave and sloping caudally with the extension of the cranially adjacent vertebra overlapping the underside of the vertebra caudal to it. These synarthrotic articulations between vertebral bodies in the cervical region of primates thus are saddle shaped and interlocking, virtually preventing much twisting and turning between adjacent vertebrae. There often is a sagittal ridge medially on the ventral aspect of the vertebral bodies.

In primates the cervical region is the most flexible section of the presacral spine with the ability to freely bend back, forth, and laterally.

The Atlas of *Tarsius*

Morphologically, the atlas of tarsiers is remarkably unremarkable. It appears that the first two cervical vertebrae of primates are all very similar, which in turn means that their basic morphology is unusually well suited to their function.

The one feature that makes the atlas stand out is the fact that it is unusually large. In *Tarsius* the atlas is the widest of all vertebrae. It exceeds the width of the second cervical vertebra, the axis, by a factor of two (see figs. 5.2 and 5.3).

The transverse processes are pierced by inconspicuous craniocaudally open foramina at their roots, and directly above these, in the dorsally ascending arches, somewhat larger foramina perforate the arches in a lateral direction. The cranially directed surface areas of the articular facets with the skull are large, positioned on the lower third of the vertebral ascending arches, and strongly turned up in a dorsal direction to accommodate and literally clasp the occipital condyles on either side of the foramen magnum. On the ventral arch, a deep median indentation carries the articular facets, which cradle the dens of the second cervical vertebra, the axis, from below and on its sides. This articular groove is not closed off dorsally by a bony sheet and thus is not separated from the neural canal in any of the tarsiers that we have been able to inspect, contra the description of Hill (1955).

Paired, caudally directed articulations with the body of the axis are located next to the ventral articular groove cradling the dens of the axis. This is the articulation between atlas and axis where most of the rotary movements of head plus atlas against the spine take place. The cranially directed and comparatively large articulations between the atlas and the occipital of the skull exclusively permit ventral and dorsal bending movements and cannot carry out any rotary movements at all. As in many primates, the atlas lacks a spinous process.

The Axis of Tarsius

As mentioned above, in tarsiers' second cervical vertebra, the axis is only about half as wide as the atlas. Overall, it very much resembles the axis of most primates. The dens is directed dorsally in a cranial direction, in an angle of 43 degrees against the long axis of the vertebral body (fig. 5.4a).

This angulation causes the axis to be tilted dorsally against the dorsoventral axis of the atlas, thus positioning the spine dorsally relative to the cranium. This is most interesting as it does not reflect the fact that tarsiers tend to have an upright posture, and that they spend about 80% of their time sitting or clinging upright and resting, according to Grand and Lorenz (1968) (see also Ankel, 1967; Ankel-Simons, 2000).

The zygapophyses between axis and third cervical vertebra are slanted, not unlike those between the postzygapophyses of the seventh cervical and the prezygapophyses of the first thoracic vertebra (fig. 5.4a, large arrows). The articular facets in front are positioned laterally on the vertebral body of the axis directly adjacent to the root of the dens. The slight spinous process is V shaped or Y shaped, with the V opening in a caudal direction.

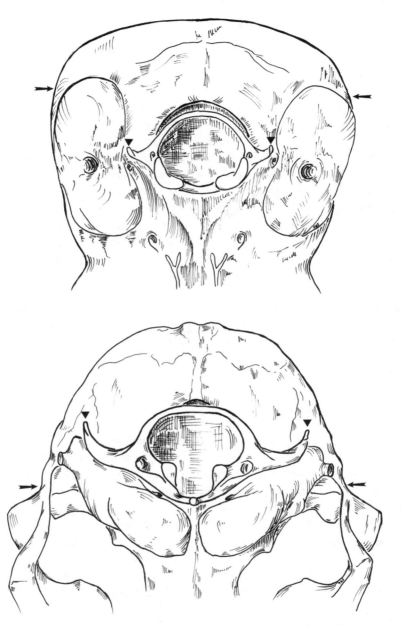

Figure 5.2. Comparison of the back of the skull base and superimposed atlas of two primate species with approximately the same body size. (Top) The New World monkey and smallest of anthropoids, *Cebuella pygmaea*. (Bottom) *Tarsius syrichta*. Note the comparatively larger foramen magnum of *Tarsius*. Width across skull base at arrows measures 2.0 cm. Transverse process ends of atlas indicated by black triangles.

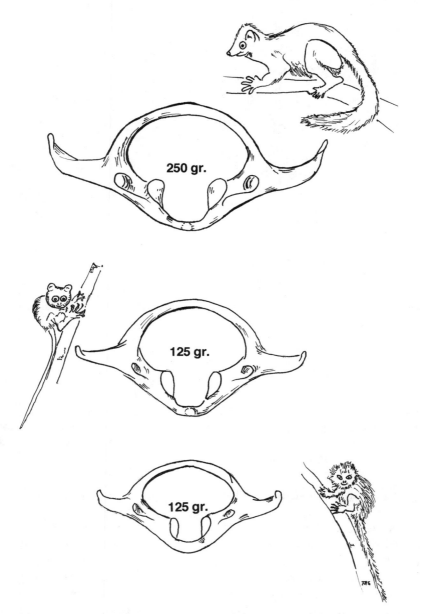

Figure 5.3. Comparison of body size and size of atlas of three different primates. *Mirza* (top) is approximately twice as large as genus *Tarsius* (middle) and genus *Cebuella* (bottom). Note the different size of the atlas; the atlas of *Tarsius* is comparatively the largest in correlation to the body weight of the three genera.

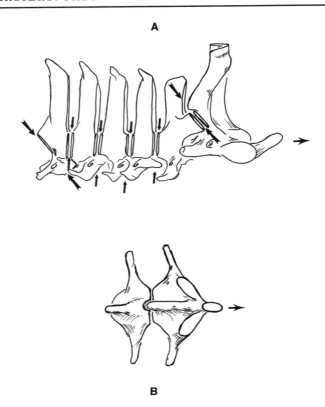

Figure 5.4. (A) Cervical region of *Tarsius,* seen from lateral, cranial to the right. Note the different positions of the zygapophyses. (B) Ventral view of the axis and first cervical vertebra of *Tarsius* showing the pseudo-articular connection between the vertebral bodies of the cervical region (cranial to the right).

The Cervical Vertebrae Three through Seven of *Tarsius*

The vertebral bodies are keeled ventromedially and equipped with a small groove on their cranio-ventral rim that accommodates the pointedly tongue-like elongated caudal ends of the vertebral body in front (fig. 5.4b). This arrangement provides a pseudoarticular connection between vertebral bodies on the bottom of the vertebral centra. Cervical vertebrae 3 through 5 have very short spines on the dorsal vertebral arch. Cervicals 6 and 7 have slightly cranially directed spinous processes that smoothly blend over to the position of the spinous processes of the caudally adjoining thoracic region. The transverse processes are slanted caudolaterally and slightly ventrally. They are not very prominent and craniocaudally perforated by comparatively large foramina. Within the cervical region the overall breadth of vertebrae 3 through 7 rapidly increases from the rather small axis.

Despite this fact, the diameters of the vertebral bodies remain small (about a third of the total width of the vertebrae), and they are triangular in crosssection. The apex of the triangle points ventrally and represents the keel along the ventral aspect of the vertebral body. Vertebrae 3 through 7 are craniocaudally short. In cervical vertebrae 3 through 7 the articular surfaces of the zygapophyses are positioned as if lying on a transverse plane that is transverse and perpendicular to the long axis of the spine, with the articular surfaces not being slanted as they are in other primates (fig. 5.4, small arrows; see also Ankel-Simons, 2000).

The seventh cervical vertebra is transitional in the position of the intervertebral articulations between the cervical and thoracic region (fig. 5.4, large arrows) that are positioned laterally on the ascending arches of the neural canal, not above the vertebral centers. This arrangement is unlike that of any other extant primate, with the angled position of these articulations preventing any rotary displacement between cervical vertebrae. In tarsiers the pseudo "ball and socket" articulations between vertebral bodies 2 through 7 described above provide the pivot for such turning movements between the cervicals. The last cervical vertebra adopts the articular position of the zygapophyses in the thoracic region, facing dorsally and ventrally, respectively.

The Thoracic Region of Primates

The seventh cervical vertebra is the transitional vertebra between the cervical and thoracic regions and is commonly regarded to be the last cervical vertebra. It has postzygapophyses in the position of the adjacent thoracic region. In primates, the number of thoracic vertebrae varies between eleven and seventeen, with an average number of twelve (Schultz, 1961). They are almost shingle-like, overlapping each other, and the articular facets of the zygapophyses are tucked underneath the arch, meeting the cranially directed associated articulations of the caudally adjacent vertebra. These articulations are positioned as if lying on the circumference of a circle with a center that is located somewhere near the midpoint of the cranial or caudal end of the vertebral bodies. Usually, however, this center, even though close to the midpoint of the cranial or caudal surface of the vertebral body, does not coincide with this midpoint of the vertebral body's ends, thus clearly making any circular movement between thoracic vertebrae impossible (Ankel-Simons, 2000: 252).

As the neural arches and vertebral bodies gradually become craniocaudally longer, the spinous processes become more prominent and are usually directed caudally. The thoracic region is the most complicated region of the primate axial skeleton as it is the region supporting the ribs that, together with the sternum, create the trunk. The vertebral bodies are short and often approximately heartshaped when seen from the cranial and caudal ends.

The first most cranially located ribs are laterally attached by means of a short cartilage to the large cranial part of the sternum, the so-called manubrium. This articulation is positioned posterior to that of the clavicles with the manubrium. The first rib has only one articulation on its dorsal end and articulates laterally with the body of the first thoracic vertebra. Two articulations exist on the dorsal end of all so-called true ribs; the capitulum costae, on the very end of the rib, straddles the intervertebral disk and attaches laterally to the dorsocranial and dorsocaudal aspect of adjacent vertebral bodies.

The second place of articulation is located slightly laterally and occurs between the tuberculum costae on the rib's dorsal aspect and the underside of the transverse processes of the caudally adjoining vertebra. The last most caudally positioned two or three ribs usually lose their direct cartilaginous connection with the sternum as well as the tubercular connection with the transverse process of the caudally adjoining vertebra. The change of zygapophyseal position from the thoracic position to the lumbar position in many primate genera occurs within the rib-bearing part of the thoracic region, so the last few ribs belong to vertebrae that have articular processes that are positioned in the manner characteristic for the lumbar region, not the thoracic region.

This fact has led researchers to count the number of thoracic versus lumbar vertebrae according to the position of the zygapophyses, not according to the presence and absence of ribs (see also Shapiro, 1993, and references therein). The difference of counting the number of thoracic versus lumbar vertebrae according to articulations as compared to the presence or absence of ribs is advocated as the functionally sensible way of counting these vertebrae. When counting in this manner, the number of thoracic vertebrae diminishes slightly as the "functional" lumbar region is intellectually elongated in cranial direction, supposedly increasing the leaping abilities of quadrupedally running and jumping primates. The vertebra that is transitional between the thoracic and the lumbar articular position is usually also the vertebra of directional change of the spinous processes. The spinous processes tend to be directed caudally in the thoracic region and cranially in the lumbar region.

Unambiguous factual documentation of this supposed functional advantage of "articular elongation of the lumbar region," however, has never been published.

The overall shape of the trunk is partially created by the vital organs that it protects (such as, for example, the constantly contracting heart, the contracting and expanding lungs, and their influence on the position of the ribs during breathing). All the following interdependently variable features ultimately determine the shape of the thorax and the thoracic cavity: trunk shape, which is the combined result of the length, location, and degree of torsion of the ribs; the length of the transverse processes; the length and breadth of the vertebral bodies; the thickness of intervertebral disks; and the consequent distance between head and tubercle of the ribs, their width, and the width of the sternum.

Mobility of the thoracic region is severely restricted and mostly involves slight possibility of forward, backward, and lateral bending. Also, the "rib cage" has the ability to slightly change shape that is directly correlated to the breathing activity of the lungs. Overall, the thoracic portion of the spine is the least flexible of the presacral regions.

The Thoracic Region of *Tarsius*

According to Schultz (1961), tarsiers usually have thirteen vertebrae in the thoracic region according to the presence of ribs. In contrast, the functional way of counting thoracic versus lumbar vertebrae uses the position of the zygapophysal articulations. The transitional vertebra, according to zygapophyseal position, is usually the eleventh thoracic. Thus, functionally there are eleven thoracic vertebrae and accordingly eight lumbar vertebrae. The first thoracic vertebra is almost as wide as the unusually large atlas in tarsiers. Both the atlas and the first thoracic vertebra are individually wider than the sacrum.

The second thoracic vertebra is about as wide as the sacrum, while the caudally following vertebrae diminish gradually in width. Medially, on the roof of the neural canal, the thoracic vertebrae carry spinal processes that are slanted caudally and overlap the roof of the neural canal of the caudally following vertebrae. Further caudally, the spines shorten craniocaudally and the spine of the transitional vertebra is directed straight dorsally. The following two thoracic spines are directed cranially like those of the lumbar region. The first seven (of thirteen) thoracic vertebrae are exceptional in the extraordinary lateral breadth of their transverse processes (diapophyses), which connect with the long vertebral portion of the unusual ribs. The first seven ribs do not have separate capitula and tubercula costae; they have one, long articular facet that is appressed to the underside of the elongated transverse processes. These ribs are then sharply angled ventrally (fig. 5.5; scale is 5 mm) and the proximal articular area of the rib and its body form a sharp angle of about 80 degrees between each other.

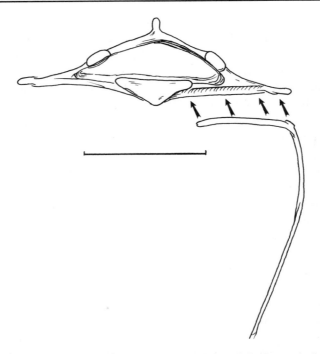

Figure 5.5. Cranial aspect of the third thoracic vertebra and rib. Arrows indicate the extraordinary length of the single articulation between the two. Scale length is 5 mm.

The cartilaginous part of the first ribs is comparatively long. The segments of the sternum (breastbone) are somewhat longer than wide but similar to both prosimian and anthropoid primates of similar size. They remain unfused, even in rather old tarsiers. Lateral width of the thoracic vertebral bodies is comparatively narrow, their craniocaudal length is short, the dorsoventral height is low, and all three dimensions increase slightly as one moves caudally.

Interestingly, the transverse processes that are so extraordinarily long laterally, and are directed more or less perpendicular to the spine's long axis in the first eight or so vertebrae, shorten laterally from then on and elongate in a caudal direction. At the same time, they increasingly become affixed to the sides of the neural arches only slightly above the upper edge of the vertebral body. In the last three or four thoracic vertebrae (thoracic vertebrae 10, 11, 12, and 13), they elongate into caudally pointing accessory processes that tightly embrace the prezygapophyses of the caudally following vertebrae from lateral. They only continue in the first two or three lumbar vertebrae, while commonly in many primates they are characteristic for the lumbar region only and are not found in thoracic vertebrae.

The articular facets are more or less incorporated into the roof of the neural canal, not projecting, and overlap one another like shingles. The prezygapophyses are actually situated on the root of the transverse processes, strutting under the roof of the neural canal of the cranially preceding vertebra. The roof of the neural canal of the thoracic vertebrae does not leave any open spaces between adjacent vertebrae.

The transitional vertebra between thoracic zygapophysis position and lumbar position is the eleventh thoracic vertebra of thirteen. The last two thoracic vertebral bodies and the adjoining first lumbar vertebral bodies widen and, to a lesser extent, gradually get longer.

The Lumbar Region of Primates

The average number of lumbar vertebrae (vertebrae without ribs) in primates varies between four and ten, averaging six (Schultz, 1961). The vertebral bodies continue to gradually become longer and wider toward the middle of the lumbar region, where the longest presacral vertebral body is usually located. From here, the length of the bodies begins to gradually diminish in a caudal direction. The bodies' cranial and caudal ends are flat. The zygapophyses of the lumbar vertebrae are positioned as if their articular surfaces were on the circumference of a small circle whose center is located dorsal to the neural arch. Often the prezygapophyses are tightly embraced from below by strong accessory bony spines that originate from the lower caudal end of the preceding vertebra's neural arch (or pedicle in human anatomy).

In the lumbar region these accessory bony processes virtually lock the articulations between vertebrae into place and make any movement between vertebrae almost impossible. The spinous processes are strong and craniocaudally long in the lumbar region. In some primates, the caudal ends of the craniocaudally long spinous processes are split into two, and they embrace the cranial aspect of the spinous process of the following vertebra, thus preventing any dorsal bending (extension) between adjacent vertebrae. Transverse processes of varying strength and direction characteristically originate either from the lower part of the pedicles or the upper lateral aspect of the vertebral bodies. On their ventral aspect the bodies of lumbar vertebrae can be sagittally keeled or flat. Mobility of the lumbar region is usually restricted to flexing forward and bending sideways.

The Lumbar Region of *Tarsius*

Schultz (1991) reports six lumbar vertebrae for genus *Tarsius*. However, eight exist if the thoracic vertebrae are counted according to the position of zygapophyses, not the presence of ribs (Washburn and Buettner-Janusch 1952; Erikson, 1963; Shapiro, 1993; Shapiro and Simons, 2000).

The lumbar vertebrae are characterized by prominent spines that are positioned cranially on the roof of the neural canal, point cranially, and reach over the roof of the neural canal of the cranially adjacent vertebrae. The articular processes in the neural arch are located at the lateral edge of the rather flat dorsal roof of the neural canal. They are positioned almost parallel to the dorsal surface of the vertebral bodies and are slightly slanted dorsally. Here, the neural arches are tightly adjoined to each other between vertebrae, and no openings exist between adjacent vertebrae. In the more proximal lumbar vertebrae, the transverse processes are thin and narrow bony ridges alongside the vertebral bodies. Only in the last three lumbar vertebrae do they gradually strengthen, become more prominent laterally, and are slightly ventrolaterally inclined and elongated cranially into a pointed tip.

In the lumbar region the lateral ascending parts of the neural canal are straight and turn inward at an almost 90 degree angle, producing a square outline of the neural canal of the lumbar vertebrae when one looks at them from the cranial or caudal aspect. The roof of the neural arch lies flat over the vertebral body and has a slitlike opening medially on its caudal end. The caudally following spinal process fits snugly into this opening. The vertebral bodies are slightly wider than long, and ventrally not keeled but unusually flat.

The Sacral Region of Primates

In primates the sacral region is a combination of several vertebrae fused together and is also the area of connection of the axial skeleton with the hindlimbs. The most common number of sacral vertebrae in primates is three, but this number can increase to nine in some lorises and up to seven or eight in apes and humans, probably in connection with the linked reduction of caudal (tail) vertebrae to only a few (Schultz, 1961).

Generally, the first two sacral vertebrae have transverse processes transformed for the attachment to the spine of the ilium of the pelvis. This "auricular surface" varies widely among primates in its surface relief and intensity of pelvic connection (Leutenegger, 1974). All intrasacral articulations are tightly fused together, as are often the sacral spinal processes. Between the transverse portions of adjacent vertebrae, dorsoventral foramina for the passage of branches of the sacral nerves and blood vessels are found. The pre- and postzygapophyses are usually unchanged from the position typical for the lumbar region.

The Sacrum of *Tarsius*

The sacrum is usually a combination of three, very rarely two, vertebrae (Schultz, 1961). The spinous processes are separate from each other, craniocaudally short, and usually pointing more or less straight dorsally. The zygapophyses are fused together, represented by small hooklike bony extensions and on their bases are perforated by dorsoventral foramina. The auricular area for the connection with the ilium of the pelvis is usually composed by the first two vertebral units and is not particularly large. The sacroiliac surfaces do not ossify (Grand and Lorenz, 1968). The neural canal index (Ankel, 1965) is like that of the New World monkey *Saimiri*, indicating a long tail that is not actively involved in locomotor activities. This also means that the diameter of the neural canal only slightly diminishes between the cranial and caudal openings.

The Caudal Region of Primates

Ankel (1962) separated the tail vertebrae of long-tailed primates into two subsections: (1) the first caudal region combined of all caudal vertebrae with neural arches and (2) the second caudal region consisting of all following vertebrae that are basically nothing but vertebral bodies and their intervertebral disks. The first caudal region is also characterized by the presence of ventral arches, the so-called chevron bones or hemapophyses. These are V-shaped craniocaudally short clasps, which straddle the caudal artery and are located ventrally under the cranial end of the vertebral bodies. They often extend farther distally into the second caudal region than the neural arches of the first region. Chevron bones are not fused to the tail vertebrae and therefore are usually lost when osteological specimens are prepared. Their presence or absence is most reliably ascertained with X-rays.

The number of caudal vertebrae varies widely among different primate groups—according to Schultz (1961), between one and thirty five with an average number of about twenty five in long-tailed primates. The caudal vertebrae of the first caudal region not only have neural arches with bony articulations on both ends, but also spinous and transverse processes of different length, width, and direction. The articular facets are generally positioned as those of the lumbar region and, while the length of the tail's vertebral bodies initially increases gradually, the width usually decreases, and the distinctiveness of the bony protuberances gradually diminishes from one vertebra to the next. The caudal vertebrae reach maximum length and diameter, then diminish caudally to the last and smallest one.

The Caudal Region of *Tarsius*

The caudal region has a comparatively short first region (Ankel, 1962) and only five to six vertebra with neural arches. The transverse processes are strongly developed on the first three caudal vertebrae. They are directed straight laterally. The spinous processes are not very prominent and straight. The craniocaudal length of the vertebral bodies increases slowly while the mediolateral width diminishes slowly. The vertebral bodies of the first few caudal vertebrae are ventrally flat, not keeled. The zygapophyses between the first four vertebrae are positioned like those of any other primate.

The length of the vertebral bodies increases more rapidly after the neural canal has disappeared, while the width decreases rapidly; the vertebral bodies are long, thin, and round in diameter, with both ends slightly wider than the middle of the vertebrae. There are chevron bones that are not very large, but they continue far backwards into the region. The longest vertebra is usually the eighth, and it is followed by three or four vertebrae of equal length that gradually become thinner. The length of the vertebral bodies decreases very gradually, and toward the distal end of the tail the vertebrae become very thin but remain comparatively long and continue to have round diameters.

Conclusions

Compared to other primate axial skeletons, the spine of two *Tarsius* species have many unique features that we have not seen in other extant primates. We do not know, however, whether the features are similar to those in the other tarsier species.

Our observations show that the axial skeleton of *Tarsius bancanus* and *T. syrichta* are morphologically unique in the cervical region (with the exception of the axis), and the thoracic, lumbar, and caudal regions. The morphology of the sacrum is unremarkable and similar to generalized mammal sacra.

The atlas shows a morphology identical with the first cervical vertebrae of all extant primates. This vertebra is, however, unusually large (fig. 5.3). We hypothesize that this remarkable size is very likely related to the unusual way tarsiers are able to turn their heads 180 degrees in either direction, from left to right. Even though the only possible movement between the atlas and occipital condyles of the skull is ventrally and dorsally as in nodding, it is the atlas that moves together with the head in the rotary movement around the dens of the axis. The caudally adjoining axis is of the usual dimensions for tarsier-sized primates, much smaller than the preceding atlas. This astonishing size of the atlas likely facilitates lateral turning movements between the skull, together with the atlas around the dens of the axis, in a more efficient and forceful manner than would be possible in other primates where the atlas is comparatively small.

The unusual size of the first cervical vertebra in accordance with the similar width of the first three thoracic vertebrae indicates that the muscles straddling these areas must be particularly well developed (particularly the musculus obliquus capitis inferior between atlas and the spinous process of the axis, the anterior lateral vertebral muscle, the longus capitis, and the posterior splenius cervicis).

The unusual position of the zygapophyses in cervical vertebrae 3 to 7 can be interpreted as being involved in the unique ability of tarsiers to look straight backwards, moving the head 180 degrees both to the right and to the left from the direct forward position. This extreme mobility of tarsiers is comparable only to owls and ostriches but not known from any other mammal. We believe that this ability can partly be explained by the exceptional arrangement of the articulations between these vertebrae (compare fig. 5.4a).

In contrast, the fact that the dens of the second vertebra—the axis—is angled just like in quadrupedal primates shows that the habitual upright posture of tarsiers is not reflected in the axis morphology as it is in truly upright (*Homo*) and partially upright (*Gorilla*) primates. These hominoids locomote habitually in upright positions, a fact that is morphologically reflected in axial dens and anterior zyapophysis position; the dens points straight upward in *Homo* and is only slightly angled in *Gorilla*. These postural differences are also mirrored in the position of the cranial zygapophyses with the atlas in these large genera. The prezygapophyses are accordingly flat and totally perpendicular to the body's long axis in the former, and only slightly inclined caudolaterally against the long axis of the spine and body in the latter. In tarsiers, as in all quadrupedal primates, these articulations are steeply slanted and thus deeply inserted into the embracing articulations of the atlas.

The shape and articulation of tarsier ribs and associated thoracic vertebrae are totally distinctive and unlike those of any other primate. The sternum is similar to other small-bodied primates. Several features associated with the lumbar region are unusual. The accessory processes that already formed in the last three thoracic vertebrae and continue to the first three of nine (counted functionally) lumbar vertebrae are unusual for primates, as they are usually found between the front of the lumbar region and the sacrum—not in the thoracic region.

Also unique (in all three dimensions) are the comparatively small, ventrally flat lumbar vertebral bodies (fig. 5.6). The tightly interlocking features of the neural canal in all lumbar vertebrae prevent much mobility in

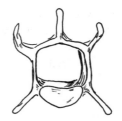

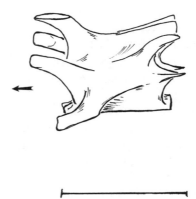

Figure 5.6. Lumbar vertebra in cranial view (top) and lateral view (bottom: cranial to the left).

this region. This observation lends support for the hypothesis formulated by Shapiro and Simons (2002) that in vertically leaping primates the lumbar region should be short and stiff. There seems to be a contradiction between the supposed "functional elongation" of the lumbar region cranially into the thoracic region in leaping primates, including *Tarsius* (Fleagle, pers. com.), and the argument that vertical leapers have short, immobile lumbar regions, unless this functional elongation is restricted to small, quadrupedally running and leaping primates.

The sacrum is very similar to other small-bodied primates and, astonishingly, does not, in its morphology, reflect the outstanding salutatory locomotion of tarsiers. The caudal region is similar to that of other long-tailed primates, even though the presence of four equally long longest-tail vertebrae and the overall length and fragility of the distal-tail vertebrae are not like other small primates.

So where do tarsiers' spines fit in the order of extant primates? They are unlike any other haplorhine or strepsirhine primate. Their spinal morphology is unique and deserves further study. For example, like the unique karyotype of genus *Tarsius*, the morphology of their axial skeleton does not resemble the norm among primates. It also, for the most part, does not reflect the unique saltatory locomotion and mostly upright posture of these outstanding little primates. It places them squarely into a unique group of their own among primates and mammals.

Acknowledgments

We would like to thank the two senior editors of this volume for inviting us to contribute our brief morphological account. We also appreciate their helpful comments and suggestions. Our gratitude also goes to John Fleagle for his thoughtful review and helpful suggestions. In the era of genomics and molecular biology, comparative morphology could be regarded by some as old-fashioned. However, genomics and molecular biology can be totally understood only in the context and within the framework of in-depth morphological knowledge. This is Duke University Primate Center Publication #732.

References

Ankel F. 1962. Vergleichende Untersuchungen über die Skelettmorphologie des Greifschwanzes südamerikanischer Affen (Platyrrhina). Z Morphol Ökologie der Tiere 52: 131–170.

Ankel F. 1965. Der Canalis Sacralis als Indikator für die Länge der Caudalregion der Primaten. Folia Primatol 3: 263–276.

Ankel F. 1967. Morphologie der Wirbelsäule und Brustkorb. Primatologia: Handbook of Primatology IV, 4, 1–120. Basel: Karger.

Ankel-Simons F. 2000. Primate anatomy: an introduction. San Diego: Academic Press.

Cartmill M. 1985. Climbing. In Hildebrand, M, Bramble M, Liem KL, Wake DB, editors, Functional vertebrate morphology, 73–88. Cambridge, Mass.: Belknap Press of Harvard University Press.

Erikson GE. 1963. Brachiation in the New World monkeys and in Anthropoid apes. Proc Zoo Soc Lond 10: 135–164.

Grand TI, Lorenz R. 1968. Functional analysis of the hip joint in *Tarsius bancanus* (Horsfield, 1821) and *Tarsius syrichta* (Linnaeus, 1758). Folia Primatol 9: 161–181.

Hill WCO. 1955. Primates, comparative anatomy and taxonomy, Vol. 2: Haplorhini. Edinburgh, UK: Edinburgh University Press.

Leutenegger W. 1974. Functional aspects of pelvic morphology in simian primates. J Hum Evol 3: 207–219.

Miller JP. 1996. Miniature vertebrates. Oxford, UK: Oxford University Press.

Schultz AH. 1961. Vertebral column and thorax. Primatologia: Handbook of Primatology IV, 5, 1–66. Basel: Karger.

Shapiro L. 1993. Functional morphology of the vertebral column in Primates. In Gebo D, editor, Postcranial adaptation in nonhuman primates, 121–149. DeKalb: University of Northern Illinois Press.

Shapiro L, Simons C. 2002. Functional aspects of strepsirrhine lumbar vertebral bodies and spinous processes. J Hum Evol 42: 753–783.

Sprankel H. 1965. Untersuchungen an *Tarsius*. I. Morphologie des Schwanzes nebst ethologischen Bemerkungen. Folia Primatol 3: 135–188.

Washburn SL, Buettner-Janusch J. 1952. The definition of thoracic and lumbar vertebrae. Am J Phys Anthro 10: 251.

CHAPTER 6

Phylogenetic Position of Tarsiers within the Order Primates: Evidence from γ-Globin DNA Sequences

Carla M. Meireles, John Czelusniak, Scott L. Page,
Derek E. Wildman, and Morris Goodman

The extant order Primates, with 233 species (Groves, 1993), consists of four major monophyletic groups (suborders) (table 6.1): Loriformes (Asian lorises, African galagos or bush babies, and pottos), Lemuriformes (Malagasy lemurs and sifakas), Tarsiiformes (Indonesian and Philippine tarsiers), and Anthropoidea (New World monkeys, Old World monkeys, apes, and humans). Loriformes and Lemuriformes have been classified as members of the semiorder Strepsirrhini, while Tarsiiformes and Anthropoidea have been classified as members of the semiorder Haplorhini (Goodman et al., 1998, 1999).

In units of millions of years ago or Ma (Mega annum, millions of years before present), the ages of the haplorhine-strepsirhine, tarsier-anthropoid, and loriform-lemuriform divergence nodes have been dated from fossil evidence and local molecular clock analyses to about 63, 58, and 50 Ma, respectively (Gingerich and Uhen, 1994; Goodman et al., 1998, 1999). The small-brained primates—i.e., the loriforms, lemuriforms, and tarsiiforms— traditionally are collectively referred to as prosimians, while the larger-brained primates, the anthropoids, are called simians. Anthropoidea (the simians) divides into the infraorders Platyrrhini for ceboids (New World monkeys), and Catarrhini for cercopithecids (Old World monkeys), and hominids (apes and humans) (Goodman et al., 1998, 1999).

The extant tarsiers are all members of one living genus, *Tarsius,* in the family Tarsiidae (Groves, 1993; Goodman et al., 1999) represented by five species (table 6.1). One tarsier species is found in the Philippines (*T. syrichta*), while four species (*T. bancanus, T. dianae, T. spectrum,* and *T. pumilus*) are found in Indonesia (Groves, 1993).

Tarsiers share features from both strepsirhines and anthropoids, and also maintain characteristics unique to themselves. They are small, weighing only 113–142 grams, and like many strepsirhines, they are nocturnal and

Table 6.1. Classification of the Major Monophyletic Groups of the Order Primates

Semiorder Strepsirrhini
    Suborder Loriformes (lorises, bush babies, and pottos)
    Suborder Lemuriformes (lemurs and sifakas)
Semiorder Haplorhini
    Suborder Tarsiiformes (Indonesian and Philippine tarsiers)
        Family Tarsiidae
          Tarsius
            T. syrichta (Philippine tarsier)
            T. bancanus (western tarsier)
            T. dianae (Dian's tarsier)
            T. spectrum (spectral tarsier)
            T. pumilus (pygmy tarsier)
    Suborder Anthropoidea
    Infraorder Platyrrhini (New World monkeys)
    Infraorder Catarrhini (Old World monkeys, apes, and humans)

SOURCE: According to Goodman et al. (1998, 1999).
NOTES: Species indicated only for the extant tarsier group. Common names for extant species of tarsiers are followed in accordance with Rowe (1996).

have grooming claws and a bicornuate uterus (Rowe, 1996). Like anthropoids, they do not have a reflective layer in their eyes, and the females present the same monthly sexual swellings (Fleagle, 1999). Unique among primates, tarsiers have only two rather than four incisors in their lower jaw (Rowe, 1996). Their relatives in the fossil record are found going back to the Eocene epoch, from 54 to 36 Ma (Fleagle, 1999).

The phylogenetic position and taxonomic status of tarsiers within the order primates are in dispute (Martin, 1990, 1993; Shoshani et al., 1996). Tarsiers have been treated as the closest sister group of loriforms (Schwartz and Tattersall, 1987; Jaworski, 1995), of strepsirhines (lemuriforms and loriforms) (Napier and Napier, 1985), of anthropoids (Martin, 1990, 1993), of an assemblage grouping anthropoids and strepsirhines (Schwartz, 1978; Gingerich, 1981), and even of humans (Jones, 1920). The position of the tarsiers in relation to the anthropoids and strepsirhines has been debated for the last century. Most traditional morphological studies have usually grouped *Tarsius* with the strepsirhine primates to the exclusion of anthropoids, thereby forming the group Prosimii (Napier and Napier, 1967, 1985). However, other studies have suggested that tarsiers are more closely related to the anthropoids to the exclusion of strepsirhines, thus forming the clade Haplorhini (Aiello, 1986; Pocock, 1918; Ross et al., 1998). Because morphological data have been ambiguous regarding the phylogenetic position of tarsiers, many studies have focused on molecular sequence, and more recently, transposition markers, in order to resolve the subordinal relationships among primates.

A relatively large body of sequence data relevant to the tarsier question has accumulated in the past twenty years. All five nuclear loci in the β-globin gene cluster support a Haplorhini-Strepsirrhini division over a Prosimii-Anthropoidea grouping (Goodman et al., 1998; Koop et al., 1989; Page and Goodman, 2001; Porter et al., 1995; this study). Other studies using sequence data that support the integrity of the haplorhine clade include cytochrome *c* oxidase subunit IV (Wildman et al., submitted) and involucrin (Dijan and Green 1991). Jaworski (1995) and more recently Murphy et al. (2001) have presented nuclear data sets that support the prosimian arrangement of subordinal primate clades. Jaworski's αA-crystallin data are problematic because the monophyly of Primates is not supported. Indeed, prosimians group with bats, rodents, and tree shrews to the exclusion of anthropoids. The Murphy et al. study used portions of fifteen nuclear and three mtDNA genes with primers derived from conserved coding regions and showed support for a prosimian clade in the analysis. However, the Murphy data did not consistently support the monophyly of the primates, with flying lemurs (Dermoptera) being shown as sister taxon to the anthropoids in some analyses. This result is likely due to homoplasy and possibly selection because the data were nuclear-coding sequences. Therefore, there are no nuclear DNA sequence studies that unambiguously support a prosimian primate clade.

Mitochondrial data present a different picture. Beginning with Hayasaka et al. (1988), and including later work by Hasegawa et al. (1990), Andrews et al. (1998), and McNiff and Allard (1998), mtDNA have not often supported the haplorhine grouping of tarsiers and anthropoids. Instead, they have supported either a prosimian clade (Hasegawa et al. 1990; Hayasaka et al. 1988), or a nonprimate position for *Tarsius* (Andrews et al., 1998; McNiff and Allard, 1998) However, these mtDNA results are potentially positively misleading because the relatively rapid mutation rate of mitochondrial compared to nuclear DNA makes the mitochondrial trees more subject to the effects of homoplasy. This error is likely to be compounded due to the "long branches attract" problem associated with phylogenies inferred among taxa that have short internodes but long terminal branches that span long evolutionary time periods (Felsenstein, 1978). Hayasaka et al. (1988:637) point out this problem with their own data and suggest that "phylogenetic relationships between distantly related species are less reliable than those between closely related species [*when mtDNA data are examined*]." Depending on the analysis employed, the Hayasaka et al., McNiff and Allard, Murphy et al., and Andrews et al. data all reject the hypothesis of primate monophyly. Therefore, it appears that the mtDNA data are either misleading or that Primates is not a monophyletic taxon (we support the position that primates are monophyletic).

Interestingly, one sequence data set from the nuclear-encoded *BC200* small nonmessenger RNA gene supports the grouping of the Anthropoidea and Strepsirrhini to the exclusion of tarsiers (Kuryshev et al., 2001). This arrangement was not seen in any of the 118 markers recently examined by Schmitz et al. (2001).

Another promising source of molecular data that informs studies of primate phylogeny is transposition markers. These markers are transposable elements (TEs) that are inserted in the genomes of organisms. TEs are powerful markers because insertions present in more than one taxon are necessarily synapomorphic because transposon integrations are not lost in genomes (Shedlock and Okada, 2000). Zietkiewicz et al. (1999) examined a type of short interspersed repetitive element, or SINE, known as Alu markers. Alu markers are primate-specific repetitive elements believed to have originated approximately 80 million years ago (Kapitonov and Jurka, 1996). The Zietkiewicz study showed that there was phenetic similarity between Alu markers present in tarsiers and anthropoids. This phenetic similarity was not seen between tarsiers and strepsirhines.

The recent study by Schmitz et al. (2001) offers the most compelling evidence to date that tarsiers are the sister taxon to anthropoids. For this study they examined 118 Alu markers and showed that fourteen loci grouped tarsiers with anthropoids based on fragment size similarity. Subsequent sequence analysis showed that of these fourteen loci, only three were orthologous, the rest being either independent transpositions on the orthologous introns or nonspecific polymerase chain reaction (pcr) products. This result shows the necessity of sequence analysis in studies of transposition markers. Finally, repeat elements present in the *BC200* locus also support grouping tarsiers in the Haplorhini (Kuryshev et al., 2001).

There is now considerable protein and DNA sequence evidence that favors sister-grouping tarsiers with anthropoids (Bonner et al., 1980; De-Jong and Goodman, 1988; Koop et al., 1989; Dijan and Green, 1991; Bailey et al., 1992; Porter et al., 1995, 1997a,b), i.e., the grouping together of Tarsiiformes and Anthropoidea in the semiorder Haplorhini (Goodman et al., 1998, 1999). Thus, most recent evidence offers support for the phylogenetic integrity of the haplorhine clade over the prosimian clade. The majority of noncoding nuclear and transposition data supports grouping tarsiers with the anthropoids. Some mtDNA and coding nuclear data suggest a sister group relationship between tarsiers and strepsirhines, but these data are problematic, and many of the gene trees that support a prosimian clade do not support a monophyletic Primates order. While the molecular evidence is at present equivocal, evidence that does exist predicts that loci examined in the future will also support the clade Haplorhini.

In this study we use the recently sequenced γ-globin gene of *Tarsius bancanus* (the Western tarsier) (Meireles et al., 1999a), along with the previously sequenced γ-globin gene of *Tarsius syrichta* (the Philippine tarsier) (Koop et al., 1989; Hayasaka et al., 1993) to reexamine the phylogenetic position of tarsiers. We compared by cladistic methods these tarsier sequences to orthologous sequences from strepsirhines (bushbaby, brown lemur, dwarf lemur) and anthropoids (New World monkeys, Old World monkeys, apes, human) and a nonprimate mammal (domestic rabbit) (table 6.2).

Table 6.2. Species Examined and Sources of the Nucleotide DNA Sequences

Taxon	Common Name	Source
Primates		
HOMINIDS		
Homo sapiens	human	Bailey et al., 1992
Pan paniscus	pygmy chimpanzee	Bailey et al., 1992
Pan troglodytes	chimpanzee	Bailey et al., 1992
Gorilla gorilla	western lowland gorilla	Bailey et al., 1992
Pongo pygmaeus	Borneo orangutan	Bailey et al., 1992
Hylobates lar	white-handed gibbon	Bailey et al., 1992
CERCOPITHECIDS		
Macaca mulatta	rhesus macaque	Bailey et al., 1992
Macaca nigra	Celebes macaque	Bailey et al., 1992
Macaca nemestrina	pig-tailed macaque	Page et al., 1999
Papio hamadryas cynocephalus	yellow baboon	Page et al., 1999
Theropithecus gelada	gelada baboon	Page et al., 1999
Lophocebus aterrimus	black mangabey	Page et al., 1999
Mandrillus sphinx	mandrill	Page et al., 1999
Mandrillus leucophaeus	drill	Page et al., 1999
Cercocebus galeritus	Tana River mangabey	Page et al., 1999
Erythrocebus patas	patas monkey	Page et al., 1999
Chlorocebus aethiops	vervet monkey	Page et al., 1999
Cercocebus cephus	mustached guenon	Page et al., 1999
Colobus sp.	colobus	Page et al., 1999
Colobus guereza	Abyssinian colobus	Page et al., 1999
Nasalis larvatus	proboscis monkey	Page et al., 1999
Trachypithecus obscurus	dusky leaf monkey	Page et al., 1999
PLATYRRHINES		
Ateles geoffroyi	black-handed spider monkey	Meireles et al., 1999b

continues

Table 6.2. *Continued*

Taxon	Common Name	Source
Ateles paniscus	black spider monkey	Meireles et al., 1999b
Brachyteles arachnoides	woolly spider monkey or muriqui	Meireles et al., 1999b
Lagothrix lagotricha	woolly monkey	Meireles et al., 1999b
Alouatta seniculus	red howler	Meireles et al., 1999b
Alouatta belzebul	red-handed howler	Meireles et al., 1999b
Alouatta caraya	black-and-gold howler	Meireles et al., 1999b
Cebus albifrons	white-fronted capuchin	Hayasaka et al., 1993
TARSIERS		
Tarsius syrichta	Philippine tarsier	Koop et al., 1989a; Hayasaka et al., 1993; Meireles et al., 1999a
Tarsius bancanus	western tarsier	Meireles et al., 1999a
STREPSIRHINES		
Otolemur crassicaudatus	thick-tailed greater bush baby	Tagle et al., 1988, 1992
Eulemur fulvus	brown lemur	Harris et al., 1986
Cheirogaleus medius	fat-tailed dwarf lemur	Harris et al., 1986
Nonprimate mammal		
Oryctogalus cuniculus	domestic rabbit	Margot et al., 1989

NOTES: Species and common names for all extant primates are from Rowe (1996). It may be noted that in the phylogenetic classification of Goodman et al. (1998, 1999), in which the age of a clade determines the rank of the taxon representing that clade, the two species of *Pan* are treated as a subgenus of *Homo,* i.e., *Homo* (*Pan*). Similarly, *Theropithecus* and *Lophocebus* are treated as subgenera of *Papio,* i.e., *Papio* (*Theropithecus*) and *Papio* (*Lophocebus*), and the two species of *Mandrillus* are treated as a subgenus of *Cercocebus,* i.e., *Cercocebus* (*Mandrillus*).

Knowledge of the evolutionary history of the nuclear genomic locus that contains either one or two γ-globin genes guided us in aligning the γ nucleotide sequences. The early primates ancestral to strepsirhines and haplorhines apparently had only a single γ-globin gene, as is the case in rabbits (Margot et al., 1989) and in nonanthropoid primates such as tarsiers (Koop et al., 1989; Hayasaka et al., 1993; Meireles et al., 1999a), lemurs (Harris et al., 1986), and bush babies or galagos (Tagle et al., 1988). However, in an early simian ancestor of platyrrhines and catarrhines, a tandem duplication of a 5 kilobase (kb) DNA genomic fragment containing the γ-globin gene produced the paired genes of the anthropoid γ locus (Fitch et al., 1991). The tandem duplication resulted from an unequal crossover between two homologous truncated L1 repetitive line elements, one (L1a) that was upstream of the ancestral simian γ gene and the other (L1b) that was downstream of it. This 10 kb locus (L1a-γ-L1ba-γ-L1b) consists mainly of noncoding sequences flanking each of the two γ genes. Each gene from its initiation codon to its termination codon has 438 base pairs (bp) of coding sequence distributed over three exons and roughly 1 kb (1000 bp) of noncoding sequence distributed in the two introns separating the three exons. These flanking and intron noncoding γ sequences evolve at a relatively rapid rate and have provided in previous studies a well-resolved picture of cladistic relationships among primate lineages (Bailey et al., 1992; Czelusniak and Goodman, 1995; Meireles et al., 1999b; Page et al., 1999).

The main sequence alignment employed spanned the 10 kb of this L1b-γ^1-L1ba-γ^2-Lb locus in anthropoid primates, although not all the anthropoids listed in table 6.2 were completely sequenced over this locus. Nevertheless, the minimum amount of sequence determined for an anthropoid species was about 2 kb, with the average amount determined per species being about 8 kb. In the nonanthropoid species, as mentioned above, the γ locus contains only one γ-globin gene. Each nonanthropoid species was represented by at least a 2 kb sequence that spanned the γ-globin gene proper region and for the rabbit, bush baby, and two tarsiers, in each an additional 2–3 kb consisting of upstream and downstream noncoding sequences flanking the gene proper (exonic, intronic) sequences. After preparing data sets in which these γ sequences were aligned against one another, we used the maximum parsimony (MP) and maximum likelihood (ML) algorithms in PAUP to construct phylogenetic trees from the sequences. When the nonanthropoid γ sequences were aligned against the anthropoid γ^2 sequences, the tarsiers grouped with strepsirhines, but very weakly so. However, when the nonanthropoid γ sequences were aligned against the anthropoid γ^1 sequences, the tarsiers grouped very strongly with anthropoids. Moreover, when the homologous anthropoid γ^1 and γ^2 sequences were first aligned against each other (a 5 kb long paralogous alignment), and then the nonanthropoid γ sequences were aligned against these paralogous γ^1 and γ^2 sequences, all anthropoid γ^1 and γ^2 sequences first joined each other, and then the tarsier γ sequences strongly joined this anthropoid clade.

The main data set consisted of aligned γ-globin nucleotide sequences from the thirty six species listed in table 6.2. The full sequence alignment for this data set can be accessed through the Internet at http://cmmg.biosci .wayne.edu/lgross/primates.html. As already pointed out, the alignment employed spans the 10 kb region of the anthropoid L1a-γ^1-L1ba-γ^2-Lb locus, with the 4.5 kb region of the nonanthropoid γ locus aligned against the γ^1 portion of the anthropoid locus. This data set was suitable for investigating the phylogenetic position of tarsiers within the order Primates in relation to the anthropoids and strepsirhines. However, we felt that portions of the anthropoid γ sequences might cause incorrect groupings among the anthropoids. These were the portions with a history of gene conversions between γ^1 and γ^2 genes. Typically, for each of the two paralogously related γ genes, the main segment with such a history spanned a 2 kb region extending from about 0.6 kb upstream of the initiation codon to about a 0.1 kb just 3′ of the termination codon. The sequences that had a history of gene conversions could yield incorrect phylogenetic groupings among the anthropoid species lineages, if in the gene conversions (nonreciprocal recombinations), γ^1 sequence replaced γ^2 sequence in some species whereas γ^2 replaced γ^1 in other species. Thus, we ran the PAUP MP and ML programs not only on the full sequence alignment but also on this alignment after deleting from each anthropoid sequence the main segment in which gene conversions occurred (fig. 6.1).

This procedure ensured that we had a data set of wholly orthologous sequences that could accurately track the branching apart of species lineages. Moreover, to calculate times of branching by the local molecular clock model (Goodman, 1986; Porter et al., 1997; Goodman et al., 1998), we wanted the γ data to consist of orthologous sequences that were both unconverted (without a history of gene conversions) and noncoding. To create this data set from the full alignment, we not only deleted the exons and other portions of the anthropoid sequences with a history of gene conversions, but we also deleted the exons of the nonanthropoid γ genes. The percentages of sequence change along the branches of the ML tree (fig. 6.2) were then used for the local molecular clock calculations (see below). The branching topology of this ML tree was essentially the same as that of the MP tree.

Figure 6.1. (facing page) Maximum parsimony (MP) tree found for thirty-six aligned γ-globin nucleotide DNA sequences. Numbers above the lines to branch points are bootstrap percentage (BP) values obtained for 1000 replicates with the full tandem γ^1–γ^2 alignment. Numbers below these lines are BP values obtained for 1000 replicates with this alignment after deleting the portions of the γ^1 and γ^2 sequences with a history of gene conversions. The three-letter species abbreviations are as follows: Ocu, *Oryctogalus cuniculus;* Hsa, *Homo sapiens;* Ppa, *Pan paniscus;* Ptr, *Pan troglodytes;* Ggo, *Gorilla gorilla;* Ppy, *Pongo pygmaeus;* Hla, *Hylobates lar;* Mmu, *Macaca mulatta;* Mni, *Macaca nigra;* Mne, *Macaca nemestrina;* Pcy, *Papio hamadryas cynocephalus;* Tge, *Theropithecus gelada;* Lat, *Lophocebus aterrimus;* Msp, *Mandrillus sphinx;* Mle, *Mandrillus leucophaeus;* Cga, *Cercocebus galeritus;* Cae, *Chlorocebus aethiops;* Cce, *Cercopithecus cephus;* Epa, *Erythrocebus patas;* Col, *Colobus* sp.; Cgu, *Colobus guereza;* Nla, *Nasalis larvatus;* Tob, *Trachypithecus obscurus;* Cal, *Cebus albifrons;* Age, *Ateles geoffroyi;* Apa, *Ateles paniscus;* Lla, *Lagothrix lagotricha;* Bar, *Brachyteles arachnoides;* Aca, *Alouatta caraya;* Abe, *Alouatta belzebul;* Ase, *Alouatta seniculus;* Tsy, *Tarsius syrichta;* Tba, *Tarsius bancanus;* Ocr, *Otolemur crassicaudatus;* Efu, *Eulemur fulvus;* Cme, *Cheirogaleus medius.*

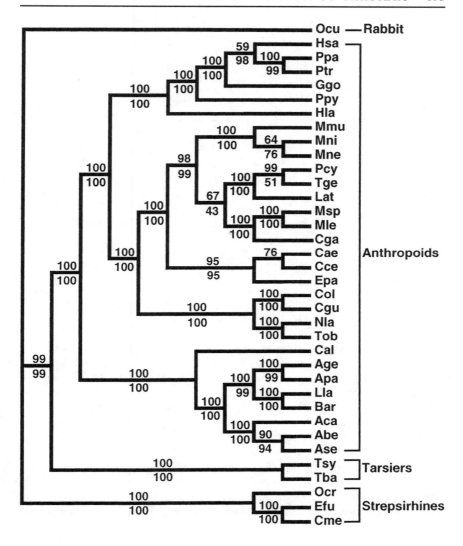

Figure 6.1 is the MP tree found for the full sequence alignment of the main data set. The numbers above the lines to branch points are the bootstrap percentage (BP) values found with this alignment. A BP value of 95 or above is viewed as very strong parsimony support for the clade (monophyletic group) specified by the line to the branch point. Note the very strong support for Haplorhini (99 BP value), the tarsier-anthropoid grouping, and also for *Tarsius* (100 BP value), the *T. syrichta-T. bancanus* grouping. After deleting those portions of the anthropoid sequences with a history of gene conversions, the MP tree found with one exception had the

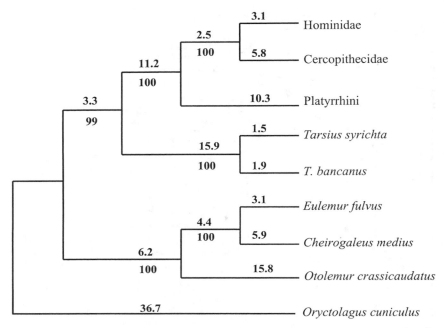

Figure 6.2. The MP tree as well as ML tree found after deleting from the full tandem γ^1–γ^2 alignment the anthropoid sequence portions with a history of gene conversions. For MP analyses, the branch-and-bound search procedure was used. For ML analyses, the heuristic search procedure was used with the following parameters: substitution model set at transition/transversion ratio = 2; the HKY (Hasegawa-Kishano-Yano) two-parameter model variant for unequal base frequencies; empirical base frequencies; starting branch lengths obtained using the Rogers-Swofford approximation method; substitution rates were set to conform to a gamma distribution; and molecular clock was not enforced. The numbers above lines (those to branch points and also those to the present-day sequences) are percentages of sequence change found by the ML algorithm. For the 6, 16, and 8 hominid, cercopithecid and platyrrhine sequences, respectively, in this data set, the percentage of sequence change on the line to Hominidae is the average of the lineages to the six hominid sequences; to Cercopithecidae is the average of the lineages to the sixteen cercopithecid sequences; and to Platyrrhini is the average of the lineages to the eight platyrrhine sequences. The numbers below lines to branch points are BP values obtained for 1000 heuristic replicates by MP analysis.

same branching topology as shown in figure 6.1. The exception occurred within one of the Old World monkey branches; instead of *Chlorocebus aethiops* (Cae) grouping first with *Cercopithecus cephus* (Cce) as shown in figure 6.1, it grouped first with *Erythrocebus patas* (Epa). The numbers below the lines to branch points are the BP values for the clades in the MP tree found after deleting from the main data set the sequence portions with a history of gene conversions. Again note the very strong support for Haplorhini (99 BP value) and for *Tarsius* (100 BP value). Further analyses were conducted using the data set in which the 5 kb long paralogously related γ^1 and γ^2 sequences were treated as separated sequences, and these separated anthropoid γ^1 and anthropoid γ^2 sequences along with the tarsier, strepsirhine, and rabbit γ sequences were all aligned against one another. This MP tree found for this data set again provided very strong support for Haplorhini (100 BP value), the sister grouping of tarsiers and anthropoids, and for *Tarsius* (100 BP value), the *T. syrichta-T. bancanus* grouping.

Previously, with a dataset of forty five noncoding ε-globin gene sequences (Porter et al., 1997b) and with paleontologically inferred reference dates of 63 Ma and 40 Ma for the haplorhine-strepsirhine and platyrrhine-catarrhine branch points, respectively, local molecular clock calculations yielded a date of 58 Ma for the tarsier-anthropoid branch point (Goodman et al., 1998). Using the 63 Ma and 40 Ma reference dates and the percentages of sequence change along the branches (lines) connecting the haplorhine-strepsirhine, tarsier-anthropoid, and platyrrhine-catarrhine branch points in figure 6.2 (numbers above the lines), we obtained with these unconverted, noncoding γ sequences the same local molecular clock estimate of 58 Ma for the tarsier-anthropoid branch point. On setting the local molecular clock with this 58 Ma date, the percentages of sequence change from the tarsier-anthropoid to the *T. syrichta-T. bancanus* branch-point, and from the latter branch-point to each of the two tarsier species, yielded the date of 5.6 Ma for the last common ancestor (LCA) of *T. syrichta-T. bancanus*.

The reason that we estimated branch times by the model of local molecular clocks rather than by the model of a global clock is that rates of molecular evolution can vary from one lineage to another. For example, as evident from the data in figure 6.2, the rate of noncoding sequence change is almost twice as fast in loriforms (bush baby, i.e., *Otolemur crassicaudatus*) than in lemuriforms (*Eulemur, Cheirogaleus*), and in the latter it is almost twice as fast in the dwarf lemur (*Cheirogaleus medius*) as in the brown lemur (*Eulemur fulvus*). Similarly, the noncoding rate is faster in cercopithecids than in hominids. If we accept the paleontologically inferred dates of 63 Ma and 40 Ma for the haplorhine-strepsirhine and platyrrhine-catarrhine branch points, respectively, then the rate in the early primate stem-haplorhine, stem-anthropoid lineage is 6.9×10^{-9} per nucleotide position per year, which is three to four times faster than the average platyrrhine-catarrhine rate of 1.7×10^{-9}. The estimated rate at which tarsier species diverge from each other is 3×10^{-9}. At such a relatively fast divergence rate, it should be possible to resolve sister-group relationships among the five recognized tarsier species (table 6.1) once genomic DNAs become available from all these species. The question of whether among the four disputed subspecies of *Tarsius spectrum* the Sanghe Island tarsiers (*T. s. sangirensis*) should be treated as a separate species (Rowe, 1996) could also be addressed.

Taxonomic Implications

In view of the relatively small time interval (5 million years) estimated to separate the tarsier-anthropoid branch point at 58 Ma from the haplorhine-strepsirhine branch-point at 63 Ma, it is not surprising that disputes exist as to whether the fossil evidence on the first true primates, the omomyids and adapids, supports their respective placement into haplorhine and strepsirhine branches (Gingerich, 1986, 1993; Martin, 1990; Shoshani et al., 1996; Kay et al., 1997). Nevertheless, cladistic analysis of extant primate lineages at both morphological (Shoshani et al., 1996; Goodman et al., 1998) and molecular (Goodman et al., 1998; this study) levels support the branching apart of the early true primates (the LCA of all extant primates) into a tarsier-anthropoid or haplorhine clade and a loriform-lemuriform or strepsirhine clade. Despite this substantial evidence that Tarsiiformes is the sister group of Anthropoidea, followers of the Simpson-Mayr school of evolutionary taxonomy employ the idea of grades of evolutionary advancement to place tarsiers along with plesiadapids, omomyids, adapids, lemuriforms, and loriforms, i.e., all extinct and extant small-brained primates, into Prosimii, the suborder presumed to be at the lower or less advanced grade. In turn, the platyrrhines and catarrhines, i.e., the larger-brained primates, are placed in Anthropoidea, the suborder at the presumed higher or more advanced grade. In contrast to this traditional gradistic division of the order, those who follow the Hennigian school of phylogenetic systematics (now called cladistics) and who accept the evidence that Tarsiiformes is the sister group of Anthropoidea and that Loriformes is the sister group of Lemuriforms (e.g., Goodman et al., 1998; and this paper) divide Primates into Haplorhini (the tarsier-anthropoid clade) and Strepsirrhini (the loriform-lemuriform clade).

Classifying organisms according to presumed grades of evolutionary advancement may be criticized for perpetrating in taxonomy the pre-Darwinian metaphysical idea of the *scala naturae* or, as Lovejoy (1936) termed it, "the great chain of being." Rather than introduce into taxonomy such a subjective anthropocentric view of nature, we prefer a strictly genealogical taxonomy in which all taxa represent real clades produced by evolution and in which the clades represented by higher-ranked taxa are of older age than those represented by lower-ranked taxa. By providing this objective phylogenetic framework for investigating evolutionary processes, the strictly genealogical taxonomy should prove beneficial for the science of systematics.

Summary and Conclusions

The results of maximum parsimony analysis of aligned γ-globin DNA sequences from thirty five primate species consisting of thirty anthropoids (twenty two catarrhines and eight platyrrhines), three strepsirhines, and two tarsiers (*Tarsius bancanus* and *Tarsius syrichta*), and rabbit, as the outgroup, demonstrate that tarsiers are the sister group of anthropoids. This agrees with the taxonomic grouping of the primate suborders Tarsiiformes and Anthropoidea into the semiorder Haplorhini and is strongly supported in the maximum parsimony tree by the bootstrap value of 99%. The sister-grouping of tarsiers and anthropoids, rather than either tarsiers and strepsirhines or anthropoids and strepsirhines, also agrees with previous molecular studies using different DNA nucleotide sequences. Local molecular clock calculations using a paleontological inferred reference date of 63 Ma (Mega annum or millions of years before the present) for the last common ancestor (LCA) of all living primates (strepsirhines and haplorhines) placed the date for the LCA of haplorhines (tarsiers and anthropoids) at 58 Ma. In turn, the date for the LCA of the two tarsiers species is 5.6 Ma.

Acknowledgments

This research was supported from the following granting agencies: National Science Foundation (INT 9602913) and National Institutes of Health (HL 33940).

References

Aiello LC. 1986. The relationships of the tarsiiformes: a review of the case for Haplorhini. In Wood A, Martin LB, Andrews PJ, editors, Major topics in human and primate evolution, 47–65. Cambridge, UK: Cambridge University Press.

Andrews TD, Jermiin LS, Easteal S. 1998. Accelerated evolution of cytochrome b in simian primates: adaptive evolution in concert with other mitochondrial proteins? J Mol Evol 47: 249–57.

Bailey WJ, Hayasaka K, Skinner CG, Kehoe S, Sieu LC, Slightom JL, Goodman M. 1992. Reexamination of the African hominoid trichotomy with additional sequences from the primate β-globin gene cluster. Mol Phylogenet Evol 1: 97–135.

Bonner TI, Heinemann R, Todaro GJ. 1980. Evolution of DNA sequences has been retarded in Malagasy primates. Nature 286: 420–423.

Czelusniak J, Goodman M. 1995. Hominoid phylogeny estimated by model selection using goodness of fit significance tests. Mol Phylogenet Evol 4: 283–290.

DeJong WW, Goodman M. 1988. Anthropoid affinities of *Tarsius* supported by lens αA-crystallin sequences. J Hum Evol 17: 575–582.

Dijan P, Green H. 1991. Involucrin gene at tarsioids and other primates: alternatives in evolution of the segment of repeats. Proc Natl Acad Sci USA 88: 5321–5325.

Fitch DHA, Bailey WJ, Tagle DA, Goodman M, Sieu L, Slightom JL. 1991. Duplication of the γ-globin gene mediated by L1 long interspersed repetitive elements in an early ancestor of simian primates. Proc Natl Acad Sci USA 88: 7396–7400.

Felsenstein J. 1978. Cases in which parsimony and compatability methods will be positively misleading. Syst Zool 27: 401–410.

Fleagle JG. 1999. Primate adaptation and evolution, 2d ed. New York: Academic Press.

Gingerich PD. 1981. Early cenozoic omomyidae and the evolutionary history of Tarsiiform primates. J Hum Evol 10: 345–374.

Gingerich PD. 1986. Early Eocene *Cantius torresi*—oldest primate of modern aspect from North America. Nature 320: 319–321.

Gingerich PD. 1993. Early Eocene *Teilhardina brandti:* oldest Omomyid primate from North America. Contributions from the Museum of Paleontology, University of Michigan, vol. 28, 321–326.

Gingerich PD, Uhen MD. 1994. Time of origin of primates. J Hum Evol 27: 443–445.

Goodman M. 1986. Molecular evidence on the ape subfamily Homininae. In Gershowitz H, Rugknagel DR, Tashian RE, editors, Evolutionary perspectives and the new genetics, 121–132. New York: A R Liss.

Goodman M, Page SL, Meireles CM, Czelusniak J. 1999. Primate phylogeny and classification elucidated at the molecular level. In Wasser SP, editor, Evolutionary theory and processes: modern perspectives, 193–211. Dordrecht: Kluwer Academic Publishers.

Goodman M, Porter CA, Czelusniak J, Page SL, Schneider H, Shoshani J, Gunnell G, Groves CP. 1998. Toward a phylogenetic classification of primates based on DNA evidence complemented by fossil evidence. Mol Phylogenet Evol 9: 585–598.

Groves CP. 1993. Order primates. In Wilson DE, Reeder DM, editors, Mammal species of the world: a taxonomic and geographic reference, 243–277. Washington, DC: Smithsonian Institution Press.

Harris S, Thackeray JR, Jeffreys AJ, Weiss ML. 1986. Nucleotide sequence analysis of the lemur β-globin gene family: evidence for major rate fluctuations in globin polypeptide evolution. Mol Biol Evol 3: 465–484.

Hasegawa M, Kishino H, Hayasaka K, Horai S. 1990. Mitochondrial DNA evolution in primates: transition rate has been extremely low in the lemur. J Mol Evol 31: 113–21.

Hayasaka K, Gojobori T, Horai S. 1988. Molecular phylogeny and evolution of primate mitochondrial DNA. Mol Biol Evol 5: 626–44.

Hayasaka K, Skinner CG, Goodman M, Slightom JL. 1993. The γ-globin genes and their flanking sequences in primates: findings with nucleotide sequences of capuchin monkey and tarsier. Genomics 18: 20–28.

Jaworski CJ. 1995. A reassessment of mammalian αA-crystallin sequences using

DNA sequencing: implications for anthropoid affinities of tarsier. J Mol Evol 41: 901–908.

Jones FW. 1920. Discussion on the zoological position and affinities of *Tarsius*. Proc Zool Soc Lond 1920: 491–949.

Kapitonov V, Jurka J. 1996. The age of Alu subfamilies. J Mol Evol 42: 59–65.

Kay RF, Ross C, Williams BA. 1997. Anthropoid origins. Science 275: 797–804.

Koop BF, Tagle DA, Goodman M, Slightom JL. 1989. A molecular view of primate phylogeny and important systematic and evolutionary questions. Mol Biol Evol 6: 580–612.

Kuryshev VY, Skryabin BV, Kremerskothen J, Jurka J, Brosius J. 2001. Birth of a gene: locus of neuronal BC200 snmRNA in three prosimians and human BC200 pseudogenes as archives of change in the Anthropoidea lineage. J Mol Biol 309: 1049–66.

Lovejoy, AO. 1936. The great chain of being: a study of the history of an idea. Cambridge, HA: Harvard University Press.

McNiff BE, Allard MW. 1991. A test of Archonta monophyly and the phylogenetic utility of the mitochondrial gene 12S rRNA. Am J Phys Anthro 107: 225–241.

Margot JB, Demers GW, Hardison RC. 1989. Complete nucleotide sequence of the rabbit β-like globin gene cluster: analysis of intergenic sequences and comparison with the human β-like globin gene cluster. J Mol Biol 205: 15–40.

Martin RD. 1990. Primate origins and evolution: a phylogenetic reconstruction. London: Chapman and Hall.

Martin RD. 1993. Primate origins: plugging the gaps. Nature 363: 223–234.

Meireles CM, Czelusniak J, Goodman M. 1999a. The *Tarsius* γ-globin gene: pseudogene or active gene? Mol Phylogenet Evol 13: 434–439.

Meireles CM, Czelusniak J, Schneider MPC, Muniz JAPC, Brigido MC, Ferreira HS, Goodman M. 1999b. Molecular phylogeny of Ateline New World monkeys (Platyrrhini, Atelinae) based on γ-globin gene sequences: evidence that *Brachyteles* is the sister group of *Lagothrix*. Mol Phylogenet Evol 12: 10–30.

Murphy WJ, Eizirik E, Johnson WE, Zhang YP, Ryder OA, O'Brien SJ. 2001. Molecular phylogenetics and the origins of placental mammals. Nature 409: 614–618.

Napier JR, Napier PH. 1967. A handbook of living primates. London: Academic Press.

Napier JR, Napier PH. 1985. The natural history of the primates. Cambridge, MA: MIT Press.

Page SL, Chiu C-H, Goodman, M. 1999. Molecular phylogeny of Old World monkeys (Cercopithecidae) as inferred from γ-globin DNA sequences. Mol Phylogenet Evol 13: 348–359.

Page SL, Goodman M. 2001. Catarrhine phylogeny: noncoding DNA evidence for a diphyletic origin of the mangabeys and for a human-chimpanzee clade. Mol Phylogenet Evol 18: 14–25.

Pocock RI. 1918. On the external characters of the lemurs and of *Tarsius*. Proc Zool Soc Lond 55: 19–53.

Porter CA, Czelusniak J, Schneider H, Schneider MPC, Sampaio I, Goodman M.

1997a. Sequences of the primate ε-globin gene: implications for systematics of the marmosets and other New World primates. Gene 205: 59–71.

Porter CA, Page SL, Czelusniak J, Schneider H, Schneider MPC, Sampaio I, Goodman M. 1997b. Phylogeny and evolution of selected primates as determined by sequences of the ε-globin locus and 5′ flanking regions. Int J Primatol 18: 261–295.

Porter CA, Sampaio I, Schneider H, Schneider MPC, Czelusniak J, Goodman M. 1995. Evidence on primate phylogeny from ε-globin gene sequences and flanking regions. J Mol Evol 40: 30–55.

Ross C, Williams B, Kay RF. 1998. Phylogenetic analysis of anthropoid relationships. J Hum Evol 35: 221–306.

Rowe N. 1996. The pictorial guide to the living primates. East Hampton, NY: Pegonias Press.

Schmitz J, Ohme M, Zischler H. 2001. SINE insertions in cladistic analyses and the phylogenetic affiliations of *Tarsius bancanus* to other primates. Genetics 157: 777–784.

Schwartz JH. 1978. If *Tarsius* is not a prosimian, is it a haplorhine? In Chivers DJ, Joysey KA, editors, Recent advances in primatology, vol. 3, 195–202. London: Academic Press.

Schwartz JH, Tattersall I. 1987. Tarsiers, adapids and the integrity of Strepsirrhini. J Hum Evol 16: 23–40.

Shedlock AM, Okada N. 2000. SINE insertions: powerful tools for molecular systematics. Bioessays 22: 148–60.

Shoshani J, Groves CP, Simons EL, Gunnell GF. 1996. Primate phylogeny: morphological vs. molecular results. Mol Phylogenet Evol 5: 102–154.

Tagle DA, Koop BF, Goodman M, Slightom JL, Hess DL, Jones RT. 1988. Embryonic ε and γ globin genes of a prosimian primate (*Galago crassicaudatus:*) nucleotide and amino acid sequences, developmental regulation and phylogenetic footprints. J Mol Biol 203: 439–455.

Tagle DA, Stanhope MJ, Siemieniak, Benson P, Goodman M, Slightom JL. 1992. The β globin gene cluster of the prosimian primate *Galago crassicaudatus:* nucleotide sequence determination of the 41-kb cluster and comparative sequence analyses. Genomics 13: 741–760.

Zietkiewicz E, Richer C, Labuda D. 1999. Phylogenetic affinities of tarsier in the context of primate alu repeats. Mol Phylogenet Evol 11: 77–83.

CHAPTER 7

The Phylogenetic Position
of Genus *Tarsius:*
Whose Side Are You On?

Anne D. Yoder

The accurate placement of the genus *Tarsius* (commonly known as the tarsier) is the fundamental key to deriving a phylogenetic classification for the order Primates. Depending on where the tarsier is placed with respect to the tooth-combed primates (lemurs and lorises) and the anthropoid primates (monkeys and apes), there are three alternative classificatory schemes whereby the primates are divided into two named subclades (fig. 7.1). If tarsiers are grouped with the lemurs and lorises, then the division is between Prosimii and Anthropoidea (fig. 7.1a); if they are grouped with monkeys and apes, then the division is between Strepsirrhini and Haplorhini (fig. 7.1b). Alternatively, if tarsiers are considered sister to all other living primates, then the division is between Tarsiiformes and Simiolemuriformes (fig. 7.1c). Each of these classificatory schemes has proponents in the literature and is upheld by varying degrees of morphological character support, as will be reviewed below. The primary purpose of this chapter, however, is to consider the genetic data as they are currently known and to determine if these data show convincing support for any one of these three phylogenetic hypotheses.

History

The systematics of tarsiers relative to other primates has been controversial from the time that biologists first began to attempt systematic arrangements of life on Earth. Indeed, early systematists did not even recognize genus *Tarsius* as a primate, instead classifying it as either a jerboa or an opossum (Buffon, 1765; Linnaeus, 1767–70). Much later, in 1916, a ferment of controversy was aroused when F. Wood Jones published his surprising "tarsioid theory of human origins." This theory was sufficiently unconventional and disturbing to prompt a symposium of the Zoological Society of London in 1919 in which the sole purpose was to discuss the relationship of *Tarsius* to other primates. The findings of this symposium were inconclusive. In the ensuing years, hypotheses of the relationship of tarsiers to other living primates can be distilled into three major schemes, as shown in figure 7.1. It was Gregory

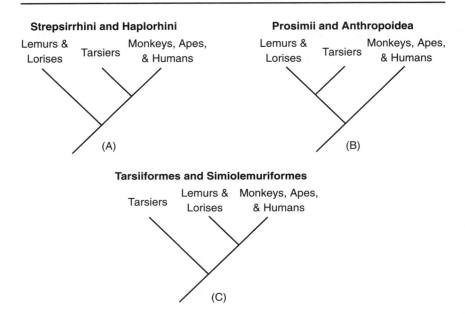

Figure 7.1. Three alternative hypotheses of higher-level primate interrelationships.

(1915) who first placed tarsiers with lemurs and lorises into the suborder Prosimii. This classification has received support from a number of more recent authorities on primate evolution (e.g., Le Gros Clark, 1971; Simons, 1972; Simpson, 1945) and has been widely accepted by the larger scientific community as the assumed primate classification (e.g., Kim and Takenaka, 2000; Wu et al., 2000). In temporal parallel, the classification proposed by Pocock (1918), in which *Tarsius* is grouped with monkeys and apes in the order Haplorhini, has also received support from a number of prominent primatologists (e.g., Aiello, 1986; Hill, 1953; Ross, 1994; Szalay and Delson, 1979), though it has received far less recognition from the general scientific community. The latter classificatory scheme is distinguished from the previous by its explicitly character-based perspective. Finally, the third scheme has been proposed more recently than either the prosimian or the haplorhine classifications of *Tarsius* (Gingerich, 1978; Schwartz 1978a, b), but has been largely ignored by most primate systematists.

Many of the controversies related to these opposing classifications derive from arguments founded on fossil data and the hypothesized relationships of tarsiers to extinct Paleocene and Eocene fossil lineages. It is my goal to avoid these arguments as much as possible, leaving the arduous task of interpreting the fossil record to the paleontologists, as it should be. Further discussion of these issues and the fossil data supporting the various viewpoints can be found in a number of excellent primary and secondary reviews (e.g., Beard, 1998; Beard et al., 1988; Beard et al., 1991; Ross et al., 1998). It is my task in this chapter to review the neontological data as they pertain to the phylogenetic affinities of the genus *Tarsius*, both morphological (although this is not a novel endeavor; see especially Aiello, 1986; Cartmill, 1982, 1994; Ross, 1994) and genetic.

Somatic Character Evidence for Tarsier Affinities

As mentioned above, Gregory (1915) was one of the first to classify tarsiers with lemurs and lorises into a single suborder Lemuroidea, within which three "series" were identified to individually recognize the distinctions among lemurs, lorises, and tarsiers. Simpson (1945) followed suit with his definitive classification of the mammals, explicitly following Gregory's arrangement though renaming the suborder Prosimii and identifying the three series as the infraorders Lemuriformes, Lorisiformes, and Tarsiiformes. Review of these two works, however, makes it clear that the rationale for the unified grouping of the three infraorders related more to the perceived "primitiveness" of the "lower primates" and to perceptions of relatedness to Eocene fossil lineages than to any fixed idea of character support for their historical unity. Indeed, both men remarked on the large morphological gaps between tarsiers and the other prosimians, with Gregory even noting that the similarities of the tarsier and anthropoid cranial arterial "condition has been derived from" (p. 429) the lemuriform pattern. He goes on to say that the contrast between lemuriforms and tarsiiforms, and their presumed fossil antecedents, is so great as to "warrant us in looking for the common stem form of the Lemuroidea in the Paleocene or even earlier."

Although Simpson (1945) formalized the prosimian-anthropoid classification, Le Gros Clark (1971) is largely responsible for its information content. Le Gros Clark established the perception of the primate order as a taxon that showed progressive evolutionary trends rather than unique defining characteristics. In this view, the order began with tree shrews and culminated with humans, showing trends for increasingly enlarged brains, perfected vision, reduced olfaction, and grasping extremities. Accordingly, the discrimination of two major subdivisions within the order was somewhat arbitrary and based primarily on the distinction of two evolutionary grades: lower primates and higher primates. Le Gros Clark further concluded that the fossil record showed that early tarsioids had developed their peculiar specializations to an advanced degree in the early Eocene, and thus they could not have "provided a basis for the subsequent evolution of the Anthropoidea in which such specializations are absent" (p. 332).

The plesiotarsiiform-simiolemuriform classification was originally conceived by Gingerich (1974) and was based on the stratophenetic linkage of tarsiers to Eocene omomyines and microchoerines, and these taxa to Paleocene plesiodapiforms. The temporal connection of these primate groups was further strengthened by stratigraphic proximity, dental similarities, and certain aspects of middle-ear morphology. The plesiotarsiiform half of the order was later revised to exclude plesiadapiforms with the name changed to Tarsiiformes to reflect the new taxon composition of the suborder (Gingerich, 1981). By and large, however, either arrangement is more a reflection of Gingerich's vehement belief that the "linking between primitive anthropoids and Eocene adapoids is one of the strongest in all of primate phylogeny" (1978: 253) more than it is a reflection of tarsier's separateness. The fact that Schwartz's work on dental homologies (Schwartz, 1978a, b) supported Gingerich's initial views has done little to buffer them from the disregard of other systematists. The most serious problem with Gingerich's hypothesis in its original formulation is that it requires independent evolution of the lemur- and tarsierlike euprimate lineages from a preplesiadapiform. Given that it is now well accepted that plesiadapiforms were not stem primates (Beard, 1990; Beard and Wang, 1991; Kay et al., 1992), Gingerich's evolutionary scenario requires convergent evolution of three of the most significant defining features of euprimates: replacement of claws by nails on at least some digits, enlargement and forward rotation of the orbits, and a postorbital bar.

Clearly, the previous two classificatory schemes both recognize and depend on the fossil record for their foundation. The strepsirhine-haplorhine classification, on the other hand, relies almost entirely on neontological data. Pocock (1918) was the first to assert that the dissimilarity of the structure of the nose and upper lip of *Tarsius* compared to lemuriforms "severs" their potential phylogenetic connection, bringing tarsiers closer in line with anthropoids instead. Indeed, the word strepsirhine is derived from the primitive retention in lemurs and lorises of a naked, moist rhinarium that exhibits a lateral cleft between the medial and lateral nasal processes. In tarsiers and anthropoids, on the other hand, the rhinarium is covered with dry, hairy skin, and the lateral cleft is completely fused (Hill, 1953).

Although proposed by Pocock (1918) and adopted by Hill (1953), the strepsirhine-haplorhine classification has gained most of its authority since the introduction of cladistic methods. Cladistically inclined primatologists have spent much of their time seeking to discover synapomorphies that ally tarsiers either with lemuriforms or with anthropoids. Cartmill (1982: 281) described the pursuit as complicated, at best, due to the difficulty of the problem: "[T]he phylogenetic histories of some organisms have been so muddled by convergence, parallelism, and evolutionary reversal that many false phylogenies can be supported. . . . Unfortunately for primate biologists, *Tarsius* may well be such an organism."

Despite the difficulty of the problem, an impressive list of putative haplorhine synapomorphies have been presented in the literature (table 7.1). Even so, doubt persists, particularly as the distribution of these characters among Oligocene anthropoids implies that many of them must be convergently derived in living tarsiers and anthropoids (Simons and Rasmussen, 1989). Continued application of phylogenetic methods to the analysis of morphological character data has unfortunately been unable to resolve the question. Although Kay et al. (1997), employing the most extensive morphological data set to date, found support for the strepsirhine/haplorhine dichotomy, subsequent analysis by the same authors identified some uncertainty in their previous conclusions (Ross et al., 1998). This latter study found that when the data were analyzed as individual cranial and postcranial data partitions, the former supported a monophyletic Haplorhini, whereas the latter supported a monophyletic Prosimii.

Genetic Data

In cases of phylogenetic uncertainty relating to extensive morphological homoplasy, it is often the case that genetic data can resolve the problem. Unlike morphological data, molecular characters (i.e., DNA sequence data) and character states are obvious and not prone to subjective interpretation. Given that the alignment is accurate (which, in certain cases, can be quite problematic, though that issue will not be addressed here), molecular characters are usually defined as the individual sites along the DNA strand, with the states defined as the individual nucleotides, A, C, G, and T. Thus, except in cases of alignment uncertainty, subjectivity and potential investigator bias are completely removed from the process of defining characters and their states. This should therefore alleviate much of the controversy that has surrounded morphological analysis of tarsier affinities. The earliest attempts to apply genetic data to phylogenetic questions are best described as genetic distance data, such as immunodiffusion (Goodman and Moore, 1971) and DNA hybridization measures (Sibley and Ahlquist, 1984). Both of these methods are "indirect," however, in that they summarize the differences and similarities among taxa into a single pairwise measure. Thus, they cannot distinguish homology from similarity, nor provide information about direction of character change. The small number of shared derived characters between two closely related taxa can easily be swamped by the much larger number of shared primitive characters. Clearly then, neither of these methods can be considered very sensitive, though their utility for providing a rapid assessment of genomic-level similarities holds some appeal.

Genetic distance methods have been essentially replaced by character-based methods derived from DNA sequence data. These data permit the investigator to employ traditional cladistic methods (such as parsimony) to genetic data, much as they have been applied to morphological data. One of the greatest advantages of DNA sequence data, however, has been the development of likelihood methods for phylogenetic analysis. Likelihood methods have been developed for incorporating knowledge of molecular evolutionary properties, such as transition/transversion rate ratio bias, codon bias, and site-specific rate variation, directly into the model and thus the analysis, giving them far more power than traditional parsimony analysis (Felsenstein, 1973; Yang et al., 1995). Such models are currently not employed for the analysis of morphological data, though ongoing work promises breakthroughs for this application (Lewis, 2001). Finally, due to the enormity of the potential data set (i.e., the millions of individual base pairs contained in the combined mitochondrial and nuclear genomes), there are virtually limitless characters from which to choose. Potentially then, systematists interested in resolving the placement of genus *Tarsius* could objectively employ multiple molecular data sets, using a variety of optimality criteria (i.e., maximum parsimony, maximum likelihood, and minimum evolution) (Swofford et al., 1996) and thus discover the hypothesis that is maximally supported by the data and the methods. Indeed, molecular phylogeneticists and other geneticists have gathered and analyzed an enormous amount of DNA sequence data; but most unfortunately for those interested in resolving the tarsier enigma, these data and analyses leave us as uncertain as have the morphological studies.

Another form of genetic data has recently emerged as potentially useful for such difficult phylogenetic questions. These genetic characters, usually referred to as SINEs (short interspersed nuclear elements), are nonautonomous transposable elements that are "interspersed" throughout the nuclear genome. These elements vary in length from about 150–500 base pairs and, according to some authorities, can be considered as ideal phylogenetic characters. The main portion of a SINE is presumably nonfunctional and thus, hypothetically, is a neutral phylogenetic marker. Also, there is a strong assumption that these characters integrate irreversibly into the nuclear genome. Although it is conceivable that a specific SINE could be deleted from a genome subsequent to integration, it is hard to imagine a mechanism whereby this could happen invisibly. Because a signature repeat is formed that flanks both ends of a SINE sequence upon integration, subsequent deletion would likely leave some portion of the signature flanking sequence, or, if the deletion was larger, would also remove a portion of the surrounding ancestral sequence. In either case, the deletion would be detectable in a comparison of related sequences. Thus, polarity can also be determined in that the shared absence of a SINE is virtually certain to be the ancestral state. Finally, due to the enormous size of the typical mammalian genome, SINE integrations observed at homologous positions are regarded as characters with infinitesimal probabilities of convergence or parallelism. For all of these reasons, some workers are convinced that SINEs are the ideal phylogenetic characters. Unfortunately, however, SINEs appear to have accumulated in the genome through multiple waves of fixation. Thus, the nonlinear accumulation and relatively small number of integration events limit their use for de novo phylogenetic reconstruction. It has been alternatively argued, therefore, that they are ideal for testing the veracity of competing hypotheses, such as might be the case for the limited number of phylogenetic hypotheses that have been proposed for tarsier affinities (Schmitz et al., 2001).

Genetic Studies Supporting Haplorhini

At first glance, the genetic evidence for tarsier's placement with the anthropoid primates appears to be overwhelming. Over the past twenty or so years, there have been many, many publications of genetic data that support a monophyletic Haplorhini (Baba et al., 1982; Bailey et al., 1991; de Jong and Goodman, 1988; Koop et al., 1989a, b; Porter et al., 1995; Stanhope et al., 1993). These accumulating studies, along with the extensive list of putative morphological synapomorphies (table 7.1), have served to solidify support for the strepsirhine/haplorhine division among primate systematists. A close inspection of these studies, however, shows that they are all taken from the nuclear genome, and that over 50% of them are from the globin family of genes. Although the first observation is not of particular concern—the nuclear genome is by far the most extensive set of potential genetic data sets—the second observation is of some consequence. Because a substantial number of these studies derive from the globin family, they cannot be said to be independent samples of phylogenetic history. To draw a morphological analogy, this is equivalent to basing a primate phylogeny strictly on dental characters, paying no attention to other aspects of the overall anatomy of the organisms under investigation.

Even so, advocates of the strepsirhine/haplorhine division can take heart in at least two new studies that have sampled nuclear genetic characters outside of the globin family. Both studies found support for tarsier's placement with the Anthropoidea in Alu repeat data (Schmitz et al., 2001; Zietkiewicz et al., 1999). Alu repeats are the largest family of SINEs in the primate genome, with more than half a million copies dispersed throughout the genome. The Zietkiewicz et al. (1999) study employed a maximum likelihood approach to examine the phylogenetic relationships among Alu

Table 7.1. Putative Haplorhine Synapomorphies

Characters	Primary reference(s)
Dry, hairy rhinarium and fused lateral cleft of nasal processes	Pocock (1918)
Characteristics of intracranial blood supply	Szalay (1975); Rosenberger and Szalay (1980); Cartmill et al. (1981); MacPhee and Cartmill (1986)
Bony contributions to postorbital spetum	Pocock (1918); Hershkovitz (1974); Cartmill (1980)
Patterns of craniogenesis	Starck (1975)
Mode of placentation	Hubrecht (1897); Pocock (1918); Luckett (1974, 1975)
Absence of tapetum lucidum	Martin (1973)
Presence of fovea centralis	Wolin and Massopust (1970)
Reduced olfactory bulbs	Rosenberger and Szalay (1980)
Loss of subtympanic recess beneath the ectotympanic	Cartmill et al. (1981)
Anterior accessory chamber of auditory bulla	Cartmill and Kay (1978)
Annular bridge of auditory bulla absent	MacPhee and Cartmill (1986)
Apical interorbital septum	Cave (1973)
Loss of olfactory recess	Cave (1973)
Delayed puberty	Groves (1986); Shoshani et al. (1996)
Separation of foramen rotundum and orbital fissure	Shoshani et al. (1996)
Sperm morphology	Robson et al. (1997)
Relative size of brain components	Joffe and Dunbar (1998)
Retinal photoreceptor types	Hendrickson et al. (2000)

sequences taken from tarsier, human, lemur, sifaka, and galago. In all analyses, tarsier and human sequences form a clade that excludes the strepsirhines sampled by the study. This phylogenetic result was also substantiated by similarities in Alu RNA secondary structure in humans and tarsiers. The Schmitz et al. (2001) study reached a similar conclusion, though these authors were careful to point out that only three of the 118 Alu markers examined supported this relationship. This latter observation emphasizes the point that even if there is a unique evolutionary branch that unites tarsiers and anthropoids, it is short indeed.

Genetic Studies Supporting Prosimii

Genetic support for one of the alternative arrangements, the prosimian/anthropoid division, has received much less impressive support from genetic data. Although there are two mtDNA studies that appear to provide robust support for tarsier's placement with lemurs and lorises (Hasegawa et al., 1990; Hayasaka et al., 1988), these two studies are based on a single data set, one that spans the ND3&4 region of the mitochondrial genome. Moreover, a subsequent investigation found that there was a taxon sampling bias in the Hasegawa et al. (1990) study that had a strong effect on the estimation of molecular evolutionary parameters (Yang and Yoder, 1999), raising the possibility that the phylogenetic results could be similarly biased. Thus, until very recently, proponents of a monophyletic Prosimii could find very little support in molecular phylogenetic studies.

This appeared to change rather dramatically with the publication of one of the most comprehensive molecular data sets to have ever been generated for the investigation of mammalian relationships (Murphy et al., 2001). This study of nearly 10,000 base pairs of nuclear and mitochondrial DNA, drawn from fifteen independent genetic regions for sixty four placental mammals, reached the conclusion that "lemur (Strepsirrhini) and tarsier (Tarsiiformes) were found to be sister taxa (bootstrap $\geq 80\%$) that were separated from anthropoids by a deep divergence" (p. 615). Given that this is one of the most extensive data sets ever to have been applied to the problem of tarsier affinities (albeit indirectly applied, in that this was not an original concern of the study), it momentarily appeared that the problem had been solved. Subsequent analysis of this data set, however, has shown that there are a number of flaws in the study (Yoder and Huelsenbeck, 2001). First, nearly 40% of the complete data set is missing for genus *Tarsius;* and second, there is an extreme taxon sampling bias in that only one strepsirhine (genus *Lemur*) is represented. The authors of the subsequent study have thus reexamined the original data set in an effort to control for missing data, test for the effects of taxon sampling bias, and, finally, to investigate the influence of various character partitions on the phylogenetic results. With regard to missing data effects, a combined data set for which all missing data were excluded was found to be supportive of the results of the original study. As before, a monophyletic Prosimii was recovered with strong statistical support. Alterations in the taxon sample were also found not to effect the original conclusions. When the data were analyzed by partition (coding nuclear versus noncoding nuclear versus mitochondrial), however, it was discovered that all three potential resolutions (fig. 7.1) of tarsier affinities could be recovered, and each with strong statistical support. Clearly, this latter result casts doubt on the confidence that we might place in the conclusion of a monophyletic Prosimii. And, as a final blow to the perceived power of the original study, its authors have found by using likelihood ratio tests that the tarsier plus strepsirhine topology cannot be shown to be significantly preferred to alternative topologies (Murphy and Eizirik, pers. com.).

Why Can't We Decide?

Situations of extreme phylogenetic uncertainty, such as seems to be the case with tarsier affinities, often relate to problems of short internal branches confounded by long external branches. In other words, it can happen that speciation and cladogenesis occur very rapidly and are then followed by a long period of independent evolution for the resulting lineages. To reconstruct these internal events is quite problematic for any data set, but most especially for molecular data due to the effects of saturation along internal branches and convergent evolution among terminal branches (Felsenstein, 1978; Huelsenbeck, 1997). This is almost certainly the situation faced at the deepest region of primate phylogeny wherein the initial divergences among strepsirhines, tarsiers, and anthropoids occurred (fig. 7.2). There are numerous genetic data sets that support the contention that the strepsirhine-tarsier-anthropoid is a virtual trichotomy, with tarsiers being so derived as to be almost unresolvable as primates (Andrews et al., 1998; Dutrillaux and Rumpler, 1988; Hayasaka et al., 1993; Inoue-Murayama et al., 1998; Jaworski, 1995; Mai, 1985; Sarich and Cronin, 1976). It is possible that these events occurred in such a short geological period that the amount of data required to solve the problem may not be worth the effort and expense involved in generating the data. Given the debate in the systematics community over the relative advantages of large character samples versus large taxon samples (Graybeal, 1998; Hillis, 1996; Kim, 1996, 1998; Purvis and Quicke, 1997; Rannala et al., 1998; Yang and Goldman, 1997), it seems that the only way that the problem can ever be "resolved" will be to generate a character set on the order of the Murphy et al. (2001) study for a taxon sample that can be considered comprehensive for primates. Given the effort involved with such a study, many might ask if it is worth our while.

In examining figure 7.2, many might ask, "Who cares what happened in such a short interval of primate evolutionary history?" I would argue that the issue is far from trivial. It is important not only for determining a phylogenetically consistent classification for primates, but also for investigating the relationship between morphological evolution and temporal constraint. If we determine that tarsiers *do* share a unique evolutionary history with monkeys and apes, then we also know that many of the presumed haplorhine synapomorphies evolved within a remarkably short geological interval. If, on the other hand, an alternative topology is found to be more accurate, then the presumed synapomorphies instead become sympleisiomorphies or parallelisms, thereby altering our perception of the ancestral primate condition and/or the evolutionary pressures applied to the primate condition. Thus, as we continue to investigate tarsier's placement among the primates, we are actually holding a magnifying glass to that essential period of evolu-

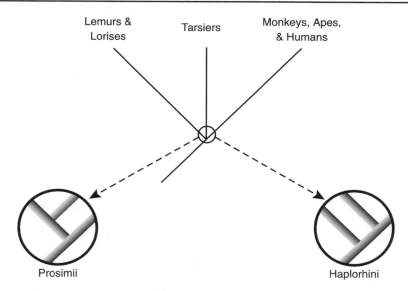

Figure 7.2. Schematic diagram illustrating problem of short internal branches within deepest region of primate phylogeny.

tionary history wherein the defining characteristics of the primate clade were being established.

References

Aiello LC. 1986. The relationships of the tarsiiformes: a review of the case for the Haplorhini. In Wood BA, Martin LB, PJ Andrews PJ, editors, Major topics in primate and human evolution, 47–65. Cambridge, UK: Cambridge University Press.

Andrews TD, Jermiin LS, Easteal S. 1998. Accelerated evolution of cytochrome b in simian primates: adaptive evolution in concert with other mitochondrial proteins? J Mol Evol 47: 249–257.

Baba ML, Weiss ML, Goodman SM, Czelusniak J.1982. The case of tarsier hemoglobin. Syst Zool 31: 156–165.

Bailey WJ, Fitch DHA, Tagle DA, Czelusniak J, Slightom JL, Goodman M. 1991. Molecular evolution of the yh-globin gene locus: gibbon phylogeny and the hominoid slowdown. Mol Biol Evol 8: 155–184.

Beard KC. 1990. Gliding behavior and palaeoecology of the alleged primate family Paromomyidae Mammalia, Dermoptera. Nature 345: 340–341.

Beard KC. 1998. A new genus of Tarsiidae Mammalia: primates from the middle eocene of Shanxi Province, China, with notes on the historical biogeography of tarsiers. Bull Carnegie Mus Nat Hist 34: 260–277.

Beard KC, Wang B. 1991. Phylogenetic and biogeographic significance of the tarsiiform primate *Asiomomys changbaicus* from the Eocene of Jilin Province, People's Republic of China. Am J Phys Anthro 85: 159–166.

Beard KC, Dagosto M, Gebo DL, Godinot M. 1988. Interrelationships among primate higher taxa. Nature 331: 712–714.

Beard KC, Krishtalka L, Stucky RK. 1991. First skulls of the early Eocene primate *Shoshonius cooperi* and the anthropoid-tarsier dichotomy. Nature 349: 64–67.

Buffon GL. 1765. Histoire Naturelle, Generale et Particuliere. Paris: L'imprimerie du Roi.

Cartmill M. 1980. Morphology, function, and evolution of the anthropoid postorbital septum. In Ciochon RL, Chiarelli AB, editors, Evolutionary biology of the New World monkeys and continental drift, 243–274. New York: Plenum Press.

Cartmill M. 1982. Assessing tarsier affinities: is anatomical description phylogenetically neutral? Geobios Mem Spec 6: 279–287.

Cartmill M. 1994. Anatomy, antinomies, and the problem of anthropoid origins. In Fleagle JG, Kay RF, editors, Anthropoid origins, 549–566. New York: Plenum Press.

Cartmill M, Kay RF. 1978. Cranio-dental morphology, tarsier affinities, and primate suborders. In Chivers DJ, Joysey KA, editors, Recent advances in primatology, 205–214. New York: Academic Press.

Cartmill M, MacPhee RDE, Simons EL. 1981. Anatomy of the temporal bone in early anthropoids, with remarks on the problem of anthropoid origins. Am J Phys Anthropol 56: 3–21.

Cave AJE. 1973. The primate nasal fossa. Biol J Linnean Soc 5: 377–387.

de Jong WW, Goodman SM. 1988. Anthropoid affinities of *Tarsius* supported by lens alpha-a-crystallin sequences. J Hum Evol 17: 575–582.

Dutrillaux B, Rumpler Y. 1988. Absence of chromosomal similarities between tarsiers *Tarsius syrichta* and other primates. Folia Primatol 50: 130–133.

Felsenstein J. 1973. Maximum likelihood and minimum steps methods for estimating evolutionary trees from data on discrete characters. Syst Zool 22: 240–249.

Felsenstein J. 1978. Cases in which parsimony or compatibility methods will be positively misleading. Syst Zool 27: 401–410.

Gingerich PD. 1974. Cranial anatomy and evolution of early Tertiary Plesiadapidae Mammalia, Primates. Ph.D. dissertation, Yale University, New Haven.

Gingerich PD. 1978. Phylogeny reconstruction and the phylogenetic position of *Tarsius*. In Chivers DJ, Joysey KA, editors, Recent advances in primatology, 249–255. New York: Academic Press.

Gingerich PD. 1981. Early Cenozoic Omomyidae and the evolutionary history of tarsiiform primates. J Hum Evol 10: 345–374.

Goodman M, Moore GW. 1971. Immunodiffusion systematics of the primates: I. The Catarrhini. Syst Zool 20: 19–62.

Graybeal A. 1998. Is it better to add taxa or characters to a difficult phylogenetic problem? Syst Biol 47: 9–17.

Gregory WK. 1915. On the classification and phylogeny of the Lemuroidea. Bull Geol Soc Am 26: 426–446.

Groves CP. 1986. Systematics of the great apes. In Swindler DR, Erwin J, editors, Comparative primate biology: systematics, evolution, and anatomy, 187–217. New York: A.R. Liss.

Hasegawa M, Kishino H, Hayasaka K, Horai S. 1990. Mitochondrial DNA evolution in primates: transition rate has been extremely low in the lemur. J Mol Evol 31: 113–121.

Hayasaka K, Gojobori T, Horai S. 1988. Molecular phylogeny and evolution of primate mitochondrial DNA. Mol Biol Evol 5: 626–644.

Hayasaka K, Skinner CG, Goodman M, Slightom JL. 1993. The gamma-globin genes and their flanking sequences in primates: findings with nucleotide sequences of capuchin monkey and tarsier. Genomics 18: 20–28.

Hendrickson A, Djajadi HR, Nakamura L, Possin DE, Sajuthi D. 2000. Nocturnal tarsier retina has both short and long/medium-wavelength cones in an unusual topography. J Comp Neurol 424: 718–730.

Hershkovitz P. 1974. The ectotympanic bone and origin of higher primates. Folia Primatol 22: 247–242.

Hill WCO. 1953. Primates, Strepsirrhini. Edinburgh, UK; Edinburgh University Press.

Hillis DM. 1996. Inferring complex phylogenies. Nature 383: 130–131.

Hubrecht AAW. 1897. The descent of the primates. New York: Charles Scribner's Sons.

Huelsenbeck JP. 1997. Is the Felsenstein zone a fly trap? Syst Biol 46: 69–74.

Inoue-Murayama M, Takenaka O, Murayama Y. 1998. Origin and divergence of tandem repeats of primate D4 dopamine receptor genes. Primates 39: 217–224.

Jaworski CJ. 1995. A reassessment of mammalian aA-Crystallin sequences using DNA sequencing: implications for anthropoid affinities of tarsier. J Mol Evol 41: 901–908.

Joffe TH, Dunbar RIM. 1998. Tarsier brain component composition and its implications for systematics. Primates 39: 211–216.

Kay RF, Ross C, Williams BA. 1997. Anthropoid origins. Science 275: 797–804.

Kay RF, Thewissen JGM, Yoder AD. 1992. Cranial anatomy of *Ignacious graybullianus* and the affinities of the Plesiadapiformes. Am J Phys Anthro 89: 477–498.

Kim HS, Takenaka O. 2000. Evolution of the X-linked zinc finger gene and the Y-linked zinc finger gene in primates. Mol Cells 10: 512–518.

Kim J. 1996. General inconsistency conditions for maximum parsimony: effects of branch lengths and increasing numbers of taxa. Syst Biol 45: 363–374.

Kim J. 1998. What do we know about the performance of estimators for large phylogenies? Trends Ecol Evol 13: 25–26.

Koop BF, Siemaniak D, Slightom JL, Goodman M, Dunbart J, Wright P, Simons E. 1989a. *Tarsius* delta- and beta-globin genes: conversions, evolution, and systematic implications. J Biol Chem 264: 68–79.

Koop BF, Tagle DA, Goodman M, Slightom JL. 1989b. A molecular view of primate phylogeny and important systematic and evolutionary questions. Mol Biol Evol 6: 580–612.

Le Gros Clark WE. 1971. The antecedents of man. Edinburgh, UK: Edinburgh University Press.

Lewis PO. 2001. A likelihood approach to estimating phylogeny from discrete morphological character data. Syst Biol 50: 913–925.

Linnaeus CA. 1767–70. Systema naturae. Vienna: J. T. Trattner.

Luckett WP. 1974. Comparative development and evolution of the placenta in primates. In Luckett WP, editors, Reproductive biology of the primates, 142–234. New York: Plenum Press.

Luckett WP. 1975. Ontogeny of the fetal membranes and placenta: their bearing on primate phylogeny. In Luckett WP, Szalay FS, editors, Phylogeny of the primates, 157–182. New York: Plenum Press.

MacPhee RDE, Cartmill M. 1986. Basicranial structures and primate systematics. In Swindler DR, Erwin J, editors, Comparative primate biology, 219–275. New York: Alan R. Liss.

Mai LL. 1985. Chromosomes and the taxonomic status of the genus *Tarsius*—preliminary results. J Hum Evol 14: 229–240.

Martin RD. 1973. Comparative anatomy and primate systematics. Symp Zool Soc Lond 33: 301–337.

Murphy WJ, Eizirik E, Johnson WE, Zhang YP, Ryder OA, O'Brien SJ. 2001. Molecular phylogenetics and the origins of placental mammals. Nature 409: 614–618.

Pocock RI. 1918. On the external characters of the lemurs and of *Tarsius*. Proc Zool Soc Lond 1918: 19–53.

Porter CA, Sampaio I, Schneider H, Schneider MPC, Czelusniak J, Goodman M. 1995. Evidence on primate phylogeny from e-globin gene sequences and flanking regions. J Mol Evol 40: 30–55.

Purvis A, Quicke DLJ. 1997. Building phylogenies: are the big easy? Trends Ecol Evol 12: 49–50.

Rannala B, Huelsenbeck JP, Yang Z, Nielsen R. 1998. Taxon sampling and the accuracy of large phylogenies. Syst Biol 47: 702–710.

Robson SK, Rouse GW, Pettigrew JD. 1997. Sperm ultrastructure of *Tarsius bancanus* Tarsiidae, Primates: implications for primate phylogeny and the use of sperm in systematics. Acta Zool 78: 269–278.

Rosenberger AL, Szalay FS. 1980. On the tarsiiform origins of Anthropoidea. In Ciochon RL, Chiarelli AB, editors, Evolutionary biology of the New World monkeys and continental drift, 139–157. New York: Plenum.

Ross C. 1994. The craniofacial evidence for anthropoid and tarsier relationships. In Fleagle JG, Kay RF, editors, Anthropoid origins, 469–547. New York: Plenum Press.

Ross C, Williams B, Kay RF. 1998. Phylogenetic analysis of anthropoid relationships. J Hum Evol 35: 221–306.

Sarich VM, Cronin JE. 1976. Molecular systematics of the primates. In Goodman M, Tashian RE, editors, Molecular athropology, 141–170. New York: Plenum.

Schmitz J, Ohme M, Zischler H. 2001. SINE insertions in cladistic analyses and the phylogenetic affiliations of *Tarsius bancanus* to other primates. Genetics 157: 777–784.

Schwartz JH. 1978a. Dental development, homologies, and primate phylogeny. Evol Theor 4: 1–32.

Schwartz JH. 1978b. If *Tarsius* is not a prosimian, is it a haplorhine? In Chivers DJ, Joysey KA, editors, Recent advances in primatology, 195–202. New York: Academic Press.

Shoshani J, Groves CP, Simons EL, Gunnell GF. 1996. Primate phylogeny: morphological vs. molecular results. Mol Phylogenet Evol 5: 102–154.

Sibley CG, Ahlquist JE. 1984. The phylogeny of the hominoid primates, as indicated by Dna-Dna hybridization. J Mol Evol 20: 2–15.

Simons EL. 1972. Primate evolution: an introduction to man's place in nature. New York: Macmillan.

Simons EL, Rasmussen DT. 1989. Cranial morphology of *Aegyptopithecus* and *Tarsius* and the question of the tarsier-anthropoidean clade. Am J Phys Anthropol 79: 1–23.

Simpson GG. 1945. The principles of classification and a classification of the mammals. Bull Am Mus Nat Hist 85: 1–350.

Stanhope MJ, Tagle DA, Shivji MS, Hattori M, Sakaki Y, Slightom JL, Goodman M. 1993. Multiple L1 progenitors in prosimian primates: phylogenetic evidence from ORF1 sequences. J Mol Evol 37: 179–189.

Starck D. 1975 The development of the chondrocranium in primates. In Luckett WP, Szalay FS, editors, Phylogeny of the primates, 127–156. New York: Plenum Press.

Swofford DL, Olsen GJ, Waddel PJ, Hillis DM. 1996. Phylogenetic inference. In Hillis DM, Moritz C, Mable BK, editors, Molecular systematics, 407–514. Sunderland, Mass: Sinauer.

Szalay FS. 1975. Phylogeny of primate higher taxa: the basicranial evidence. In Luckett WP, Szalay FS, editors, Phylogeny of the primates, 91–126. New York: Plenum Press.

Szalay FS, Delson E. 1979. Evolutionary history of the primates. New York: Academic Press.

Wolin L, Massopust L. 1970. Morphology of the primate retina. In Noback CR, Montagna W, editors, The primate brain, 1–27. New York: Appleton-Century-Crofts.

Wu CWH, Bichot NP, Kaas JH. 2000. Converging evidence from microstimulation, architecture, and connections for multiple motor areas in the frontal and cingulate cortex of prosimian primates. J Comp Neurol 423: 140–177.

Yang Z, Goldman N. 1997. Are big trees indeed easy? Trends Ecol Evol 12: 357.

Yang Z, Goldman N, Friday A. 1995. Maximum likelihood trees from DNA sequences: a peculiar statistical estimation problem. Syst Biol 44: 384–399.

Yang Z, Yoder AD. 1999. Estimation of the transition/transversion rate bias and species sampling. J Mol Evol 48: 274–283.

Yoder AD, Huelsenbeck JP. 2001. Testing the effects of methods, characters, and taxa on a persistent problem in primate phylogeny. Knoxville, TN: Joint Meetings of SSB, ASN, and SSE.

Zietkiewicz E, Richer C, Labuda D. 1999. Phylogenetic affinities of tarsier in the context of primate Alu repeats. Mol Phylogenet Evol 11: 77–83.

PART 2

PRESENT:
Taxonomy, Behaviorial Ecology, and Vocalizations

CHAPTER 8

The Tarsiers of Sulawesi

Colin P. Groves

The distinction of species of tarsiers living on the island of Sulawesi has become controversial in recent decades. The tarsiers of Sulawesi and neighboring islands have traditionally been assigned to a single species, *Tarsius spectrum* (Pallas, 1778). Hill (1953a) argued that the type locality of Pallas's *Lemur spectrum,* supposed to have come from Ambon, was actually Macassar (now Ujung Pandang, the capital of Sulawesi Selatan province, Indonesia), because he gave "the native name as *podje,* by which term the creature is known in Macassar." The evidence that the name designated a Sulawesi tarsier was regarded by Musser and Dagosto (1987) as "actually rather slim," and indeed the description of its tail, as naked with a tufted tip, does tend to suggest *T. bancanus* rather than a Sulawesian tarsier. But ever since the writing of Hill (1953a), the name *spectrum* has been applied to a Sulawesian tarsier, and it would be most confusing to alter this assessment now. I would therefore accept Hill's restriction of the type locality to Macassar. It is interesting, however, that Meyer (1896–97) quoted information from Teijsmann that the name *podi* is used in P. Salayar, and M. Shekelle (pers. com.) informs me that variants of the same word are used in many other places in and around Sulawesi, for example in the Togian Islands.

As recounted by Hill (1953a) and Musser and Dagosto (1987), another tarsier from Macassar was described by Fischer (1804) as *Tarsius fuscus,* inadvertently renamed *Tarsius fuscomanus* by E. Geoffroy Saint Hilaire (1812), and redescribed as *T. fischeri* by Burmeister (1846). Meyer (1896–97), used the name *T. spectrum* for Bornean tarsiers, as was usual in his day, and called the Sulawesian tarsier, which he knew from Minahasa, Bonthain (= G. Lompobatang), Tonkean, P. Peleng and P. Salayar, *T. fuscus.* He described a new species, *T. sangirensis,* from the Sangihe Islands (P. Siao and P. Sangihe Besar), distinguishable by its less extensively haired tail and less-haired tarsus. The tail of a Salayar specimen in Dresden was assigned to *T. fuscus* but is "weniger behaart" (less hairy), although this, he thought, was perhaps because the specimen was still young; but Groves (1998) found that a Salayar specimen in the Natural History Museum, London, likewise stands out by its less-haired tail.

The Sulawesi tarsier, still being called *T. fuscus,* was divided into subspecies by Miller and Hollister (1921). Using the name *Tarsius fuscus fuscus* for specimens from northeastern Sulawesi, they distinguished a new sub-species *Tarsius fuscus dentatus* (type locality Labua Sore, north of Parigi, central Sulawesi) as more grayish with a longer tail and larger skull. At the same time they described a new Sulawesian species, *Tarsius pumilus,* also from the highlands of central Sulawesi; as shown by Musser and Dagosto (1987), this is a valid species, but only the type specimen actually belongs to it, the two referred specimens (from Gimpu) being actually young examples of the common lowland species, and so should have been included by Miller and Hollister (1921) in their *T. f. dentatus.*

The first author to comment explicitly on these divisions was Sody (1949), who recognized *T. f. dentatus,* though finding the race "very weak": one of his north Sulawesi specimens was grayer than any from central Sulawesi, the relative tail length overlapped, and only the tooth row lengths seemed to differ (on the evidence of very few specimens). He also recognized *sangirensis,* but as a subspecies ("a very good race") of *T. fuscus,* and described a new subspecies, *Tarsius fuscus pelengensis* (from P. Peleng), "a very poor race," with a slightly smaller tooth row but slightly larger skull than mainland forms.

Hill (1953b) described for the first time in detail the differences in the tail between the Western, Sulawesian, and Philippine tarsiers, alluded to earlier by such authors as Meyer (1896–97). Hill (1953a) argued for the applicability of the name *spectrum* to the Sulawesi tarsier and keyed out the subspecies (exclusive of the very small-sized *T. s. pumilus*) as follows:

a. Underparts buff: *T. s. sangirensis*
b. Underparts white
   a'. Dorsal pelage browner.
      a". Skull smaller, with larger tooth row: *T. s. spectrum*
      b'. Skull larger, with shorter tooth row: *T. s. pelengensis*
   b'. Dorsal pelage grayer: *T. s. dentatus.*

Two odd aspects of this arrangement call for comment. The first is that he continued to maintain the subspecies *dentatus* on the basis of color, despite Sody's (1949) strictures; the second is that, although he had argued for Macassar (in the far south of Sulawesi) as the type locality of *spectrum,* his nominotypical subspecies was based entirely on specimens from the far north!

The three species *T. spectrum, T. syrichta,* and *T. bancanus* were described and clearly differentiated by Niemitz (1974), who also tested the validity of the described subspecies one by one, on the characters ascribed to them by their describers and by later commentators (especially Sody, 1949). The subspecies *dentatus* and *pelengensis* were rejected as differing feebly if at all. *T. s. pumilus* was retained because of its small size, although the ascribed localities could not be found by him and questions remained about its relationship to other *T. spectrum.* The validity of *T. s. sangirensis* remained questionable, because the hairiness of the tarsus was not part of Niemitz's study protocol, but the claimed lesser hairiness of the tail was refuted. Later the same author (Niemitz, 1984a) definitively sank *pelengensis* and proposed that *pumilus,* which he had encountered at Marena in the Lore Lindu Nature Reserve, be raised to specific rank.

The first field observations of Sulawesi tarsiers were published by MacKinnon and MacKinnon (1980). In the present context, the most significant findings of this study were that the territorial calls differed in three different regions where tarsiers were observed: from the northeastern tip (nearly?) as far west as Gorontalo (the Manado form); from Gorontalo as far west as Tanjung Panjang, on the south coast at about 121°50'E (the Gorontalo form); and near Palu, in Central Sulawesi (the Palu form). The behavior of the Manado and Gorontalo forms also differed slightly.

Niemitz (1984b) made further recordings of tarsier vocalizations. Those made at Dumoga, in Minahasa, were like MacKinnon and MacKinnon's (1980) Minahasa form; those recorded at Marena, in the mountains near Gimpu, in central Sulawesi, were quite different and seemed, though the sonagrams were admittedly unsatisfactory, to be more like their Palu form. The Marena tarsiers, because of their montane habitat and proximity to Gimpu, from where the two paratypes of Miller and Hollister's (1921) *Tarsius pumilus* had come, were ascribed by Niemitz (1984b) to *T. pumilus.* In fact, as noted above, the Gimpu specinmens were actually not that species at all, but fall within the range of variation of *T. spectrum sensu lato.* Gimpu is some 90 km SSE of Palu.

Reexamination of Miller and Hollister's *Tarsius pumilus* by Musser and Dagosto (1987) concluded that this is a thoroughly distinct species. Most of the distinctive features had actually been missed by Miller and Hollister (1921), who in addition (as already mentioned) had misidentified two juvenile *T. spectrum* (*sensu lato*) as examples of *T. pumilus.* Musser and Dagosto (1987) reviewed the taxonomy of the genus *Tarsius,* and to some extent that of *T. spectrum.* They provisionally retained *T. s. sangirensis* on the basis of its reduced caudal and tarsal hairiness on average, and possibly on the lack of the white postauricular spot. They tended to accept the greater tail length and skull length of *T. s. dentatus,* but noted that sample sizes are small. Likewise, the larger size and smaller upper teeth of *T. s. pelengensis* seemed to distinguish it. Their Table 4 indicates also that the lower teeth of Peleng, Minahasa, and central Sulawesi tarsiers are large compared to those from the southern and southeastern arms and from the northern arm west of Gorontalo. Feiler (1990) recorded the skull characters of a Sangihe specimen and supported its status as a separate species.

Niemitz et al. (1991) described the first new taxon within *Tarsius* for over forty years, the first new species for seventy years: *Tarsius dianae*, based on (at least?) two specimens from Kamarora, 1°10′S, 120°09′E, at the northern boundary of Lore Lindu National Park in central Sulawesi. These were contrasted to (at least?) three specimens, ascribed to *T. spectrum*, from G. Tangoko Batuangus Reserve on the far northeastern tip of Sulawesi. *T. dianae* had a more conspicuous black paranasal spot; white on lip; no brown tones on the hip, thigh, or lateral side of knee; a bare patch at base of ear; a medium cleft down nasal septum; a smaller, less symmetrical eye; and darker tail, digits and nails. The vocalizations were different from any of MacKinnon and MacKinnon's three types, including that from Palu, only 25 km to the west.

A preliminary revision was presented by Groves (1998), based in part on multivariate analyses of cranial, dental, and external features. Specimens from Sangihe and from Salayar separated so cleanly from the rest that there seemed no doubt that they represented different species; specimens from Peleng could "equally well" be a further species or a subspecies of *T. spectrum*. Special attention was paid to a specimen that showed most of the conditions ascribed by Niemitz et al. (1991) to *T. dianae;* in the craniometric analysis, it fell close to, but outside, the range of central Sulawesi tarsiers. I take this opportunity to confess that the locality of Rorokan was, in the 1998 analysis, mistakenly assigned to central Sulawesi; it is in fact in Minahasa, near the tip of the northern peninsula; and the *dianae*-like specimen, USNM 83967, is from Rorakan. Thus, the analysis of mainland specimens needs to be rerun.

In a recent field study, focusing on vocalizations, Shekelle et al. (1997) made recordings, took hair samples, and made morphological observations on tarsiers at approximately 100 km intervals along the northern peninsula, from Tangkoko Batuangus Reserve in the far northeast to Lore Lindu National Park in central Sulawesi; as well as on Great Sangihe Island and in the Togian Islands. Morphologically, they could not confirm that *T. dianae* is distinct from other Sulawesi tarsiers, while those on Great Sangihe Island were highly distinct. Although they did not give any sonograms, nor actually describe the vocalizations, they noted that those on Sangihe were distinct from the rest; in Manahasa, from the tip west to Suwawa (123°10′E), there was a single call type, in agreement with MacKinnon and MacKinnon (1980); there were unique calls at each of three southern coastal localities, Libuo (approx. 122°10′E), Sejoli (approx. 121°15′E) and Tinombo (120°16′E); a different call type again was found at Marantole and Kamarora, respectively 20 km north and 50 km south of Parigi, Kamarora being the type locality of *dianae* and Matantole, close to Labua Sore, the type locality of *dentatus;* and they found a final call type on two of the Togian Islands. Highly significant were the results of playback experiments. Tarsiers duetted in response only to their own call type, exhibiting "negative responses" (jumping about, ear twitching) to foreign calls; the Togian tarsiers responded to everyone else's vocalizations, but no one else responded to theirs. Shekelle et al. (1997) concluded that each call denoted a separate species, and the Togian call type was the most primitive. Given that the same call type occurred at both Marantole and Kamarora, they concluded that *dianae* is probably a synonym of *dentatus*.

Nietsch and Kopp (1998), extending a previous preliminary report by Nietsch and Niemitz (1993), reported on their own studies of vocalizations at several sites. Those recorded at Kamarora, Lake Poso, Ampana, and Morowali (all in central Sulawesi) were attributed to *T. dianae;* those at Palu, Gorontalo, and Manado were referred to as *T. spectrum,* although it was agreed that there were differences between tarsiers at the three localities; those recorded at Marena, farther south in central Sulawesi, were considered to be *T. pumilus* (as earlier identified by Niemitz, 1984b); and a population from Wakai, in the Togian group, was different yet again (Nietsch and Niemitz, 1993). Nietsch and Kopp (1998) illustrated and described the spectrograms of *T. dianae, T. spectrum,* and Togian tarsiers; and they reported the results of playback experiments, like those of Shekelle et al., but this time the recordings from the wild were replayed to a group of *T. spectrum* (from Minahasa) in captivity; their results reinforced the now-growing conviction that the northern, central, and Togian tarsiers represent three different species. (Gursky [1998] ascribed a species name, *T. togianensis,* to Nietsch, but this name does not appear, as far as I can trace, in any of the latter's publications, and in the absence of any description it rates as a *nomen nudum*). It is finally worth reiterating the conclusion from Groves (1998), that Sulawesi tarsiers are strongly distinct from those of western Indonesia and the Philippines, and it might be helpful to separate them generically.

Materials and Methods

Marian Dagosto very kindly sent me measurements of skulls in the American Museum of Natural History and the United States National Museum. Her measurements of the USNM skulls correspond closely to my own, so I have confidently used her measurements from the AMNH, which I have not studied in person; the tooth measurements, however, could not be exactly duplicated. Not all the measurements used by Groves (1998) are represented, so I have here used more limited data than in the Groves (1998) paper. The measurements used, in different combinations in different analyses, are as follows:

sk, Greatest skull length

bi, Biorbital breadth

orb, Breadth of a single orbit

bl, Bulla length

bu, Bulla breadth

pal, Widest palate breadth

ut, Upper tooth row length (including canine)

lo, Lower tooth row length (including canine)

areaumo, Crown area of M^1

areaumt, Crown area of M^2

arealmo, Crown area of M_1

arealmt, Crown area of M_2

The groups finally used were as follows:

1. Sangihe
2. Peleng
3. Manado, including Manembo-nembo, Kalabat, Rurukan, Tonsea Lama, and Minahasa
4. Molengkapoti
5. Bumbulan
6. Tolitoli
7. Parigi, including Labua Sore
8. Palopo, including Malili
9. Southeast, including only Wawo
10. South, including only Lombasang

These localities, and those for the vocalization data, are depicted in figure 8.1. The sample is much the same as in Groves (1998) but correctly sorted this time, and further divided into groups corresponding, as far as possible, to the call types discriminated by MacKinnon and MacKinnon (1980) and Shekelle et al. (1997). Group 3 corresponds to MacKinnon and MacKinnon's Manado group, and covers Shekelle et al.'s sites 2 to 5. Group 5 is within the range of MacKinnon and MacKinnon's Gorontalo group, and originates from very close to Shekelle et al.'s site 6. Group 7 covers the same range as Shekelle et al.'s sites 9 and 10.

On the other hand, the affiliation of Group 6 is uncertain. There is no guarantee that it falls within MacKinnon and MacKinnon's Gorontalo group, or that it can be associated with either of Shekelle et al.'s sites 7 or 8. All the localities, at least those west of Minahasa, where both teams made recordings, are on the Tomini Gulf coast, while Molengkapoti and Tolitoli (Groups 4 and 6 of the present study) are on the northern (Celebes Sea) coast.

As in Groves (1998), I compared the samples by both univariate and

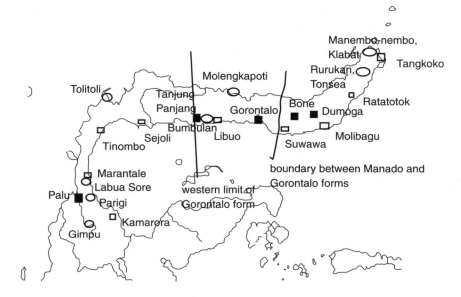

Figure 8.1. Locations of northern and central Sulawesi samples used in this study (open circles), and of vocalization localities of MacKinnon and MacKinnon (1980) and Shekelle et al. (1997).

multivariate methods, and made observations on external features. Re-arranging the groupings, with larger data sets, enabled new insights into interrelationships.

Results

Univariate comparisons

Table 8.1 shows that skulls from Sangihe and Peleng average considerably larger than those from the mainland of Sulawesi; those from Parigi are the next largest. For their size, Sangihe has narrow bullae while Peleng has relatively small orbits and short tooth rows, although the crown areas of the upper molars (but not of the lowers) are large.

Among the mainland samples, Bumbulan is very similar to Manado, its narrower palate, smaller upper first and lower second molars, and larger upper second molars having slight differences. Compared to both of these, Tolitoli has both shorter and narrower bullae, a very narrow palate, and small molar crown areas except for M^1. The Parigi sample—the main sample from the central area—averages larger in all dimensions of skull size than those from the northern arm, and the molar crown areas are relatively larger still. The two from Palopo/Malili are distinguished by the long bulla,

Table 8.1. Means and Standard Deviations, in Tenths of a Millimeter, for Samples of Tarsiers in This Study

Location	Stats	Variables											
		Skull	Borb	Orb	BL	BU	PAL	UT	LO	AREAUMO	AREAUMT	AREAMO	AREALMT
Sanglhe	Mean	381.25	300.25	161.00	116.75	56.50	148.50	131.50	141.25				
	N	4	4	4	4	4	4	4	4				
	SD	12.82	7.68	4.32	1.71	1.73	6.03	3.11	3.30				
Peleng	Mean	380.31	291.23	156.92	116.00	68.92	145.42	126.62	136.00	82.89	49.39	77.53	54.00
	N	26	26	26	26	26	26	26	26	19	18	19	18
	SD	8.74	9.14	4.48	3.19	4.38	5.06	2.21	2.98	3.97	5.14	3.78	4.26
Manado	Mean	365.79	288.26	156.00	111.35	61.10	145.35	127.89	134.70	77.42	47.73	78.50	53.42
	N	19	19	20	20	20	20	18	20	12	11	12	12
	SD	6.27	7.94	4.71	3.10	4.29	3.87	3.39	4.89	4.42	3.82	6.13	5.23
Bumbulan	Mean	384.00	286.40	155.71	111.38	62.00	141.67	125.57	131.29	74.71	49.29	78.57	50.14
	N	4	5	7	8	8	6	7	7	7	7	7	7
	SD	7.12	7.96	3.82	2.33	2.78	3.67	2.99	4.89	5.25	3.25	3.46	4.67
Tolitoli	Mean	363.25	286.25	157.75	107.50	57.25	136.75	123.50	129.25	74.25	45.00	72.50	46.75
	N	4	4	4	4	4	4	4	4	4	4	4	4
	SD	11.59	6.70	5.19	2.38	3.95	4.35	.58	.50	2.99	2.58	2.52	3.77
Parigi	Mean	375.00	294.67	162.00	112.67	63.00	146.25	130.00	136.25	87.50	54.75	80.25	57.25
	N	3	3	3	3	3	4	4	4	4	4	4	4
	SD	7.00	3.51	4.36	3.51	2.65	3.77	2.71	5.12	8.19	3.86	9.46	4.27
Palopo	Mean	369.50	297.00	158.50	121.00	63.50	150.50	130.00	139.50				
	N	2	2	2	2	2	2	2	2				
	SD	.71	9.90	2.12	7.07	6.36	.71	5.66	9.19				
South	Mean	364.50	285.50	155.00	110.60	65.50	141.00	122.40	129.00	71.60	44.00	72.80	49.00
	N	4	4	4	5	4	3	5	5	5	4	5	4
	SD	7.68	8.23	5.72	3.21	2.89	4.58	2.88	3.54	4.28	1.63	2.68	3.56
Total	Mean	372.64	290.25	156.97	113.39	63.99	144.75	126.93	134.72	79.06	48.62	77.25	52.57
	N	66	67	70	72	71	69	70	72	51	48	51	49
	SD	10.82	8.70	4.63	4.19	5.69	5.07	3.43	4.89	6.43	4.71	5.13	5.07
Molengkapoti (N = 1)		361	295	160	115	66	140	121	123	81.4	48.3	81.4	55.2
Southeast (N = 1)		367	296	161	112	62	139	125	130	79.2	52.8	77.7	57.6

wide palate, and long lower tooth row. The "South" sample, from the southern arm, are small like those from the northern arm, with a relatively wide bulla, short tooth row (not unlike Tolitoli), and very small molar crown areas, especially M1 in both upper and lower jaws.

The single Molengkapoti specimen is like no other. Small in overall size like others from the northern arm, it has big orbits, large bullae, narrow palate (like Tolitoli) and short tooth rows but large crown areas, especially the first molars. The single "Southeast" specimen, from Wawo, is likewise fairly small, between Palopo and "South," but with larger crown areas than the latter.

Multivariate Comparisons

In all the multivariate cranial comparisons, Sangihe has been omitted, and Peleng has been excluded from all but one; this is because I wanted to focus on the mainland groups. Only three of the analyses are depicted here.

In figure 8.2, based on sk, bi, orb, bl, bu, ut, and lo, Bumbulan, Parigi, and Southeast are not distinguished from Manado, and South is distinguished on average only, but the three Tolitoli specimens are (just) outside the range of the others, the Molengkapoti specimen is more distinct, and the two Palopo specimens are quite different from the rest and from each other. The first Discriminant Function largely contrasts sk (negative) with bi and bl (positive); the second, sk and lo (positive) with bu (negative).

Figure 8.3, in which pal takes the place of the tooth row length measure-

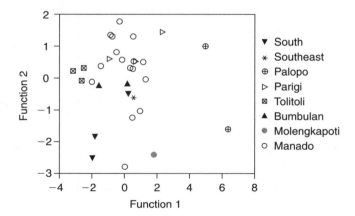

Figure 8.2. Discriminant analysis of tarsier samples, using skull length, biorbital breadth, orbit breadth, bulla length, bulla breadth, upper tooth row length, and lower tooth row length. Plot of first and second discriminant functions. DF1 accounts for 74% of the total variance, DF2 for 14%.

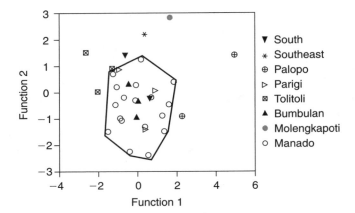

Figure 8.3. Discriminant analysis of tarsier samples, using skull length, biorbital breadth, orbit breadth, bulla length, bulla breadth, and palate breadth. Plot of first and second discriminant functions. DF1 accounts for 56% of the total variance, DF2 for 28%. The polygon encloses the Manado sample.

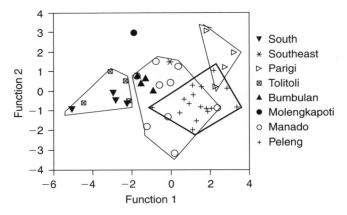

Figure 8.4. Discriminant analysis of tarsier samples, using upper tooth row length, area of M^1, area of M^2, lower tooth row length, area of M_1, and area of M_2. Plot of first and second discriminant functions. DF1 accounts for 73% of the total variance, DF2 for 15%. Polygons surround the ranges of Peleng and of South plus Tolitoli (solid), Manado (dotted), and Parigi (dashed).

ments, depicts much the same relationships as figure 8.2, except that South is less distinct, but the Southeast specimen is outside the range of Manado. Molengkapoti, Palopo, and (less strikingly) Tolitoli again stand out. Exactly as in the previous analysis, DF1 contrasts sk (negative) with bi and bl (positive), but DF2 contrasts pal (strongly negative) with bi and bl (positive).

Figure 8.4 is based entirely on dental measurements (ut, lo, arealmo, arealmt, areaumo, areaumt). South, Manado, and Parigi this time form three

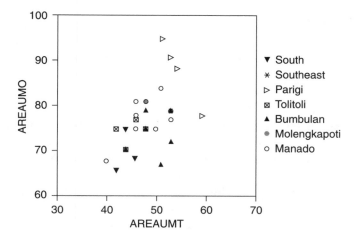

Figure 8.5. Plot of area of upper first molar (AREAUMO) against that of upper second molar (AREAUMT).

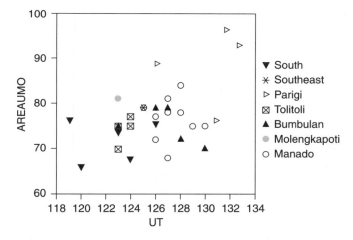

Figure 8.6. Plot of upper tooth row length (UT) against area of upper first molar (AREAUMO).

quite separate clusters, not overlapping. Oddly, Tolitoli falls in the South polygon, while Bumbulan and the Southeast specimen are within the Manado polygon. Molengkapoti is again quite separate. For comparison, Peleng has been included; it overlaps both Manado and Parigi. DF1 is largely a size component, in which lo, arealmt, and areaumo are strongly positive, but the other variables are almost unweighted; DF2 contrasts ut and areaumt (positive) with lo (heavily negative). Figure 8.5 plots area pf upper first molar against the area of the upper second molar, while figure 8.6 plots the area of the tooth row against the area of the upper first molar.

External Features

Recall that I have seen no skins from South or Southeast samples. The comparisons are, therefore, entirely among samples from North and Central Sulawesi, and the insular samples (table 8.2).

The basic distinction is between Sangihe, Peleng, and the mainland samples. Sangihe is yellower, less mottled; Peleng is redder; mainland specimens are gray-buff. In Sangihe the tarsus hair is very sparse, while in Peleng, it seems to be unusually dense. The pelage of the Sangihe specimens is finer, less woolly than most (not all) others. In both Sangihe and Peleng, the white postauricular spot is very small compared to nearly all mainland skins. On the mainland, the Molengkapoti skin immediately stands out by its rather silky pelage, and the tail is not dark like others, and has a pale friction surface. The Parigi skins are characterized by the thigh being light yellow or creamy; in others, the thigh is brown, or colored like the body. Two skins from Gimpu (identified by Miller and Hollister [1921] inexplicably as *Tarsius pumilus*)—being juveniles, their (rather fragmented) skulls were not included in the multivariate analyses—are slightly different from those from Parigi (and Labua Sore). Their tail pencils are thicker and blacker; the

Table 8.2. Distribution of Some External Phenotypic Characters in Tarsiers from Northern and Central Sulawesi

Character	Manado (N = 12)	Molengkapoti (N = 1)	Bumbulan (N = 1)	Tolitoli (N = 4)	Parigi (N = 4)
Thigh colour	Brown or as body	As body	Brown	As body or paler	Yellowish or creamy
Black paranasal spot	Usually copious	Yes	Yes	Yes	Present in 3 out of 4
White hair on lip	Usually	Yes (in middle)	Yes (on sides)	Yes	Yes, but usually more grayish
Bare subaural patch	Varies	No	Yes	No	Yes
Light postauricular spot	Usually large	Yes	Yes	Very conspicuous	Varies from conspicuous to ill-defined
Median nasal cleft	Varies	Yes	Yes	Poor or absent	Poor or absent
Lateral eye notch	Usually	No	No	No	No
Black eye rims		No	No	No or slight	No or poor
Colour of fingernails	Usually pale or slightly dark	Pale	Pale	Pale	Pale
Nails reach end of pad	No or barely	Yes	Barely	No	Barely

papillary ridges on the tail's friction surface appear to be less developed (unless this is an artifact of preservation), the light hairs on the lips are buffy rather than gray or white, and they have very prominent black eye rims. The tail hair is denser in Bumbulan and Palopo/Malili than in other samples, in which it tends to be rather sparse. The Palopo and Malili skins are rather different from each other. The Palopo skin, uniquely, lacks a postauricular spot. The Malili skin has very prominent black eye rims. Recall that the two skulls (one from Palopo, one from Malili) that made up the "Palopo" sample are far apart in the discriminant analyses as well.

The following external features are variable within samples: suppression of the black paranasal spot; presence of white hair on the lips; presence of a bare spot at the base of the ear; degree of development (and occasional absence) of the median groove down the nasal septum; lateral eye notch; darkness/paleness of fingernails; and relative tail length. Moreover, these features do not correlate with one another.

Discussion

It is clear that there is a great deal of craniodental and pelage variation within Sulawesi; much of it is rather subtle, but it does, as far as the evidence goes, seem to correlate with the divisions made on the basis of vocalizations. There is a great need for further evidence to test this proposition. One would be reluctant to collect new specimens; precise, detailed descriptions (and photographs) should be made of individuals caught by hand or in mist nets, while owl pellets might well be a source of further skull material.

There is absolutely no doubt that the Sangihe tarsier, as several authors have noted, is distinct from any on the mainland. The same is true, to only a slightly lesser extent, of the tarsier from Peleng. *Tarsius sangirensis* Meyer, 1896–97 and *T. pelengensis* Sody, 1949, should be definitively recognized as distinct species.

On the mainland, the large Manado sample demonstrates the range of variation to be expected within a restricted population. On this basis, the consistent separation of the Parigi, Tolitoli, and South skull samples in the discriminant analyses (especially the dental one, fig. 8.4) is surely significant; while the differentiation of the Parigi skins on the basis of thigh color would also probably be a diagnostic difference. The significance of the differentiation of the single specimen from Molengkapoti is less clear: Would other specimens confirm this uniqueness, or would they bridge the gap between it and the Manado or possibly the Bumbulan sample? Likewise, vocalizations should be recorded in that region to compare with those of MacKinnon and MacKinnon's (1980) Manado and Gorontalo types.

I combined the Palopo and Malili specimens, preferring a single sample of two rather than two samples of one each, but I think this was probably a poor decision, as they turn out in every analysis to be rather different from each other. Nor do they seem particularly close to South and Southeast, respectively.

There is finally the status of Niemitz et al.'s (1991) *Tarsius dianae* to consider. Shekelle et al. (1997) have already commented that they could detect no differences in vocalizations between tarsiers at Kamarora, the type of locality of *T. dianae,* and at Morantale, very close to the type locality of *T. fuscus dentatus* Miller and Hollister, 1921; moreover, there was no indication of heterogeneity at Kamarora indicating sympatry of two species. Likewise, they indicated that the morphological features attributed to *T. dianae* do not, in their opinion, consistently differentiate it. The external features (and some cranial features) that I examined were partly designed to test the *T. dianae* features. I have already noted (under Results) that there is great variability in such features as nasal cleft, subaural bare patch, paranasal black spot, and so on, all of which vary independently within all samples. I thought (Groves, 1998) that I had identified a specimen of *T. dianae* in the U.S. National Museum collection, but the specimen is from Rurukan, which I had mistakenly placed in central Sulawesi, whereas in fact it is in Minahasa, not far from Manado; as the *T. dianae* features vary independently between specimens, in both Parigi and Manado samples, it is to be expected that some of them would by chance recombine in a few individuals, hence the "*T. dianae* look-alike" in the far north.

These results show some congruence with the recent field studies of vocalizations. In particular, the observation by Nietsch and Niemitz (1993) that vocalizations from Palu all along the northern peninsula, to its tip, are in the nature of minor variants within a general type and quite different from those from most of central Sulawesi, makes sense in the light of the morphological results. Unfortunately, the exact localities from which specimens were obtained and where recordings were made are not the same, with a few exceptions, so the congruence of the two data sets cannot be tested very far.

Conclusions

The evidence to date supports the recognition of several of tarsier species on Sulawesi and its offshore islands. These include the following:

1. *Tarsius sangirensis* Meyer, 1896–97. Sangihe Island, at least Sangihe Besar.
2. *Tarsius pelengensis* Sody, 1949. P. Peleng.
3. Northern arm species or species group, with variants (species?) at intervals along the peninsula as far as the Palu valley.

4. Southern and southeastern arms species, apparently several species involved.

5. *Tarsius dentatus* Miller and Hollister, 1921. This may be the prior available name for the species from the northern part of the central land mass of Sulawesi, from Labua Sore as far east as Moruwale. Both the morphological evidence and the vocalization data of Nietsch and Niemitz (1993) and Shekelle et al. (1997) support the hypothesis that a single species exists throughout this region; as this includes Kamarora, it seems likely that *T. dianae* is a synonym of *T. dentatus.* It remains possible, however, that two species are sympatric in parts of this region — indeed, this seems to be the case at Moruwali (Gursky, 1998) — so further study is needed to confirm or refute this proposed synonymy. It should be pointed out, in addition, that I have not seen the type material of *T. dianae.*

6. The tarsier of P. Salayar (Groves, 1998).

7. The distinctive *Tarsius pumilus;* but it needs to be demonstrated for certain whether this was really the species that was recorded at Marena by Nietsch and Niemitz (1993), and whether the type locality, Rano Rano, was correctly identified by Musser and Dagosto (1987), since Myron Shekelle (in press) has suggested to me that a different Rano Rano, farther southwest in central Sulawesi, might be the correct place.

8. Finally, there are at least two "unknowns": the tarsier of Molengkapoti, known only by a museum specimen, and that of the Togian Islands, known only by vocalizations.

The outstanding problem in nomenclature concerns the type locality of *Lemur tarsier* Erxleben, 1777. Erxleben (1777) based his description on "Le Tarsier" of Buffon (1749), which seems to be that of a Sulawesi tarsier, not of a Philippine one as generally assumed (for example, by Hill, 1953a,b). As far as we know, Buffon's specimen is no longer extant. Pallas (1778) described a different specimen, and this too no longer seems to be extant. Hill (1953a) designated the type locality for Pallas's *spectrum* as Makassar, and in the absence of contrary evidence this must be accepted; Niemitz (1984a) stated that the type locality was Manado, although he did not argue against Hill's designation; and subsequent authors seem to have accepted this assumption. Eighteenth-century voyage and trading patterns make it much more likely that both Buffon's and Pallas's tarsiers were from Makassar rather than elsewhere in Sulawesi, certainly not Manado. If future study confirms that tarsiers from Makassar and Manado are distinct species, then it looks as if the prior available name for the Makassar species would be *Tarsius tarsier* (Erxleben, 1777), while the Manado species, which is by far the best studied of the Sulawesi tarsiers, remains curiously without a name.

Acknowledgments

I would first of all like to thank Marian Dagosto for sending me measurements of tarsiers in the American Museum of Natural History, New York. I thank the curators of collections in which I studied tarsiers in the 1990s: Boeadi in the Museum Zoologici Bogoriense, Bogor; Paula Jenkins in the Natural History Museum, London; and Linda Gordon in the Smithsonian Institution, Washington, D.C. For various discussions on tarsiers I would like to thank Boeadi, Guy Musser, Dick Thorington, and Doug Brandon-Jones. Most of all, I would like to thank Myron Shekelle, whose discussions have greatly broadened my understanding of tarsiers and stimulated me to go deeply into the old literature; he and I plan a series of papers on tarsier taxonomy, and I have been very careful not to preempt our findings or thought directions in any way.

References

Buffon GLL, Comte de. 1749. Histoire naturelle, générale et particulière, vol. 13. Paris: Imprimerie du Roi.

Burmeister C. 1846. Beitrage zur näheren Kenntnis der Gattung Tarsius, Berlin.

Erxleben CP. 1777. Systema regni animalis per classes, ordines, genera, species, varietates cum synonymia et historia animalium. Classis I. mammalia. Leipzig: Weygand.

Feiler A. 1990. Über die Säugetiere der Sangihe- und Talaud-Inseln—der Beitrag A. B. Meyers für ihre Erforschung. Zoologische Abhandlungen Staatliches Museum für Tierkunde Dresden 46 (4): 75–94.

Fischer. 1804.

Geoffroy-Saint Hilaire E. 1812. Suite au tableau des guadrumanes. Ann Mus Hist Natl Paris 19: 156–170.

Groves C. 1998. Systematics of tarsiers and lorises. Primates 39: 13–27.

Gursky S. 1998. Conservation status of the spectral tarsier *Tarsius spectrum:* population density and home range size. Folia Primatol 69 (suppl. 1): 191–203.

Hill WCO. 1953a. Note on the taxonomy of the genus *Tarsius*. Proc Zool Soc Lond 123: 13–16.

Hill WCO. 1953b. Cutaneous caudal specializations in *Tarsius*. Proc Zool Soc Lond 123: 17–26.

MacKinnon J, MacKinnon K. 1980. The behavior of wild Spectral Tarsiers. Int J Primatol 1: 361–379.

Meyer AB. 1896–97. Säugethiere vom Celebes- und Philippinen-Archipel. Abhandlungen der königlichen zoologischen und anthropologischen Museum Dresden 6: 1–36.

Miller GS, Hollister N. 1921. Twenty new mammals collected by Raven in Celebes. Proc Biol Soc Wash 34: 93–104.

Musser GG, Dagosto M. 1987. The identity of *Tarsius pumilus,* a pygmy species endemic to the montane mossy forests of Central Sulawesi. Am Mus Nov 2867: 1–53.

Niemitz C. 1974. Zur Biometrie der Gattung Tarsius Storr, 1780 (Tarsiiformes, Tarsiidae). Inaugural dissertation, Justus-Liebig-Universität Giessen.

Niemitz C. 1984a. Taxonomy and distribution of the genus *Tarsius* Storr, 1780. In Niemitz C, editor, Biology of tarsiers, 1–16. Stuttgart: Gustav-Fischer-Verlag.

Niemitz C. 1984b. Vocal communication of two tarsier species (*Tarsius bancanus* and *Tarsius spectrum*). In Niemitz C, editor, Biology of tarsiers, 129–141 Stuggart: Gustav-Fischer-Verlag.

Niemitz CC, Nietsch A, Warter S, Rumpler Y. 1991. *Tarsius dianae:* a new primate species from Central Sulawesi (Indonesia). Folia Primatol 56: 105–116.

Nietsch A, Kopp ML. 1998. Role of vocalization in species differentiation of Sulawesi Tarsiers. Folia Primatol 68 (suppl. 1): 371–378.

Nietsch A, Niemitz C. 1993. Diversity of Sulawesi tarsiers. Deutsche Gesellschaft fÅr Säugetierkunde 67 (Hauptversammlung): 45–46.

Pallas PS. 1778. Novae species quadrupedum e glirium ordine cum illustrationibus variis complurium ex hoc ordine animalium. Erlangen: Wolfgang Walther.

Shekelle M, Leksono SM, Ichwan LLS, Masala Y. 1997. The natural history of the tarsiers of North and Central Sulawesi. Sulawesi Primate Newsletter 4, 2: 4–11.

Sody HJV. 1949. Notes on some primates, carnivora and the babirusa from the Indo-Malaysian and Indo-Australian regions. Treubia 20: 121–190.

CHAPTER 9

Outline of the Vocal Behavior of *Tarsius spectrum:* Call Features, Associated Behaviors, and Biological Functions

Alexandra Nietsch

The vocal behavior of Sulawesi tarsiers is particularly interesting because of the relationship between signal structure and species differentiation. This study describes the acoustic properties of the vocal repertoire of *T. spectrum* as well as the associated contexts of sound emission. The data give evidence for a large vocal repertoire, organized in at least fifteen distinctive call types, most of which clearly covary with distinctive contexts. In addition I used play-back experiments to test possible functions of calls. Among the calls tested, duet calls proved to be especially good stimuli to elicit a vocal response. The results of playback experiments clearly indicate that duet calls are multifunctional displays. Animals responded to played calls in a gender- and age-dependent way. The analysis revealed that duet calls are not stereotypic displays. Rather, temporal and structural organization of calls emitted in relaxed morning duet context may differ from calls emitted during territorial context. I discuss the results with respect to the use of playback experiments in species discrimination.

The vocal behaviors of nonhuman primates routinely reveal inter- and intraspecific variation in acoustic structure. In a variety of Old and New World primates as well as in prosimians, differences in call structure were shown to be related to taxonomic differentiation (e.g., in Pongidae: Mitani and Gros-Louis, 1995; Hylobatidae: Marshall et al., 1984; Cercopithecidae: Struhsaker, 1970; Marler, 1973; Colobidae: Oates and Trocco, 1983; Callitrichidae: Snowdon, 1993; Hodun et al., 1981; Galagidae: Masters, 1991; Zimmermann et al., 1988; Lemuridae: Macedonia and Taylor, 1985). Thus, more traditional forms of taxonomic classifications may profit from the study of vocalizations by clarifying phylogenetic affinities between species, e.g., langurs: Wilson and Wilson (1975); gibbons: Haimoff et al. (1982); lion tamarins: Snowdon et al. (1986); lemurs: Macedonia and Stanger (1994); or galagos: Zimmermann (1990). The use of vocal structure as a supplementary diagnostic feature in taxonomic studies is of particular interest in the monomorphic group of Sulawesi tarsiers. Here, morphological features do not unambiguously differentiate species or subspecies. Hence, past studies disagreed on the significance of morphological differences among specimen for a taxonomic differentiation and, as a consequence, tarsier specimens from the same region were recognized as typical *T. s. spectrum*, as a different subspecies, or even as a distinct species (see Niemitz, 1977, 1984a, 1985; Musser and Dagosto, 1987; Feiler, 1990; Groves, 1998).

To date, three species—*T. spectrum, T. dianae,* and *T. pumilus*—are recognized, while a number of geographical populations, for instance the tarsiers from Greater Sangihe Island, Selayar Island, or Togian Islands, are proposed to be separate species (see Groves, 1998; Nietsch and Kopp, 1998; Shekelle et al., 1997; Groves, chapter 8). The question of how many distinctive taxa should be recognized has gained renewed attention by the evidence that vocalizations may differentiate tarsier species better and more reliably than morphological features. Vocal differences related to species differentiation have been proved for *T. dianae* and *T. spectrum* (Niemitz et al., 1991). These species can be reliably distinguished by the structure of the loud calls that mated males and females may coordinate to a joint display, termed "dueting." Other populations that have been formerly included in *T. spectrum* have acoustically distinct duet songs as well, suggesting that these populations may be distinctive on the species level, too (MacKinnon and MacKinnon, 1980; Nietsch and Niemitz, 1993). Some of these acoustically distinctive populations have not yet been morphologically investigated and are known solely by their vocal output. The Togian population is an example for clear acoustical differentiation from *T. spectrum* or *T. dianae*, whereas the evidence of morphological differentiation has yet to be established (Nietsch, 1999). Lack of support from morphological studies may evoke skepticism on the conclusions drawn from behavioral studies. However, in these cases playback experiments using duet songs as a stimulus were able to provide further evidence for taxonomical differentiation. For instance, the behavior of *T. spectrum* shows that subjects behaviorally discriminate the population differences in acoustic structure of duet calls (Nietsch, 1995; Nietsch and Kopp, 1998). When exposed to duet songs, they respond only to the playback of their own calls in a species-specific manner.

With respect to functions of duet songs in intraspecific competition, the fact that the *T. spectrum* individuals almost ignored the *T. dianae* songs—that is, they did not routinely reply vocally to the played calls and never approached the sound source—could only be explained by their biological differentiation. In a similar vein, the suggested species status of the Togian tarsiers could be furnished, since again *T. spectrum* did not respond species specifically to the calls of this population as well (Nietsch and Kopp, 1998). Furthermore, a recent study revealed that a variety of distinctive forms might exist along the northern peninsula among the population of *T. spectrum*, because groups that were separated by only 100 km did not respond to each other's duet calls (Shekelle et al., 1997). Past and recent taxonomic studies on Sulawesi tarsiers thus arrive at extremely opposing conclusions: morphological studies on the one hand provide evidence for an almost homogeneous population, splitting off into few distinct species and subspecies. On the other hand, studies based on vocal differentiation depict a scenario of a high number of distinct taxa, far beyond the number that is currently recognized.

To account for the differences in vocal characteristics among populations in the framework of current and future work on Sulawesi tarsiers, it is advisable to put our knowledge on properties of vocal repertoires, variation in vocal structure, or possible variability in response behaviors to played calls in these primates on a solid basis. Such a study should include calls from different behavioral contexts to extend the array of context homologue calls that could be potentially used for species differentiation.

In this chapter, I present data on the acoustic characteristics of the vocal repertoire as well as on the vocal behavior of *Tarsius spectrum* from Tangoko Batuangus Nature Reserve, located at the easternmost tip of the northern peninsula. This report will include a description of acoustic properties of duet calls as well as of calls from nonduet contexts, together with a description of the behavioral context of vocalizations. The aim of the study is to contribute to our knowledge of the vocal behavior of *T. spectrum* to establish a broad basis for future comparison of vocal characteristics of this species with those of other populations. In addition, I present data from playback experiments that were designed to clarify biological functions of dueting behavior in *T. spectrum*. The results allow the discussion of the use of duet calls in playback studies to elucidate issues of species differentiation in Sulawesi tarsiers.

Methods

We recorded calls of feral *T. spectrum* in Tangkoko Batuangus Nature Reserve (125°14′ E / 1°34 N), which is located at the easternmost tip of the northern peninsula (fig. 9.1). The study area was situated approximately 2.5 km from the village of Batuputih and comprised approximately 4 sq km. The area comprises a broad range of floral communities, including shoreline scrub vegetation, extended fields of the tall grass *Saccharum spontaneum* or *Imperata cylindrica*, and patches of secondary as well as primary rain forests (for detailed description of the main vegetation types of the reserve see World Wildlife Fund, Management Plan, 1980).

The recordings were obtained during nightly observation periods from

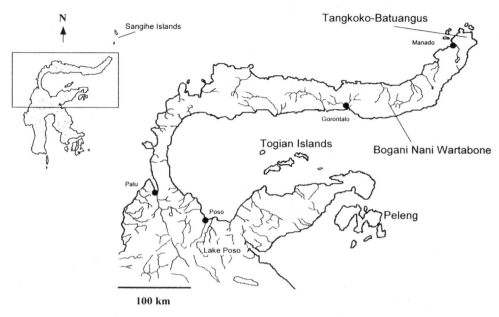

Figure 9.1. Map of central Sulawesi and the northeastern peninsula. The location of the Tangkoko-Batuangus Nature Reserve is indicated.

tarsiers living in secondary and primary forest areas from sea level up to approximately 500 miles above sea level. We recorded duet calls (N = 76; 8 pairs) at dawn between 5 and 6 A.M. (see, for details, Nietsch, 1999).

The study was conducted between November and May, that is, during the rainy season. Recordings were made during all kind of weather, including rainy nights. Because of high ambient noise level caused by heavy rainfalls some recordings of tarsier sounds did not produce clear spectrograms and were thus excluded from the analyses.

Tarsiers are members of the lower canopy and employ high frequencies in their calls (Nietsch 1999). The ecological constraints of a humid and noisy rain-forest habitat reduces the carrying distance of sounds through attenuation and reverberation (Marten et al., 1977). In addition, high frequency sounds, when transmitted through dense foliage, may experience severe degradation because of the increase in attenuation rate that occurs with increase in frequency (e.g., Marten et al., 1977). Hence, to obtain high-quality recordings of tarsier sounds, all recordings were done at close distance to calling animals. The distance between calling animals and the microphone was at most 30 m. Furthermore, several individuals of both sexes and all age classes were caught within the study area. They were housed temporarily in a large open air enclosure, allowing observations and recordings under seminatural conditions at close distances (Nietsch, 1993). Concomitant to sound recordings, we recorded the overall context of sound emission. In addition, these individuals were subjected to playback experiments (see below for procedure of experiments).

Further playback experiments were conducted with tarsiers kept in two Indonesian facilities, the Primate Research Center in Bogor and Yogyakarta Zoo. The Bogor group contained one adult male, five adult females. The Yogya group consisted of three adult males, two adult females, and a subadult male. In either group, only one pair was dueting. Crucial to the aims of my study was the identification of the population to which these animals belonged. We had promising hints that the Bogor tarsiers originated from Tangkoko-Batuangus Reserve. Tarsiers from Yogyakarta were caught in Bogani Nani Wartabone National Park (formerly known as Dumoga Bone National Park), not far from Doloduo (123°54′ E / 0°34 N). Animals from this region have been shown to exhibit the same duet call pattern as Tangkoko tarsiers, however (MacKinnon and MacKinnon, 1980; Shekelle et al., 1997). By comparing their duet calls with calls from my field study, I could confirm that the dueting individuals in Bogor and Yogyakarta were fully compatible in acoustics with the Tangkoko tarsiers. A problem arose with the proper allocation of nondueting individuals. Encouragingly, all vocalizations that I obtained from these animals were clearly congruous in acoustic structure compared to those obtained from Tangkoko tarsiers.

Recordings and observations were made using a tape recorder UHER 4000 Report Monitor, or optional, a Sony Professional Cassette-Recorder WM DC6, connected to a directional Sennheiser microphone ME 64, or optional ME 88 (frequency response of the equipment was 20–25,000 Hz for UHER 4000; 40–15,000 Hz for Sony WM DC6; 50–15,000 Hz for ME 64; 40–20,000 Hz for ME 88). For playbacks I used either one of these devices or a Sony DAT Walkman TCD-D7, connected with a 15-meter-long cable to a Sony loudspeaker.

Playback experiments were done as follows. We placed a loudspeaker close to the animals, and playback was started only if the animals had not vocalized for ten minutes. During the playbacks and five minutes afterwards, the animals were observed and the following responses were recorded:

1. Type of vocalization produced.
2. Latency of vocal responses, that is, the time elapsing between the start of the playback and a vocal response.
3. Approach to the loudspeaker.

We changed the order of trials from session to session to minimize potential habituation to the playback situation.

Real-time spectrograms of calls were made with the software package Avisoft for PC (R. Specht, Berlin; see McGregor and Holland, 1995). Analogue signals were digitized with a sampling rate of 44,100 Hz. Spectrograms were produced by 512-point FFT. To assess the acoustic structure of calls, each spectrogram was analyzed according to maximum frequency, minimum frequency, frequency range, and call duration. These variables were measured on screen with a time resolution of 5.80 ms and a frequency resolution of 111 Hz. Sound energy in calls is largely concentrated in the fundamental frequency (f 0), and harmonics are weakly expressed in the spectrograms. Therefore, the analysis refers to the fundamental frequency (f 0). If necessary, spectrograms of calls used for presentation were edited to eliminate tracings of bird and insect sounds. The duet calls were analyzed according to their total duration, regardless which sex initiated or terminated the song. Further, female and male contributions to the duet call were analyzed according to the maximum as well as minimum frequencies that could be reached. These measurements were used to calculate the frequency range of a call.

To examine whether the duet note repertoires of either sex contained distinct note types, the following procedure was used. First, all notes that were produced in a particular duet were analyzed according to the following variables: start frequency (Start-F), peak frequency (Peak-F), terminal frequency (End-F), frequency range (F-range; maximum–minimum frequency), as well as duration of a note. Second, using these five variables, a hierarchical cluster analysis was then performed on the entirety of notes made by both the male and the female.

Results

In the following three sections I will outline the vocal repertoire of *T. spectrum* by considering the acoustical features of signal patterns of juvenile and adult animals of both sexes, together with the associated behaviors of sound emission. The first section deals with the calls produced by mated tarsiers in dueting context, including information on dueting behavior, as well as a description of the acoustic properties of duet calls. The second section deals with the vocal nondueting repertoire of *T. spectrum*, both juvenile and adult. Finally, I will proceed with the results of playback experiments that I conducted to draw conclusion on possible signal functions (see section on Playback Experiments).

Duet Calls

Mated tarsier pairs engage in dueting with great regularity each morning between 5 and 6 A.M. in the vicinity of their sleeping sites. During that time I heard a particular pair calling on twenty seven mornings out of thirty consecutive all-night-observation periods. The earliest duet was heard at 4:47 A.M., the latest at 5:41 A.M. In contrast, around sunset (approximately 6 P.M.) when the animals leave their sleeping sites (tree holes or dense vines/vegetation), the same pair dueted only two times. Fifty seven out of seventy six morning duets (75%) were initiated by the female. As a rule the males responded very quickly to the call of their mates. The latency of the male's call to the first duet note of the female was between less than 1 and 15 s.

The duration of *T. spectrum* duets varies considerably. The shortest duet in my study is about 24 s, but duet songs could also last for more than 4 min (max. duration: 293 s). The average duration of duet songs is 118 s (median; 1st, 3rd quartiles: 90,168; N = 76).

The duet calls of male and female tarsiers can be easily distinguished from each other. The differences are far from subtle and comprise phonological as well as temporal features. Figure 9.2a gives an example for a section of a *T. spectrum* duet song, showing differences in the call of the male and the female in note phonology as well as overall temporal patterning of the calls. In general, the call of the female covers a higher-frequency band than the call of the male. The female song is located in the frequency band between 4.7 (4.2, 5) kHz and 13.4 (13, 13.8) kHz and the maximum frequency range of a song is 8.7 (8.4, 9.1) kHz (medians, in parentheses 1st, 3rd quartiles, N = 76). The song of the male is located between 4.5 (4.1, 4.8) kHz and 12.2 (11.7, 12.6) kHz, and the frequency range is 7.5 (7.2, 8.1) kHz (medians, in parentheses 1st, 3rd quartiles, N = 76).

Characteristically, in her song a female produces several distinctive series of notes, termed "phrases," spaced by short silent intervals (median duration: 4 s; 1st, 3rd quartiles: 3.4; 4.5 s, N = 77). Each individual phrase is characterized by an almost gradual change in note shape from the initial note toward the end of the series (see fig. 9.2a). The average duration of a phrase is 11.3 s (median; 1st, 3rd quartiles: 8.1; 24.2, N = 90). A female may utter up to twelve such phrases in her song.

The male contributes to the entire song of the female a series of almost evenly spaced notes. The average length of internote intervals in a song was, for two different males, 0.75 s (0.68, 0.83) and 0.7 s (0.63, 0.83) (median; in parenthesis 1st, 3rd quartiles; N = 102 and 93, respectively), a note delivery rate of one to two notes per second. Further, visual inspection of the note spectrograms indicates that the shape of the male notes is far less conspicuously varied when compared to the female duet contribution.

To identify distinctive note patterns within the duet call of females or males, the entire notes (N = 143) vocalized by a particular female during a duet comprising eight phrases, as well as 132 notes that her mate contributed, were subjected to a hierarchical cluster analysis. The analysis identified four clusters in the note repertoire of the female and four clusters in the reper-

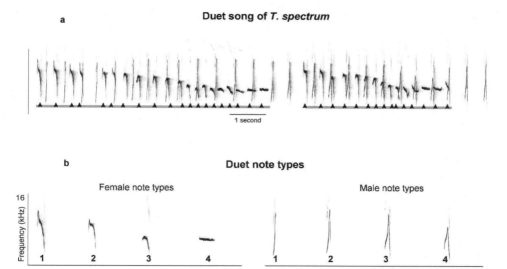

Figure 9.2. (a) Spectrogram of a section of a duet song of *T. spectrum*. Two female phrases are shown, indicated by gray bars. Female notes are indicated by black arrows. (b) Spectrograms of basic duet note types 1–4 in female (left) and male (right) duet calls.

toire of the male. Representative spectrograms of the four note types of the female as well as the note types of the male are shown in figure 9.2b. Notes within each cluster show a distinctive parameter configuration (table 9.1).

In the female, note types 1 to 4 differ from each other in Start-F, Peak-F, as well as in F-range (p < 0.01, M-W U test). Note type 4 additionally differs from all other notes in duration and End-F (p < 0.01, MW-U test). Note 1 has the highest pitch (12 to 13 kHz), followed by note 2 (10 to 11 kHz). Note 3 is considerably lower in pitch and lacks the initial steep increase in pitch toward the note peak. Note 4 is characterized by similar height of Start-F, End-F, and Peak-F, a small F-range (approximately 1 kHz) and its long duration (up to 500 ms).

Investigation of the female call reveals that all four-note types are present in four out of eight phrases. One of the remainder phrases lacks note type 1 and starts with note type 2; the other three phrases miss note type 2. However, in all eight phrases either note 1 or 2 precedes note 3, which is then followed by type 4.

The four notes of the male are less conspicuously differentiated than those of the female. Note types 1 to 3 have a similar high pitch (11–12 kHz) and similar values for height of Start-F (5–6 kHz) or F-range (6–7 kHz). However, these notes can be distinguished by their duration, which increases from note 1 to note 3, and by the height of the end frequency, which

Table 9.1. Acoustic Parameters of Duet Note Types (*Tarsius spectrum*)

Parameters	Female				Male			
	1 (N = 43)	2 (N = 10)	3 (N = 50)	4 (N = 44)	1 (N = 7)	2 (N = 6)	3 (N = 93)	4 (N = 26)
Start-F (kHz)	10*	9*	8.2*	7.5*	4.5	4.9	5.8	6.3
	(0.7)	(0.4)	(0.6)	(0.4)	(0.2)	(0.1)	(0.3)	(0.5)
Peak-F (kHz)	11.9*	10.4*	8.8*	7.7*	11.3	11.4	11.5	9.6*
	(0.7)	(0.5)	(0.7)	(0.3)	(0.1)	(0.2)	(0.4)	(0.8)
End-F (kHz)	5.4	5.2	5.4	6.9*	10.8*	9.6*	5	4.2
	(0.3)	(0.3)	(0.3)	(0.7)	(0.3)	(0.8)	(0.8)	(0.2)
F-range (kHz)	6.5*	5.2*	3.4*	0.8*	6.4	6.5	6.7	5.5*
	(0.7)	(0.5)	(0.7)	(0.5)	(0.2)	(0.2)	(0.3)	(0.9)
Duration (ms)	190	190	180	260*	60*	80*	130	130
	(20)	(20)	(20)	(100)	(10)	(10)	(20)	(10)

*Significant at $p < 0.01$, Mann-Whitney U test.

decreases from note 1 to note 3 ($p < 0.01$, MW-U test). Note type 4 differs in all parameters under consideration from notes 1 and 2, but differs from note 3 only in having a lower pitch (9–10 kHz) and a smaller F-range ($p < 0.01$, MW-U test).

Investigation of the spectrograms shows that the male initiates his call with note type 1, followed by note 2. Yet, these note types account for only 10% of all notes (N = 132) because they grade quickly into note type 3, which represents the most abundant note type in the male call and is considered to be the "typical" male duet note. Characteristically, this note starts at approximately 6 kHz, rises quickly to 11–12 kHz, and declines then to approximately 5 kHz. The mean duration of this note is 130 ms. Further inspection of the spectrograms reveals that note type 3 may change into note type 4. This change occurred at the end of the phrase, that is, when the pitch of the female notes is low. As soon as the female has finished her phrase, the male will change again to note type 3 (see fig. 9.2a), that is, he regains pitch. This may suggest that the male synchronizes his note pitch with that of the female note (Nietsch, 1999). However, in the particular song that I analyzed, the male showed this behavior only during six out of eight phrases.

As shown above, the number of phrases that a female may produce during a morning duet is highly variable, ranging between one single phrase per song and more than ten. Female songs that lack phrase structure were also recorded, however. Without exception, these were cases when two females were calling simultaneously, for instance during group encounters. During group encounters, pairs may engage in counter duets. I recorded five such incidents in the early morning in the vicinity of the sleeping site of

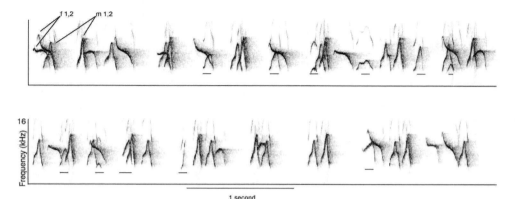

Figure 9.3. Two sections of a vocal dispute between two pairs of *T. spectrum* at the boundary of their territories. The horizontal bars indicate aberrantly modulated notes, i.e., notes that are not part of the duet note repertoires of males or females in spontaneous morning duets. F1 and m1 refer to notes emitted by the female and the male from the resident pair; f 2 and m2 are notes from the intruders.

a particular pair, when the neighbor pair came very close and both pairs started to sing. The duration of these counter duets was 240 s (median; 1st, 3rd quartiles: 112, 301.5 s; N = 5). Both females routinely produced phrases in their songs when calling alone with their mates. Yet, in this particular context, the same females failed to sing phrases in three out of the five cases. In one case, both the resident female and the female from the neighbor pair each called a single phrase. In the other case, the resident pair continued to duet after the intruders fell silent, and only then did the song of the female exhibit a "normal duet structure," that is, the female sang phrases.

Visual inspection of spectrograms of a short section of a counterduet, displayed in figure 9.3, reveals that the participating females produced notes that can be identified according to shape as *T. spectrum* female duet notes, types 1, 2, or 3, while note type 4 is lacking. In addition to "normal" duet notes, both females delivered a variety of oddly shaped notes, however. Examples for such aberrant vocalizations are indicated in the spectrogram (fig. 9.3). Seemingly, both males contributed to the song in a quite "normal" way with respect to the shape of notes as well as note spacing. However, as it is not possible to ascribe each note to a particular individual, I cannot exclude that some of the atypical notes displayed refer to a male, too.

Vocal Nonduet Repertoire: Call Types and Behavioral Contexts

The visual inspection of spectrograms of the samples from adult and juvenile *T. spectrum* of both sexes showed fifteen different call pattern types.

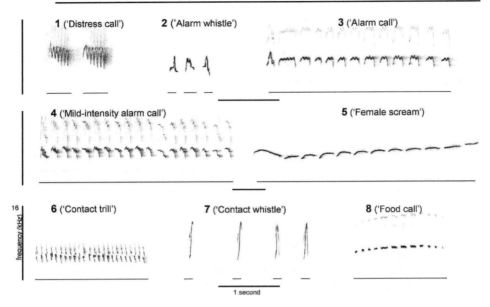

Figure 9.4. Spectrograms of vocalizations of adult *T. spectrum:* call types 1 to 8.

Most calls in *T. spectrum* are loud and piercing and carry far through the habitat, indicating a use of long-distance signaling. Soft calls were recorded as well but might be less well represented in my sample than loud calls since it was not possible to observe continuously and to record tarsiers at close proximity during the study. Examination of the vocal output of the animals revealed covariations between call types and age or sex of signaler. Further, it was possible to relate most call patterns to a particular behavioral context. In cases where the behavioral background could be fairly determined the term for any particular call type refers to the associated behavioral context. Examples for distinctive call patterns of adults or juveniles are shown in figures 9.4 and 9.5, respectively.

In distressing situations, for instance when caught and handled by the observer, animals of all ages may utter a very loud, shrill, and piercing call (call type 1, "Distress call," fig. 9.4a). In adults this trill is high-pitched and may cover a broad frequency band (table 9.2). In a similar context, infants and juveniles may vocalize as well. These calls were not recorded, however. They sounded lower in pitch than calls of adults and much softer.

When exposed to a sudden loud noise, when detecting a snake, rat, or a cat, animals may utter loud and piercing whistles (call type 2, "Alarm whistle," fig. 9.4, table 9.2). A small infant (approximately four weeks of age) gave four very high-pitched short whistles when detecting a big rat (not re-

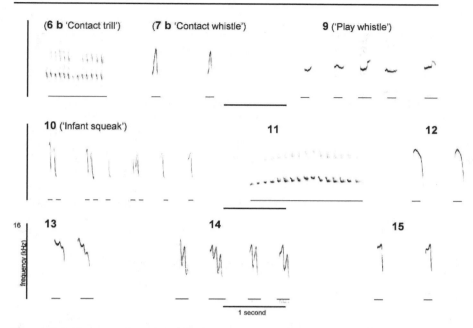

Figure 9.5. Spectrograms of vocalizations of juvenile *T. spectrum:* call types 9 to 15. Calls 6b and 7b correspond to the calls 6a and 7a in the adult repertoire (fig. 9.4).

corded). The calls of this infant sounded much higher in pitch than whistles of adults in a similar context.

Whenever animals are disturbed, for instance by the human observer or by free-ranging cattle, but also in response to distress calls of conspecifics, adults may produce two different types of loud and prolonged calls (call types 3 and 4, fig. 9.4, table 9.2, see section "Playback Experiments"). Call type 4 is less strongly related in time to a preceding event than call 3 ("alarm call"), and I therefore tentatively refer to it as "mild-intensity alarm call." Both calls do not differ much from each other in pitch and are located in a similar narrow frequency band (table 9.3). They differ noticeably in duration as well as in the number and shape of the constituting units (table 9.2), however.

Call type 5 was recorded exclusively from unpaired females when exposed to male duet vocalizations ("female scream," fig. 9.4, table 9.2). This call is very piercing, may reach considerable duration (up to 12 s) and is characterized by a high number of individual units (up to nineteen units) that are hardly frequency modulated. The context of calling suggests that this call serves a function in female advertising for mating.

When being approached by a group mate, animals of all ages may utter

Table 9.2. Adults: Acoustic Parameters of Call Types (*Tarsius spectrum*), Calls 1–8

Parameters	1 Distress call (N = 23)	2 Alarm whistle (N = 28)	3 Alarm call (N = 7)	4 Mild-intensity alarm (N = 15)	5 Female mating scream (N = 25)	6 Contact trill (N = 14)	7 Contact whistle (N = 17)	8 Food call (N = 15)
Duration (s)	0.4	124.3	1.5	2.3	6.9	1.2	70	1.3
	(0.1)	(13.9)	(0.7)	(1.7)	(2.5)	(0.2)	(10)	(0.6)
Max f (kHz)	10.2	8	7.6	7.0	8.7	5.6	12.3	7.0
	(1.4)	(0.9)	(0.05)	(0.4)	(0.6)	(0.1)	(0.3)	(0.2)
Min f (kHz)	3.4	5.2	4.4	4.6	6.5	3.1	4.2	5.4
	(0.7)	(0.5)	(0.5)	(0.8)	(0.9)	(0.5)	(0.2)	(0.6)
F-range (kHz)	6.8	2.8	3.2	2.5	2.2	2.5	7.5	1.6
	(1.5)	(0.8)	(0.5)	(0.8)	(0.9)	(0.6)	(2)	(0.6)
No. of units	5.3		9.4	12	9.7	17		9.5
	(1.8)		(3.7)	(7.2)	(4.6)	(3.5)		(3.1)

a very soft twittering sound (call type 6a, fig. 9.4, and call type 6b, fig. 9.5; "contact trill"), which is especially low in pitch (located between 3 and 5 kHz) and is constituted by a large number of extremely short units (table 9.2). This sound often preceded friendly interactions between caller and conspecifc individuals, such as huddling or grooming. It was also frequently recorded when animals were exposed to duet songs (see Nietsch and Kopp, 1998).

Animals of all ages may produce long series of loud and piercing whistles when alone, as well as when exposed to the duet songs of conspecifics (call type 7a, figs. 9.4, 9.5; "contact whistle"). A two-month-old captive infant called particularly frequently if left alone during daytime but ceased calling immediately when the human caregiver came back. The call may serve as long-distance contact or locator call, informing conspecifics about the whereabouts of the caller. Calls of adults and juveniles are very similar in acoustic structure. Interestingly, the acoustic structure of this call is very similar compared to male duet notes (compare male duet note types 1 and 2, fig. 9.2b). This may indicate that the male duet call emerges of the contact call. Likewise, it has been suggested that the gathering call of *Galago demidovii* emerges of the "tsic" call of very young infants used to establish contact with the mother. It would be most interesting to compare contact calls and male duet calls in other tarsier populations.

Call type 8 was recorded from adults when they were detecting prey items ("food call"; fig. 9.4a, table 9.2). The constituting units are very short and almost unmodulated in frequency, and the entire call covers a narrow fre-

Table 9.3. Juveniles: Acoustic Parameters of Particular Call Types (*Tarsius spectrum*), Calls 9–15

Parameters	9 Play whistle (N = 4)	10 Infant squeak (N = 21)	11 Trill (N = 12)	12 Whistle (N = 14)	14 Whistle (N = 9)	15 Whistle (N = 6)
Duration (ms)	143	48.6	1.35	154.5	177	102
	(5.4)	(13.7)	(0.4)	(25)	(13)	(6.9)
Max f (kHz)	8.6	11.7	7.4	12.6	13.5	12.5
	(0.5)	(1.2)	(0.4)	(0.5)	(0.34)	(0.5)
Min f (kHz)	6.9	4.9	5.5	5.9	5.9	6.2
	(0.4)	(1)	(0.3)	(0.5)	(0.34)	(0.6)
F-range (kHz)	1.7	6.7	1.8	6.7	7.6	6.4
	(0.8)	(1.25)	(0.5)	(0.8)	(0.6)	(1)
No of units	—	—	11.7	—	—	—
			(3.5)			

quency band (1–2 kHz; table 9.2). The hearing range to the human ear is approximately three to five meters.

During playful interactions with conspecifics, infants may utter a very soft whistle (call type 9, "play whistle"; fig. 9.5, table 9.2). If an infant is isolated, hungry, or cold, it may produce a string of faint, squeaky, or twittering sounds (call type 10, "infant squeak"; fig. 9.4b). The hearing range for this call was less than 0.5 m. This call disappeared in a hand-reared infant after the third week of age and was never recorded again from this particular individual.

Call types 11–15 were recorded from juveniles as they were exposed to duet songs. It was not possible to determine the behavioral context of calling more precisely. Yet, while calling, the juveniles tried to approach the sound source, suggesting that they are attempting to establish contact to the parental pair. Call 11 resembles in acoustic characteristics the "food call" of adults (fig. 9.4, table 9.2), but differs in the modulation of individual units. Calls 12–15 (fig. 9.5) are single-sound emissions, which may be repeated in long series with irregular spacing between individual whistles. All are very loud and piercing and exhibit a similar high pitch or F-range (for Juvenile calls I–III sec; table 9.3), and all are distinctive in shape. The diversity of call types emitted by juveniles during duet contexts in a similar situation (call types 6, 7, 11–15) leads to the question, Does the apparently consistent contextual category I set up correspond to that of the animals? It is also likely that distinct call types covary with changes during earlier phases of ontogeny, that is, within the first few months of life, and that a distinction of juveniles on a finely graded age scale may show that. For example, calls 11 and 14 were recorded exclusively from juveniles older than three months.

Table 9.4. Vocal Responses to Played Vocalizations

Stimulus	N trials	% Females (N = 7)	% Males (N = 3)	N trials	% Juveniles (N = 3)
			Number of Responses by Test Subjects (%)		
Duets	29	52	93	8	8
Male duet calls	17	82	29	3	0
Female duet calls	18	50	100	8	8
Contact whistle	10	0	0	4	0
Distress trill	4	100	100	n.t.	n.t.
Mild-intensity alarm call	3	0	0	1	0
Food trill	5	0	n.t.	n.t.	n.t.
Female mating scream	6	33	100	n.t.	n.t.

NOTE: n.t. = not tested.

Playback Experiments

A total of ninety two playback experiments was conducted with feral as well as captive adult or juvenile animals of both sexes. The animals were exposed to duet songs, or male or female duet calls. In addition, some "nonduet" calls were used in playback experiments (see table 9.4).

In general, duet vocalizations proved to be effective stimuli for males, females, and juveniles. The type as well as the frequencies of responses obtained during the course of the trials differed according to individuals as well as the type of playback. Duet songs and, in particular, female duet calls could stimulate all males to reply with duet calls in series of male duet notes. With respect to the females, in each tested group, only the female that engaged in dueting with a particular male in the early morning could give duet calls during playback sessions. If a duet song was given in response to a playback trial, this was always between this female and the particular male with whom she would normally perform duets in the morning. The "nondueting" females as well as the juveniles replied to duet vocalizations with a variety of call types, including "contact trills" or "contact whistles" (see above), but never uttered duet vocalizations. In addition, these females could emit "female screams" in response to playback of male duet calls. Further, males responded more frequently to duets or female calls than females did, whereas females responded more frequently to male duet calls than males did. Juveniles replied regularly to duet songs and female calls, but never vocalized when exposed to male calls.

A gender-related pattern in the responses of adults was also found when the responses were examined according to "who calls first." When duet songs were played, I obtained a total of twenty seven vocal responses. In twenty six trials, the male gave the first vocal response. The latency of a male's response varied between 5 and 173 s, with the average duration being 87.8 ± 56.4 s (N = 27). In females, latencies were considerable longer and varied between 35 and more than 200 s, on average 142.2 ± 50.7 s (N = 15).

When female calls were played, the first response was from a male (N = 18 trials). Latencies varied between 13 and 100 s, average duration was 45.5 ± 21.4 s (N = 18). The latencies of females again were longer than those of males, varying between 78 and 218 s, on average 196.7 ± 29.6 s (N = 9).

When male calls were played, the response behaviors differed noticeably from the other trials. In twelve trials, a "non-dueting" female gave the first response. The latency of a female response varied between 16 and 134 s and was on average 86.5 ± 43 s (N = 13). Males started calling in five out of the twelve trials and had a mean latency of 134 ± 39.5 s (range of latencies: 68 to 159 s, N = 5). They always started vocalizing after the females had called; hence, it is not clear whether they replied to the played call or to the actual female.

Playback with "non-duet" calls was far less effective in eliciting vocal responses. From the five call types used here (table 9.4), only the "female scream" and the "distress call" were responded to with vocalizations.

"Female screams" (fig. 9.4) reliably elicited duet vocalizations from males. The males replied to this call as they would do to a female duet call, that is, they coordinated their duet vocalizations with the played call and tried to approach the sound source. The response behaviors of tested males confirm a suggested function of the call in female advertising for mating.

When "distress calls" (fig. 9.4) were played, the animals approached the sound source very rapidly and stayed at close range to the speaker while uttering alarm calls (fig. 9.4) as well as "mild-intensity alarm calls" (fig. 9.4). This behavior is very similar to that described by Petter and Charles-Dominique (1979) for a number of prosimian species and might be interpreted as mobbing behavior.

Food calls (fig. 9.4) elicited a nonvocal response in animals. When calls were played, the animals immediately moved closer to the ground and scanned the floor as though they were foraging. It is suggestive to think that animals could profit from a caller's skill to detect prey if they commence foraging close to him when hearing his call. Since this particular call is very soft, only animals that are close to the caller would benefit from this behavior. How this behavior fits with the idea that tarsiers are solitary foragers needs further investigation.

Discussion

I have presented a study on the vocal behavior of *T. spectrum* carried out on feral and captive adult and juvenile subjects. The data give evidence for a large vocal repertoire, organized in categorically discrete call types, most of which clearly covary with distinct contexts.

Several studies have been conducted in the past on dueting behavior in *T. spectrum* (MacKinnon and MacKinnon, 1980; Niemitz, 1984b; Niemitz et al., 1991; Nietsch, 1999). The present study provides additional information about the acoustic properties of the duet note repertoires of both female and male *T. spectrum*. With this study I also provided data on temporal and structural coordination of calls of both sexes and showed for the first time that the structure of duet calls may vary according to context.

Comparison among Tarsier Species

Previous studies provided comparative data of duet calls of *T. spectrum* and *T. dianae* (Niemitz et al., 1991; Nietsch, 1999). These studies have shown that the differences in call structure are evident and not subtle. Distinctiveness in call pattern definitely holds for the Togian tarsier (Nietsch, 1999).

With spectrograms of sections of calls from the "Gorontalo form" and the "Palu form" MacKinnon and MacKinnon (1980) have shown that both populations exhibit clear, different call patterns. Spectrographic display of calls from other populations with divergent duet call patterns, for example the Sangihe tarsier (Shekelle et al. 1997), were not available at the time this manuscript was written. Hence, a comparison of call features is still pending. We cannot compare calls of different populations of Sulawesi tarsiers from nondueting contexts because such data are not available. We hypothesize that such data would confirm the assumption that species are vocally distinctive.

Bioacoustic studies should also shed light on how the three major groups of tarsiers—the Sulawesi tarsier group, *T. bancanus* from Borneo and Sumatra, and *T. syrichta* from the Philippines—are related taxonomically. Morphometric analysis suggested that the degree of separation among specimens of these three groups supports ideas on generic separation between the Sulawesi tarsiers on the one hand, and *T. bancanus* or *T. syrichta* on the other (Groves 1998). The special status of Sulawesi tarsiers is corroborated by the fact that neither *T. bancanus* nor *T. syrichta* performs pair duets, reflecting a fundamental difference in social systems: on the one side the nongregarious lifestyle of both *T. bancanus* and *T. syrichta,* on the other side the rather gregarious lifestyle of *T. spectrum.*

Furthermore, previous studies unequivocally claimed that vocal activity in *T. bancanus* is relatively low, except in a sexual context (Fogden 1974). Niemitz (1979, 1984b) distinguished four basic patterns of vocalizations in *T. bancanus* from Semongok, Sarawak, and confirmed that *T. bancanus* is not often heard in the wild. This finding is in contrast to Crompton and Andau (1987), who studied *T. bancanus* in Sepilok Forest, Sabah. They reported that this species frequently calls during the night. They distinguished three different types of vocalizations, one call type being used in long calling concerts that involved up to five animals at a time. They suggested that these calls were involved in male-female interactions as well as in territorial behaviors.

Information on calling behavior in *T. syrichta* is even more scanty. Wright and Simons (1984) described three different call types for males. One call ("male call I") was suggested to function as a territorial or location call. A second call was identified as "courtship call," emitted by the male when the female is in estrus (Wright et al., 1986). In a comparative study on social behaviors of *T. syrichta* and *T. bancanus,* Haring et al. (1984) reported that the "loud call" given by male *T. syrichta* was not given by male *T. bancanus.* We need more information on calling behavior in these species to clarify taxonomic relationships based on bioacoustic analysis.

Biological Functions of Dueting Behavior in T. spectrum

Many attempts have been made to explain the biological significance of duet displays in animals (reviews in von Helverson, 1980; Farabaugh, 1982; Cowlishaw, 1992). The most quoted functions of vocal dueting include those that addresses signals to extrapair conspecifics, for example advertising and defense of the territory or the pairbond (Todt, 1970; Seibt and Wickler, 1977; Hultsch and Todt, 1984; Geissmann, 1999). More recently, a role for duet calls in mate guarding has been suggested, based on the assumption that individuals participate in dueting to repel sexual rivals who are attracted to the call of a male's mate (Sonnenschein and Reyer, 1983; Levin 1996). Others proposed that duets serve for within-pair communication, such as mutual reproductive stimulation and synchronization (Todt and Hultsch, 1982), or serve in the maintenance and mutual strengthening of the pair-bonds (Wickler, 1980).

Combined analyses of duet behavior with playback experiments in *T. spectrum* strongly indicate that vocal dueting in this species is a multifunctional display. In the following I will consider three biological functions of duet behavior: the defense of the territory, mate attraction, and vocal mate guarding.

TERRITORIAL FUNCTION

Duet vocalizations in *T. spectrum* accompany group encounters and have been interpreted as behavior that serves a function in intergroup spacing or redefining boundaries (MacKinnon and MacKinnon, 1980). Given that playback of conspecific calls mimic a territorial intrusion, the vocal responses during playback can be interpreted as territorial responses. As males always respond first to the call of intruding pairs, this behavior indicates a predominance in territorial defense behavior in favor of the male sex (Mac-Kinnon and MacKinnon, 1980; Nietsch, 1993). The female may then join the male's call, as shown by a number of playback trials. With joint calling, pair mates demonstrate their readiness to conduct a joint and coordinated defense of the common territory.

Tarsiers spend more time dueting at the end of their nocturnal activity period in the vicinity of their sleeping site than calling in the early hours of the night when groups disperse and territorial intrusion by trespassing individuals is likely. Both features, their favorite time to call and the preferred location of calling, suggest a particular role of dueting in defense of the sleeping site. The possession of a suitable sleeping site may be vitally important, so tarsiers may need to advertise its occupation to prevent the takeover by nongroup conspecifics. There is evidence that tarsiers become extinct in areas that had been cleared of suitable sleeping sites, although they could still forage in this area. It has been suggested that an abundance of sleeping sites might have a greater impact on population density than other ecological factors (Leksono et al., 1997).

MATE ATTRACTION AND VOCAL MATE GUARDING

My playback experiments strongly suggest that duet calls are involved in mate attraction or mate recognition. The responses that males and (unmated) females give to the playback of calls from the opposite sex can best be explained by courting behavior.

By coordinating his call to that of the mate, an individual may signal to sexual rivals that the partner is already paired and therefore unavailable as a potential mate. If such vocal mate guarding applies to dueting behavior in tarsiers, then we should expect that the call of an individual that is dueting with a mate would elicit less strong responses from a potential mate than a solo-calling individual. Furthermore, we would expect that mates do not tolerate their partners' calling alone. Rather, individuals need to pay attention to the call of their mates and should not fail to reply to their calls.

Analysis of *T. spectrum* calling behavior in spontaneous duets and the responses obtained to playbacks strongly suggests that male tarsiers engage in vocal mate guarding, but the evidence for this behavior in females is less convincing. First, males responded with greater latencies to played pair duet songs than to played female solos. In a number of cases, males even failed to respond to duet songs but replied to all female solos—possibly because males should be more reserved if a female is accompanied by her mate.

Second, males almost never fail to join the call of their mates. Hence, female solo calls are rare events in *T. spectrum*. In contrast, we sometimes hear males call solos in the presence of their female mates. Further, the male responds to the first note of his female mate within a few seconds. The time that elapses between male and female notes when the male initiates the duet song is much longer. I did not measure this timespan in this study, but Theile (1999), with data of captive tarsiers, confirmed that a female's reply to the initiating note of her mate was significantly delayed compared to that of males. Further, females do not routinely intervene when their mates are responding to a female solo. In these cases, the male could coordinate his call in time and structure with that of the mimicked female as if the males were dueting with their mates. For example, he could decrease the pitch of notes at the end of the played phrase (Nietsch, in prep.). Females apparently do not repel other females that show interest in their male partner's call. Together, these observations indicate that intrasexual aggression in *T. spectrum* females is weak. Therefore, tarsier groups may tolerate numerous adult females but rarely more than one adult male (Nietsch, 1993; Gursky, 1995).

Use of Playback Experiments in Species Discrimination

We still continue to discover populations with divergent duet call patterns. Shekelle et al. (1997) found four new duet forms on the northern peninsula of Sulawesi along a transect circling the Tomini Bay and pointed out that even more distinct forms could be detected along the northern coast of the peninsula. We recently recorded duets in south and southeast Sulawesi that are different to the calls of *T. spectrum* or *T. dianae* (Nietsch and Burton, in prep.). Yet, as long as we lack support from morphological studies, the question remains whether the observed differences in calls are significant for species differentiation. Needless to say, the animals may base their discrimination on completely different acoustic cues than humans. Moreover, one could argue that *T. spectrum* should not be too particular about deviations from their "normal" duet call structure because the duet call structure is modifiable. Whether other populations modify their duet calls and to which degree is still a matter for investigation. Exposing tarsiers to duet calls and letting the animals themselves discriminate among different calls is the best way to evaluate the significance of acoustic differences that we ourselves can detect in the calls.

Indeed, it has been shown that tarsiers do not respond to calls from other populations in the same way as they respond to the call of conspecific animals. The results of playback experiments clearly stress the important role of duet calls in intraspecific communication. It is particularly interesting that duet calls serve a function in mate attraction or mate location. Tarsier males and females that do not respond to each other's calls most likely do not recognize each other as potential mates. Playback experiments may thus serve as a tool to identify reproductively isolated populations.

In testing for species recognition, we have to decide which kind of vocal responses should be taken into account. The quality and the quantity of vocal responses obtained in playback sessions vary according to age and sex, and most likely according to sexual arousal or motivation. In my study, pairs did not necessarily respond to conspecific duets with a joint duet. Sometimes only the male responded. Why the female in playback situations did not vocally support her mate against the intruders (or defend her mate against a sexual rival; see above) in each trial needs further investigation. Further, in my experiments it was not possible to decide on whether the female merely responded to the call of her (already vocalizing) mate or whether she responded to the mimicked intruders. Only the latter would indicate that she recognized the played calls as calls of conspecific animals. If the call of her mate is the important stimulus, her participation might depend on the quality of his calls, that is, his motivation. Careful examination of the male's call is required to detect any differences in his performances that could trigger the female's behavior.

A recent study revealed that the quality of calls given in response to played calls is affected in a similar way, as is the quality of calls in the counter duet (Theile, 1999). During playbacks, the phrase structure of female contributions may be lacking, and the female may even cease calling. Moreover, some of the vocalizations that the females emitted during playbacks were not identifiable as female duet vocalizations. Thus, in a situation when a resident pair is threatened by conspecific intruders, mimicked or not, they might fail to perform what we would consider to be a correct vocal response, that is, a neatly coordinated duet song. To sum up, if we only accept a joint and well-coordinated duet performance as a valid response to a played call in a species discrimination experiment, we would interpret the obtained response as a response given to nonconspecifc calls.

If response behaviors indicate reproductive isolation between populations, one should expect test results to be reciprocal. This is indeed the case with all tarsier populations tested to date (Shekelle et al., 1997). However, one noticeable exception is the Togian Island tarsier. Although the evidence from playback experiments suggests that no mainland tarsiers would recognize a Togian tarsier duet song as conspecific vocalization (Shekelle et al., 1997; Nietsch and Kopp, 1998), this does not hold when the Togian tarsiers are exposed to foreign calls. According to Shekelle et al. (1997), these tarsier reply to the calls of all mainland tarsiers. This phenomenon of unidirectional discrimination warrants further examination. However, it would be informative to compare the quality of Togian tarsier responses to mainland tarsier calls with those given to their own calls. Some "wrong" responses were also obtained from *T. spectrum* males when exposed to *T. dianae* or Togian tarsier duet songs. In these few cases the males responded with noticeable hesitancy and delivered a significantly shorter song (Nietsch and Kopp, 1998). The latency and intensity of a vocal response may thus serve as further important cues to determine whether the animals discriminated the played calls (Nietsch and Kopp, 1998).

My future work will include studies on vocal repertoires in different geographical populations and playback experiments. I will focus especially on studies in which potentially discriminating vocal features are varied systematically in order to examine which acoustic parameters in signals are responsible for species recognition and mate recognition in Sulawesi tarsiers.

Acknowledgments

I thank Dr. Marie-Luise Kopp and Stefanie Theile for assistance during fieldwork and for reading the manuscript. Special thanks to Dr. Henrike Hultsch, who read first drafts of the manuscript and made useful comments. I am grateful to Dr. Ir. Supraptini Mansjoer, Department of Animal Breeding and Genetics, and Prof. Dr. Dondin Sajuthi, Primate Research Centre, Bogor Agriculture University, for cooperation and permission to study the tarsiers at Bogor. Similarly, I am thankful to the director and stuff of Gembira Loka, Yokjakarta, for permission to study the tarsiers at this facility. This study was supported by LIPI; financial support was provided by DFG, Germany, and FNK, Freie Universität Berlin.

References

Cowlishaw G. 1992. Song function in gibbons. Behaviour 121: 131–153

Crompton RH, Andau MP. 1987. Ranging, activity rhythms, and sociality in free-ranging *Tarsius bancanus:* a preliminary report. Int J Primatol 8: 43–71.

Farabough SM. 1982. The ecological and social significance of duetting. In Kroodsma DE, Miller EH, Ouellet H, editors, Acoustic communication in birds, 85–124. New York: Academic Press.

Feiler A. 1990. Über die Säugetiere der Sangihe- und Talaud-Inseln—der Beitrag A. B. Meyers für ihre Erforschung. Zoologische Abhandlungen Staatliches Museum für Tierkunde Dresden 46: 75–94.

Fogden MP. 1974. A preliminary field-study of the Western tarsier, *Tarsius bancanus* Horsfield. In Martin RD, Doyle GA, Walker AC, editors, Prosimian biology, 151–165. London: Duckworth.

Geissmann T. 1999. Duet songs of the siamang, *Hylobates syndactylus:* II. Testing the pair-bonding hypothesis during a partner exchange. Behaviour 136: 1005–1039.

Groves CP. 1998. Systematics of tarsiers and loris. Primates 39: 13–27.

Gursky S. 1995. Group size and composition in the spectral tarsier, *Tarsius spectrum:* implications for social organization. Trop Diversity 3: 57–62.

Haimoff EH, Chivers DJ, Gittins SP, Whitten T. 1982. A phylogeny of gibbons (*Hylobates* spp.) based on morphological and behavioural characters. Folia Primatol 39: 213–237.

Haring DM, Wright PC, Simons EL. 1984. Social behaviors of *T. syrichta* and *T. bancanus*. Am J Phys Anthro 66: 179.

Hodun A, Snowdon CT, Soini P. 1981. Subspecific variation in the long calls of the tamarin, *Saguinus fuscicollis*. Z Tierpsychol 57: 97–110.

Hultsch H, Todt D. 1984. Spatial proximity between allies: a territorial signal tested in the monogamous duet singer *Cossypha heuglini*. Behaviour 91: 286–293.

Leksono SM, Masala Y, Shekelle M. 1997. Tarsiers and agriculture: thoughts on an integrated management plan. Sulawesi Primate News Letters 4: 11–13.

Levin RN. 1996. Song behavior and reproductive strategies in a duetting wren, *Thryothorus nigricapillus*. II. Playback experiments. Anim Behav 52: 1107–1117.

Macedonia JM, Stanger KF. 1994. Phylogeny of the Lemuridae revisited: evidence from communication signals. Folia Primatol 63: 1–43.

Macedonia J, Taylor L. 1985. Subspecific divergence in a loud call of the ruffed lemur (*V. variegata*). Am J Primatol 9: 295–304.

MacKinnon JR, MacKinnon KS. 1980. The behavior of wild spectral tarsiers. Int J Primatol 1: 361–379.

Marler P. 1973. A comparison of vocalizations of red-tailed monkeys and blue monkeys, *Cercopithecus ascanius* and *C. mitis*, in Uganda. Z Tierpsychol 33: 223–247.

Marshall JT, Sugardjito J, Markaya M. 1984. Gibbons of the lar group: Relations based on voice. In Preuschoft H, Chivers D, Brockelman W, Creel N, editors, The lesser apes: evolutionary and behavioural biology, 532–541. Edinburgh: Edinburgh University Press.

Marten, K, Quine DB, Marler P. 1977. Sound transmission and its significance for animal vocalization. II. Tropical habitats. Behav Ecol Sociobiol 2: 291–302.

Masters JC. 1991. Loud calls of *Galago crassicaudatus* and *G. garnettii* and their relation to habitat structure. Primates 32: 135–167.

McGregor PK, Holland J. 1995. Avisoft-Sonagraph Pro: A PC-program for sonagraphic analysis. V 2.1. by Raimund Specht. Anim Behav 50: 1137–1143.

Mitani JC, Gros-Louis J. 1995. Species and sex differences in the screams of chimpanzees and bonobos. Int J Primatol 16: 393–411.

Musser GG, Dagosto M. 1987. The identity of *Tarsius pumilus*, a pygmy species endemic to the montane mossy forest of Central Sulawesi. Am Mus Nov 2867: 1–53.

Niemitz C. 1977. Zur Funktionsmorphologie und Biometrie der Gattung *Tarsius* Storr, 1780. Courier Forschungsinstitut Senckenberg 25 (Frankfurt-am-Main): 1–161.

Niemitz C. 1979. Outline of the behavior of *Tarsius bancanus*. In Doyle GA, Martin RD, editors, The study of prosimian behavior, 631–660. New York: Academic Press.

Niemitz C. 1984a. Taxonomy and distribution of the genus *Tarsius* Storr, 1780. In Niemitz C, editor, Biology of tarsiers, 1–16. Stuttgart: Gustav-Fischer-Verlag.

Niemitz C. 1984b. Vocal communication of two tarsier species (*Tarsius bancanus* and *Tarsius spectrum*). In Niemitz C, editor, Biology of tarsiers, 129–141. Gustav-Fischer-Verlag.

Niemitz C. 1985. Der Koboldmaki: Evolutionsforschung an einem Primaten. Naturwissenschaftliche Rundschau 38: 43–49.

Niemitz C, Nietsch A, Warter S, Rumpler Y. 1991. *Tarsius dianae:* a new primate species from Central Sulawesi (Indonesia). Folia Primatol 56: 105–116.

Nietsch A. 1993. Beiträge zur Biologie von *Tarsius spectrum* in Sulawesi—Zur Morphometrie, Entwicklung sowie zum Verhalten unter halbfreien und unter Freilandbedingungen. Dissertation, Freie Universität Berlin.

Nietsch A. 1995. On the role of vocalisations in species differentiation of Sulawesi tarsiers. Biology and conservation of prosimians. International Conference and Workshop at the North of England Zoological Society, Chester, September 1995.

Nietsch A. 1999. Duet vocalizations among different populations of Sulawesi tarsiers. Int J Primatol 20: 567–583.

Nietsch A, Kopp ML. 1998. The role of vocalization in species differentiation of Sulawesi tarsiers. Folia Primatol 69: 371–378.

Nietsch A, Niemitz C. 1993. Diversity of Sulawesi tarsiers. Deutsche Gesellschaft für Säugetierkunde 67: 45–46.

Oates JF, Trocco TF. 1983. Taxonomy and phylogeny of black-and-white colobus monkeys. Folia Primatol 40: 83–113.

Petter JJ, Charles-Dominique P. 1979. Vocal communication in prosimians. In Doyle GA, Martin RD, editors, The study of prosimian behavior, 247–305. New York: Academic Press.

Seibt U, Wickler W. 1977. Duettieren als Revier-Anzeige bei Vögeln. Z Tierpsychol 43: 180–187.

Shekelle M, Leksono, SM, Ichwan LLS, Masala Y. 1997. The natural history of the tarsiers of north and central Sulawesi. Sulawesi Primate Newsletter 4: 4–11.

Snowdon CT. 1993. A vocal taxonomy of the Callitrichids. In Rylands AB, editor, Marmosets and tamarins: systematics, behaviour, and ecology. Oxford: Oxford University Press.

Snowdon CT, Hodun A, Rosenberger AL, Coimbra-Filho AF. 1986. Long-call structure and its relation to taxonomy in lion tamarins. Am J Primatol 11: 253–261.

Sonnenschein E, Reyer H-U. 1983. Mate-guarding and other functions of antiphonal duets in the slate-coloured boubou (*Laniarius funebris*). Z Tierpsychol 63: 112–140.

Struhsaker TT. 1970. Phylogenetic implications of some vocalizations of *Cercopithecus* monkeys. In Napier JR, Napier PH, editors, Old World monkeys: evolution, systematics and behaviour. New York: Academic Press.

Theile S. 1999. Untersuchung zur kontextabhängigen Variabilität im Duettgesang von *Tarsius spectrum* (Pallas 1778). Diploma thesis, Freie Universität Berlin.

Todt D. 1970. Die antiphonen Paargesänge des ostafrikanischen Grassängers *Cisticola hunteri prinoides* Neumann. J Ornithol 111: 332–356.

Todt D, Hultsch H. 1982. Impairment of vocal signal exchange in the monogamous duet-singer *Cossypha heuglini* (Turdidae): effects on pairbond maintenance. Z Tierpsychol 60: 265–271.

von Helversen D. 1980. Structure and function of antiphonal duets. Acta XVII Int Ornith Congr, Berlin 1978: 682–688.

Wickler W. 1980. Vocal dueting and the pair bond. Z Tierpsychol 52: 201–209, 217–226.

Wilson WL, Wilson CC. 1975. Species-specific vocalizations and the determination of phylogenetic affinities of the *Presbytis aygulamelalophus* group in Sumatra. In Kondo S, Kawai M, Ehara A, editors, Contemporary primatology. 459–463. Basel: Karger.

World Wildlife Fund. 1980. Cagar alam gunung Tangkoko dua saudara Sulawesi Utara. Management plan 1981–1986. Bogor, Indonesia.

Wright P, Simons EL. 1984. Calls of the Mindano tarsier (*Tarsius syrichta*). Am J Phys Anthro 63: 236.

Wright PC, Toyama LM, Simons EL. 1986. Courtship and copulation in *Tarsius banconus*. Folio Primatologica 46: 142–148.

Zimmermann E. 1990. Differentiation of vocalizations in bushbabies (Galaginae, Prosimie, Primates) and the significance for assessing phylogenetic relationships. Z Syst Evolution 28: 217–239.

Zimmermann E, Bearder SK, Doyle GA, Anderson AB. 1988. Variations in vocal patterns of Senegal and South African lesser bushbabies and their implications for taxonomic relationships. Folia Primatol 51: 87–105.

CHAPTER 10

Territoriality in the Spectral Tarsier, *Tarsius spectrum*

Sharon Gursky

Numerous hypotheses have been proposed to explain the presence of territoriality in animal species. Two of the primary hypotheses proposed are the defense of mates and the defense of resources. My primary goal in this paper is to attempt to elucidate the function of territoriality in a small nocturnal primate, the spectral tarsier, *Tarsius spectrum*. Although previous observations have demonstrated that the spectral tarsier is territorial, no attempts have been made to elucidate the function of territoriality in this species. The results of this long-term radio-tracking study in Tangkoko Nature Reserve in Sulawesi, Indonesia, indicate that spectral tarsiers were frequently hostile toward members of neighboring groups. In particular, spectral tarsiers were consistently hostile toward members of the same sex, but variably hostile toward members of the opposite sex. The frequency of territorial encounters statistically increased during the mating season. The frequency with which the male female pair were encountered together during nightly foraging also statistically increased during the mating season. In contrast to the predictions of the resource defense hypothesis, male female pairs did not jointly defend their territory from intruders. Similarly, a substantial decrease in the frequency of encounters during the dry season was observed, contrasting with the prediction from the resource defense hypothesis. These results suggest that spectral tarsiers defend their territories as a means of defending their mates and not to defend access to resources.

The spectral tarsier is a small nocturnal primate found only on the island of Sulawesi in Indonesia. The spectral tarsier is unusual in that it is presently the only known tarsier species that exhibits territorial behavior (MacKinnon and MacKinnon, 1980; Niemitz, 1984; Gursky, 1997, 1998). Although previous studies of this species have amply demonstrated the presence of territorial behavior, no attempts have been made to determine why spectral tarsiers exhibit territoriality.

If spectral tarsiers exhibit territoriality in order to defend their mates, then (1) it is predicted that males will be hostile toward nonresident males but not toward nonresident females. Similarly, females will be hostile toward nonresident females but not toward other males. (2) It is also predicted that the number of intergroup confrontations will increase during the mating season as individuals attempt to locate potential mates outside their ranges. (3) It is also predicted that there will be an increase in the number of intragroup encounters during the mating season since males will be mate-guarding the female.

On the other hand, if spectral tarsiers exhibit territoriality in order to defend resources from nongroup members, then (1) it is predicted that confrontations between neighboring groups will involve both males and females. Similarly, (2) it is predicted that confrontations between groups will increase during the dry season when resources are less abundant relative to the wet season. (3) It is also predicted that there will be an increase in the number of intragroup encounters during the dry season to minimize within-group competition.

In virtually all primate species, groups range over a relatively fixed area, and members of a group can be consistently found in a particular area over time. The majority of these ranges, commonly called home ranges, represent an undefended living space. However, some primate species maintain exclusive access to fixed areas. For territorial primates, the boundaries of the territory are essentially the same as for their home range, and territories do not overlap. The defense of a territory has been observed in several primate species, including *Hapalemur griseus, Indri indri, Hylobates klossii, Hylobates lar, Lemur catta, Colobus guereza, Presbytis entellus, Presbytis rubicunda, Presbytis thomasi, Leontopithecus rosalia* (Peres, 1989; Nievergelt et al., 1998; Tenaza, 1975; Palombit, 1993; Waser, 1977, 1981; Mertl-Millhollen et al., 1979; Mitani and Rodman, 1979; Kaufmann, 1983; Niemitz, 1984; Cheney, 1987; Stanford, 1991; Kinnaird, 1992; van Schaik et al., 1992; Jolly et al., 1993).

Numerous hypotheses have been proposed to explain the presence of territoriality in primate species (Hinde, 1956; Brown, 1964; Kaufmann, 1983; Cheney, 1987; Krebs and Davies, 1987). One of the primary hypotheses that has been proposed to explain the function of territoriality is the defense of resources (Brown, 1964; Kodric Brown and Brown, 1978; Mitani and Rodman, 1979; Ostfeld, 1985; Kinnaird, 1992). This hypothesis was first advanced by Brown (1964), who introduced the concept of economic defensibility of the resources present in an animal's range. According to the concept of economic defensibility, individuals will establish exclusive ranges that are defended when it is cost-efficient to defend those resources; that is, when the exclusion of other individuals from the territory has higher payoffs than ignoring the intruders. For example, when food resources are ephemeral, if individuals have to roam over a large area to obtain sufficient nutrients, then it would not pay to be territorial. Mitani and Rodman's (1979) work extended Brown's concept of economic defensibility with the development of an index of range of defensibility. The index of defendability (D) is the ratio of observed daily path length (d) to an area equal to the diameter (d′) of a circle with area equal to the home range of the animal. Their results indicated that territoriality was observed only in populations whose ranges were economically defendable, having an index of 1.0 or greater, a conclusion supporting the resource defense function of territoriality in primates.

Other researchers have suggested that the function of territoriality is the defense of mates (Stanford, 1991; van Schaik et al., 1992; Slagsvold et al., 1994). This hypothesis is based on sexual selection theory (Bateman, 1948; Trivers, 1972), which suggests that the factors that limit the reproductive success of females and males tend to be different. Specifically, female reproductive success is limited by their access to resources, while male reproductive success is limited by access to fertile mates. Females are expected to compete among one another primarily for access to resources whereas males will compete mainly for matings. For example, Stanford (1991) showed that territorial behavior in the capped langur, *Presbytis pileata,* supports the predictions of mate defense better than resource defense. In intergroup encounters, male capped langurs directed their aggression toward their group females, thus controlling access by extragroup males to the females. Similarly, van Schaik et al. (1992) also found that in six *Presbytis* species, not only were loud calls produced solely by males, but between-group antagonism was also restricted to males.

Methods

Study Site

Sulawesi (fig. 10.1), formerly known as Celebes, is a four-armed island located to the east of Borneo and northwest of Australia–New Guinea (long. 125°14′ E, lat. 1°34′ N) (Audley-Charles, 1981). Sulawesi is the eleventh-largest island in the world. It is also the largest and most central island of the biogeographical region of Wallacea, where the Australian and Asian zoogeographical regions meet. Sulawesi shows a blend of Asian and Australian elements in its fauna and flora, but also exhibits very high levels of local endemics. Throughout the island's protected areas, various species of the marsupial *Phalanger* live sympatrically with the primates *Macaca* and *Tarsius.* Of the 127 indigenous mammals, 79 (62%) are endemic, including the spectral tarsier (Musser, 1986). Endemic species include *Macaca nigra* (Celebes ape or black monkey), *Phalanger ursinus* (bear cuscus), and *Babyrousa babirussa* (babirusa). In comparison, the neighboring island of Borneo, the largest Indonesian island, has only thirty six endemic mammal species.

This study was conducted at Tangkoko-Dua Saudara Nature Reserve on the easternmost tip of the northern arm of the island of Sulawesi (long. 125°14′ E, lat. 1°34′ N). When the reserve was first formed in 1980, it was

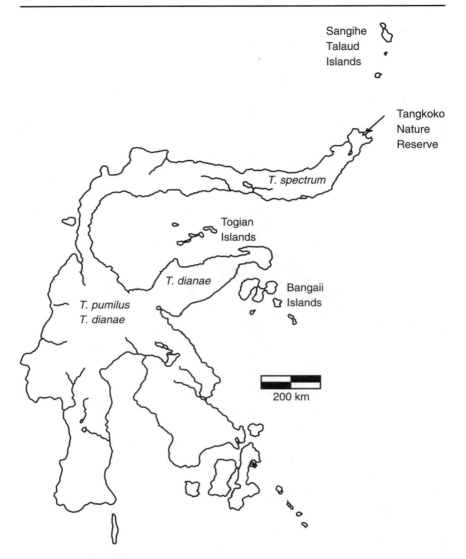

Figure 10.1. Map of Sulawesi, Indonesia.

comprised of 8867 hectares with a sea boundary of 12 km (World Wildlife Fund, 1980; Whitten et al., 1987). Based on a vegetation survey conducted by MacKinnon and MacKinnon, the reserve exhibits a full range of floral communities from sea-level coastal communities, to lowland forests, sub-montane forests to mossy cloud forests on the summits of Dua Saudara and the Tangkoko Crater (MacKinnon and MacKinnon, 1980).

In order to quantitatively characterize the habitat within the study area, a vegetation survey within a four-hectare plot was conducted (Brower et al., 1990; Gursky, 1997). One hundred and twenty-seven different tree species (all trees that were greater than one meter in height) were identified in the plot. The most common tree species as measured by the Importance Value Index (Brower et al., 1990) were *Leea indica* (Leeaceae), *Morinda citrifolia* (Pubiaceae), *Piper aduncum* (Piperaceae), *Palaquium obvatum* (Sapotaceae), *Barringtonia acutangula* (Lecythidaceae), and *Vitex quinata* (Verbenaceae) (Gursky, 1997). The total number of trees in the four-hectare plot with a diameter at breast height (dbh) greater than or equal to five cm was 3164 trees, with a mean of 791 trees per hectare. The total number of trees with a dbh greater than or equal to ten cm was 1727 trees, with a mean of 432 trees per hectare (Gursky, 1997). The majority of the reserve has been disturbed by human influence due to selective harvesting for fishing-boat production and firewood. Annual rainfall at Tangkoko Nature reserve averaged approximately 2500 mm (Gursky, 1997).

Capture and Attachment

The following procedures were used to locate individuals. Prior to dawn, my field assistant and I would stand on the periphery of a one-hectare plot. Plots were chosen randomly (following a block design) within one square kilometer of the trail system. As the tarsiers returned to their sleeping site, or at their sleeping site, they gave loud vocal calls for three to five minutes that could be heard from 300 to 400 meters (MacKinnon and MacKinnon, 1980; Niemitz, 1984). All groups that were heard vocalizing were then followed to their sleeping site. My field assistant and I then returned to the sleeping site prior to dusk to set up several mist nets in the vicinity of the sleeping site (Bibby et al., 1992). The mist nets were continually monitored for captured tarsiers.

Upon capture, individuals were placed in a cloth bag and weighed with a portable scale providing an accuracy of ±1 gram. An SM1 radio collar weighing either 3.5 or 7.0 grams (depending on the size of the battery in the radio collar) with a groove-loop was attached to the tarsier's neck by a simple folding of the thermoplastic band. In this study, thirteen individuals from six groups were observed from April 1994 to June 1995.

Data Collection

A radio receiver using 151 MHz frequency and a three-element collapsible Yagi antenna were used to determine the location of each individual radio frequency. Each individual tarsier was followed approximately every fourth night. An Indonesian student assistant and I conducted behavioral focal follows. Initially, the student and I conducted focal follows together on a single individual until approximately 99% of the data recorded by both of us were the same. At this point, I felt comfortable enough with the consistency in data recording to permit the student to conduct independent focal follows. Once each month thereafter, the student and I conducted an interobserver reliability test to determine if we were still consistent in our data recording. I found that our data recording was at least 98% during each interobserver reliability test.

The focal individual's behavior was recorded at five-minute intervals by the researcher with the aid of an ITT third-generation nightscope, moonlight, and red-filtered flashlights (Altmann, 1974). The following behaviors were recorded: foraging, feeding, resting, traveling, and socializing (i.e., scent mark, allogrooming, playing, and vocalizing) (Gursky, 1992). Definitions of all behaviors recorded are presented in table 10.1. In addition, all occurrences of insect pursuits and captures were collected continuously, ad libitum.

Table 10.1. The Definitions of the Behavioral States Recognized during This Study

Behavior	Definition
Forage	Actively searching the ground, leaves, or air for a moving prey item. This also involves ears twitching while trying to locate prey auditorally, but primarily involves active scanning behavior. Head may not be in normal position, turned from 0 degrees. Also included travel movements while foraging.
Feed	The animal is actively eating a prey item. This includes all handling time of prey such as putting the prey into the mouth.
Rest	The animal is motionless. Its ears and head are not moving. Its head is not rotated from the frontal position. Its eyes may be closed and the animal may be sleeping.
Travel	Actively moving from one support to another via various locomotor styles such as vertically clinging and leaping, quadrupedalism, and climbing. Excludes movements while actively foraging.
Social	Involves scent marking (moving the genital region against a substrate from side to side with the tail in a raised position or urinating), grooming others, vocalizing, and play grappling (running and jumping and tail pulling).
Nurse	Infant's face is located toward mother's chest in pectoral or urogenital region.
Transport	Infant is located within the mouth or on the body of an individual who is traveling.
Miscellaneous	Grooming self (scratching with the grooming claw and scent gland exudite or tooth comb, may include marking own body with scent glands), cleaning body with tongue or hands like a cat and scent gland exudate or tooth comb.

Each individual's spatial position was determined at fifteen-minute intervals using a combination of reflective flagging tape, tape measure, and compass locations. During the focal follows, reflective flagging tapes noting the individual, the day, and the time were attached to the substrate the focal individual was utilizing during the scan. The following day, two local Indonesian assistants located all flagging tapes from the previous night's focal follows and used the compass and tape measure to determine the location of each reflective flagging tape relative to the trail system in the reserve. Based on the locational data points, the actual home-range size was calculated using minimum convex polygons (Kenward, 1987; White and Garrott, 1987). To determine the distance each tarsier individual traveled per unit time, I used fifteen-minute step distances (Whitten, 1982; Altmann and Samuels, 1992; Kinnaird, 1992). Thus, distance traveled was calculated as the straight-line distance between successive fifteen-minute locations. Nightly path length was calculated as the sum of all fifteen-minute step distances each night. An intergroup encounter is defined as an instance were members of the focal group and members of another neighboring group were observed in proximity of one another.

Since the diet of spectral tarsiers is restricted to insects (Gursky, 1997, 2000), three methods of insect sampling were used to record resource abundance: (1) sweep nets, (2) pitfall traps, and (3) malaise traps (Southwood, 1993; Janzen, 1973; Muirhead-Thompson, 1991; Brower et al., 1990). These three trapping techniques were chosen since spectral tarsiers are known to consume insects from the air (malaise traps and sweep nets), from vegetation (sweep nets), as well as from the ground (pitfall traps) (Gursky, 1997, 2000). On nights when hourly insect sampling was conducted, a local PHPA counterpart (Department of Forestry) and Indonesian local assistant were responsible for sweeping the air and vegetation throughout the territory of each focal group with sweep nets hundred times (sweeps) each hour from dusk until dawn. Twenty pitfall traps were also placed in the ground, and two malaise traps were distributed throughout the tarsier group's territory. The local PHPA counterpart and Indonesian local assistant were responsible for collecting the insects from the pitfall and malaise traps each hour throughout the night. The contents from the sweep nets, malaise traps, and pitfall traps were measured (insect length) and weighed to the nearest milligram using an Ohaus digital scale. The PI and Indonesian local assistants identified each insect to their taxonomic level of order and then dried them to determine the biomass for the sample period. Data from the three sampling methods were lumped together in the analyses.

During this study, an intergroup encounter is defined as an instance where members of the focal group and members of another neighboring group were observed in proximity to one another. Encounters were considered agonistic if individuals engaged in lunging and retreating, and were vocalizing loudly, including alarm calling. Intergroup encounters were considered affiliative if non-alarm calls were emitted and no lunging and retreating behavior was exhibited.

Data Analyses

Behaviors sampled at short time intervals are often autocorrelated (Janson, 1990; Sokal and Rohlf, 1981). As a result, considering each sample as independent exaggerates sample size and biases the statistical results. I performed a chi-square contingency table analysis to determine if the data collected in this study were autocorrelated (Sokal and Rohlf, 1981). Data collected at fifteen-minute intervals were not found to be autocorrelated. All statistics follow Sokal and Rohlf (1981).

Results

Descriptive Territorial Behavior

The home range of the individual spectral tarsiers obtained during this study with the assistance of radio telemetry varied from 1.6 to 4.1 hectares, with an average size of 2.3 ha for females and 3.1 ha for males (Gursky, 1998). All groups minimally overlapped with 1–3 groups (table 10.2). Overlapping areas between neighboring groups constituted, on average, approximately 15%. Thus, the majority of each group's range was exclusively used by the resident group and only occasionally occupied by neighboring individuals. Both successive and simultaneous use of overlap areas occurred, the latter of which represent the basis for encounters.

Throughout this study, 243 agonistic intergroup encounters were observed over a period of 442 nights. This amounts to approximately 0.55 territorial disputes each night, or approximately one territorial conflict every other night. When a group member observed a stranger in its territory, it immediately gave a loud call. All the resident group members responded by congregating around the initial caller to vocalize their defense of the territory, as well as lunging and retreating at the intruder, until the intruder departed. The location of the territorial disputes was often at territory boundaries (N = 204). Sometimes, though, the dispute occurred closer to the center of the group's territory (N = 29). The average overall duration of intergroup encounters was eight minutes. Intergroup encounters occurred throughout the entire night but were unevenly distributed (fig. 10.2). Encounter rates peaked between 18:00 and 20:00 and then again shortly between 04:00 and 06:00 (Gursky, 1997).

Table 10.2. Spectral Tarsier (*Tarsius spectrum*) Home Range Estimates

Group	Group type	Home range (in hectares)	Number of neighboring groups	Percent home range overlap
C600	Monogamous	2.91	1 (E650)	0.21 ha
E650	Monogamous	2.12	2 (C600; F600)	0.41 ha (0.22 + 0.19)
F600	Polygynous	2.83	2 (E650; G850)	0.43 ha (0.24 + 0.19)
G850	Monogamous	3.45	3 (F600; G1000; J700)	0.69 ha (0.27 + 0.22 + 0.20)
G1000	Monogamous	2.28	2 (G850; J700)	0.46 ha (0.21 + 0.25)
J700	Polygynous	4.05	3 (G1000; G 850; M600)	0.73 ha (0.12 + 0.31 + 0.30)
M600	Monogamous	2.68	1 (J700)	0.33 ha

NOTES: Estimates based on radio telemetry locations and percentage overlap with neighboring groups. Data were collected at Tangkoko Nature Reserve, Sulawesi, Indonesia during 1994–95.

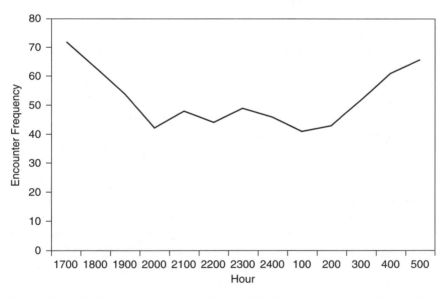

Figure 10.2. The frequency that two adult group members encountered each other during each hour of the night. Encounter is defined as less than or equal to 10 m. Three groups observed for 60 nights.

Mate Defense

During intergroup encounters when the observed intruder was a male and was encountered by the resident male (N = 87), then the resident male always (100%) immediately responded harshly by vocalizing (alarm calling) and lunging at the intruder. Often other group members would come to the area where the resident male was vocalizing. Subadult males would often contribute calls. When an intruding male was encountered by the resident female (N = 64), the resident female varied in her response. Frequently she emitted a loud call (63%), but sometimes (37%) the female emitted another softer call that could not be heard by other group members, only the intruder. When the intruder was a female (N = 43), then the resident male occasionally (26%) emitted a loud call, but more often gave another softer call (74%). This led to several extrapair copulations. However, when a female intruder was encountered by the resident female (N = 49), the resident female immediately responded harshly by vocalizing, emitting a loud call and lunging at the intruder (100%).

The mean number of territorial disputes in the nonmating season (June–October and January–March) was 12.3 per month (SD = 6.7, N = 74 disputes, N = 6 months) compared to the mean number of territorial disputes in the mating season (April–May and November–December), which was 19.8 per month (SD = 11.4, N = 169, N = 9 months). Territorial disputes were much more frequent in the mating seasons than they were during other times of the year (X^2 = 37.92, P = .0001, DF = 1).

While conducting focal follows on a single adult, interactions between the focal individual and another adult group member were observed on 1072 occasions. The mean number of intragroup encounters during this study was 2.42 encounters per night (SD = 2.7). They ranged from zero encounters to as many as eighteen within-group encounters per night. The frequency of encounters was not normally distributed (Shapiro Wilk W test, W = 0.728, P = 0.000), nor did the frequency of encounters represent a Poisson distribution (Kolmogorov-Smirnov, D = 0.241, P < .01; X^2 = 401.93, DF = 6, P = .00000). The modal duration of each intragroup encounter was approximately 4 minutes, and the mean duration was 48 minutes. There was substantial variation in the duration of intragroup encounters ranging from less than 1 minute to as long as 3 hr 12 min.

The mean number of intragroup encounters per night of observation during the mating season was 3.97 (SD = 3.8, N = 698 encounters, N = 176 nights) compared to 1.41 (SD = 1.14, N = 374 encounters, N = 266 nights) intragroup encounters per night of observation during the nonmating season. That is, group members encountered one another significantly more frequently during nightly forays in the mating season than during the non-mating season ($X^2 = 86.96$, $P = .0001$, $DF = 1$).

Resource Defense

Of the 243 intergroup territorial encounters observed throughout this study, only 63% (N = 153) of them involved joint defense of the resources by the resident male and female. When the resident male encountered an intruder (N = 73 male intruders, N = 53 female intruders) and began emitting a loud call to elicit support from his mate, the female returned her mate's call on only forty six occasions (35%). Despite how infrequently the female responded to her mate's alarm calls, she was equally likely to return her mate's call for both male and female intruders. However, when the resident female encountered an intruder (N = 69 male intruders, N = 44 female intruders), and began emitting a loud call to elicit support from her mate, the male always returned her call and immediately traveled to her location.

Resource availability, as measured according to insect biomass, is illustrated in figure 10.3. Two distinct periods of resource abundance at Tangkoko Nature Reserve can be distinguished: a period of high insect biomass (November–April) and a period of low insect biomass (May–October). These two periods are also observed when comparing the number of insects captured during each month (Gursky, 2000) and roughly corresponds to

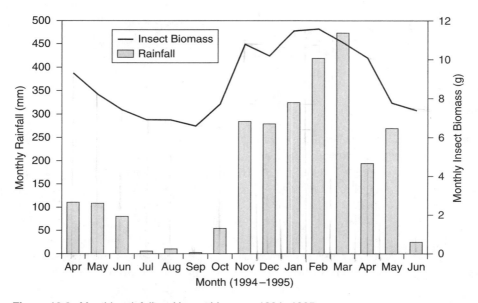

Figure 10.3. Monthly rainfall and insect biomass, 1994–1995.

the rainy season and the dry season, respectively. Consequently, it seems appropriate to use interchangeably the terms "dry season" and "period of low resource abundance," and "rainy or wet season" and "period of high resource abundance," when discussing how spectral tarsiers modify their behavior in response to seasonal resource bases.

There is a significant correlation between monthly rainfall and the number of insects captured in the traps each month (Spearman rank order $Z = 3.05$, $P = .0023$), between monthly rainfall and insect biomass each month (Spearman rank order $Z = 3.42$, $P = .0006$), and between monthly insect biomass and the number of insects captured by the tarsiers each month (Spearman rank order $Z = 3.33$, $P = .0009$).

During months of high resource abundance (December–May), only eighty one territorial disputes were observed over a period of 256 nights. In contrast, during months of low resource abundance, a total number of 162 territorial disputes were observed over a 186-night period. The mean number of territorial disputes per night of observation during the wet season was only 0.32. In contrast, the mean number of territorial disputes per night of observations during the dry season was 0.87. Territorial disputes were significantly more common during the dry season than during the wet season ($t = -13.83$, $P = .0001$, DF = 440 nights).

The mean number of intragroup encounters per night of observation during the dry season was 1.09 (SD = 0.83, N = 204 encounters, N = 186 nights) compared to 3.39 (SD = 2.15, N = 868 encounters, N = 256 nights) within-group encounters per night of observation during the wet season. This measure of the effect of seasonality on group social behavior indicates that group members encountered one another significantly less frequently during nightly forays in the dry season than during the wet season ($t = 6.32$, $P = .0001$, DF = 440).

Discussion

The results of this study provide substantial support for the mate defense function of territoriality in spectral tarsiers (table 10.3). Predictions number 2 and 3 of the mate defense hypothesis were supported. There was an increase in the number of territorial encounters, as well as intragroup encounters, during the mating season compared to the nonmating season. However, the results of prediction number 1 do not provide complete support for the mate defense hypothesis. Spectral tarsiers were not consistent in their behavioral response toward strange members of the opposite sex. Sometimes resident individuals were friendly and responded with a friendly soft call, while sometimes they responded harshly with a loud territorial defense call. This variability in response to members of the opposite sex was

Table 10.3. Summary of the Hypotheses, Predictions, Results, and Conclusions

Hypothesis	Predictions	Results
Spectral tarsiers exhibit territoriality to defend access to mates	Individuals will be hostile to members of the same sex, but not toward members of the opposite sex.	Mixed: although individuals hostile toward same sex, not consistently friendly toward opposite sex
	The number of intergroup confrontations will increase during the mating season.	Prediction supported
	The number of intragroup encounters will increase during the mating season.	Prediction supported
Spectral tarsiers exhibit territoriality to defend access to resources	Territorial confrontations will involve males and females equally.	Prediction not supported
	The number of intergroup encounters will increase during the dry season.	Prediction supported
	The number of intragroup encounters will increase during the dry season.	Prediction not supported

exhibited by both males and females. This may reflect the fact that spectral tarsiers are facultatively polygynous (Nietsch and Niemitz, 1992).

The results of this study also provide mixed support for the resource defense function of territoriality in spectral tarsiers. Both prediction 1 and prediction 3 were not supported. Male and female spectral tarsiers did not jointly encourage the removal of nonresident individuals that were encountered in the territory. Similarly, in contrast to prediction number 3, there was also a substantial decrease in the number of intragroup encounters during the dry season. Interestingly, although predictions number 1 and 3 were not supported, prediction number 2 of the resource defense model was supported. That is, there was an increase in the number of intergroup encounters (territorial disputes) during the dry season.

Overall, the results of this study indicate that the primary function of territoriality in spectral tarsiers may be mate defense and not resource defense. Nonetheless, there is some minor support that some territorial behavior stems from the defense of resources. It is interesting that territorial behavior increases so substantially during the mating season. The mating system is also concurrent with the biannual birth season (Gursky, 1992, 1997). The high rate of within-group encounters during the birth season may indicate that the function of territoriality is perhaps infanticide avoidance whereby potential intruders are kept away from vulnerable infants (Reichard and Sommer, 1997). This model has been invoked to explain female territoriality among rodents (Ebensperger 1998). Among rodents the intensity of female territoriality increases during pregnancy and decreases after the weaning of infants. Unfortunately, because the spectral tarsier mating season and birth season occur during the same months, it was not possible to distinguish between these two hypotheses. Future studies with more specific predictions distinguishing between these two hypotheses will be necessary to fully tease apart these conflicting results.

Acknowledgments

The author acknowledges that this research would not have been possible without the permission and assistance of the following organizations and people: LIPI (The Indonesian Institute of Sciences), SOSPOL, POLRI, PHPA (Manado, Bitung, Tangkoko, and Jakarta), Romon Palette, Yoppy Muskita (WWF), Jatna Supriatna, the University of Indonesia–Depok and Tigor P.N. (UNas). Thanks go to my field assistants for their help in collecting the data (Nestor, Petros, Celsius, Frans, Nolde, Nellman, Ben, and Uri). Funding for this research was provided by the Wenner Gren Foundation, National Science Foundation Grant SBR-9817705, Douroucouli Foundation, Chicago Zoological Society, L.S.B. Leakey Foundation, Primate Conservation Incorporated, and Sigma Xi.

References

Altmann JA. 1974. Observational study of behavior: sampling methods. Behaviour 49: 227–267.
Altmann J, Samuels A. 1992. Costs of maternal care: infant carrying in baboons. Behav Ecol Sociobiol 29: 391–398.
Audley-Charles M. 1981. Geological history of the region of Wallace's line. In Whitmore TC, editor, Wallace's line and plate tectonics. Oxford: Clarendon Press.
Bateman A. 1948. Intra-sexual selection in Drosophila. Heredity 2: 349–368.
Bibby R, Southwood T, Cairns P. 1992. Techniques for estimating population density in birds. New York: Academic Press.
Brower J, Zar J, von Ende C. 1990. Field and laboratory methods for general ecology. Iowa: Wm. C. Brown.
Brown J. 1964. The evolution of diversity in avian territorial systems. Wilson Bull 76: 160–169.
Cheney DL. 1987. Interactions and relationships between groups. In Smuts BB,

Cheney DL, Seyfarth RM, Wrangham RT, Struhsaker TT, editors, Primate societies, 267–281. Chicago: University of Chicago Press.

Ebensperger L. 1998. Strategies and counterstrategies to infanticide in mammals. Biol Rev 73: 321–346.

Gursky S. 1992. Behavioral observation of *Tarsius spectrum* in Sulawesi, Indonesia. Am J Phys Anthro (suppl.) 14: 85–86.

Gursky S. 1997. Modeling maternal time budgets: the impact of lactation and gestation on the behavior of the Spectral Tarsier, *Tarsius spectrum*. Ph.D. dissertation, SUNY–Stony Brook.

Gursky S. 1998. The conservation status of the Spectral Tarsier, *Tarsius spectrum*, in Sulawesi Indonesia. Folia Primatol 69: 191–203.

Gursky S. 2000. The effects of seasonality on the behavior of an insectivorous primate. Int J Primatol 21: 477–495.

Hinde R. 1956. The biological significance of the territories of birds. Ibis 98: 340–369.

Janson C. 1990. Social correlates of individual spatial choice in foraging groups of brown capuchin monkeys, *Cebus apella*. Animal Behav 40: 910–921.

Janzen D. 1973. Sweep samples of tropical foliage insects: effects of seasons, vegetation types, elevation, time of day and insularity. Ecology 54: 687–708.

Jolly A, Rasamimanana HR, Kinnaird M, O'Brien T, Crowley H, Harcourt C, Gardner S, Davidson J. 1993. Territoriality in *Lemur catta* groups during the birth season at Berenty, Madagascar. In Kappeler PM, Ganzhorn J, editors, Lemur social systems and their ecological basis, 85–109. New York: Plenum Press.

Kaufmann J. 1983. On the definitions and functions of dominance and territoriality. Biological Reviews of the Cambridge Philosophical Society 58: 1–20.

Kenward R. 1987. Wildlife radio tagging. Academic Press: New York.

Kinnaird M. 1992. Variable resource defense by the Tana river crested mangabey. Behav Ecol Sociobiol 31: 115–122.

Kodric Brown A, Brown J. 1978. Influence of economics, interspecific competition and sexual dimorphism on territoriality of migrant rufous hummingbirds. Ecology 59: 285–296.

Krebs J, Davies N. 1987. An introduction to behavioral ecology. Oxford: Blackwell Scientific.

MacKinnon J, MacKinnon K. 1980. The behavior of wild spectral tarsiers. Int J Primatol 1: 361–379.

Mertl-Millhollen A, Gustafson H, Budnitz N, Dainis K, Jolly A. 1979. Population and territory stability of the lemur catta at Berenty, Madagascar. Folia Primatol 31: 106–115.

Mitani J, Rodman P. 1979. Territoriality: the relation of ranging pattern and home range size to defendability, with an analysis of territoriality among primate species. Behav Ecol Sociobiol 5: 241–251.

Muirhead-Thompson RC. 1991. Trap responses of flying insects. New York: Academic Press.

Musser G. 1986. The mammals of Sulawesi. In Whitmore TC, editor, Biogeographical evolution of the Malay Archipelago. Oxford: Clarendon Press.

Niemitz C. 1984. The biology of tarsiers. Sttutgart: Gustav-Fischer-Verleg.

Nietsch A, Niemitz C. 1992. Indication for facultative polygamy in free ranging *Tarsius spectrum*, supported by morphometric data. Int Primatol Soc Abstracts, Strasbourg, 318.

Nievergelt C, Mutschler T, Feistner A. 1998. Group encounters and territoriality in wild Alaotran Gentle Lemurs (*Hapalemur griseus alaotrensis*). Am J Primatol 46: 251–258.

Ostfeld R. 1985. Limiting resources and territoriality in microtine rodents. Am Nat 126: 1–15.

Palombit R. 1993. Extra-pair copulations in a monogamous ape. Anim Behav 47: 721–723.

Peres C. 1989. Costs and benefits of territorial defense in wild golden lion tamarins, *Leontopithecus rosalia*. Behav Ecol Sociobiol 25: 227–233.

Reichard U, Sommer V. 1997. Group encounters in wild gibbons (*Hylobates lar*): agonism, affiliation and the concept of infanticide. Behaviour 134: 1135–1174.

Slagsvold T, Dale S, Saetree G. 1994. Dawn singing in the great tit (*Parus major*): mate attraction, mate guarding or territorial defense? Behaviour 131: 11–138.

Sokal R, Rohlf J. 1981. Biometry. New York: Freeman and Co.

Southwood T. 1993. Ecological methods with particular reference to the study of insect populations. New York: Chapman and Hall.

Stanford C. 1991. The capped langur in Bangladesh: behavioral ecology and reproductive tactics. Contrib Primatol 26.

Tenaza RR. 1975. Territory and monogamy among Kloss' Gibbons (*Hylobates klossii*) in Siberut Island, Indonesia. Folia Primatol 24: 60–80.

Trivers R. 1972. Parental investment and sexual selection. In Campbell B, editor, Sexual selection and the descent of man, 136–179. Chicago: Aldine.

van Schaik CP, Assink PR, Salafsky N. 1992. Territorial behavior in southeast Asian langurs: resource defense or mate defense? Am J Primatol 25: 233–242.

Waser P.1977. Individual recognition, intragroup cohesion and intergroup spacing: evidence from sound playback to forest monkeys. Behaviour 60: 28–41.

Waser P. 1981. Sociality or territorial defense. The influence of resource renewal. Behav Ecol Sociobiol 8: 231–237.

White G, Garrott R. 1987. Analysis of wildlife radiotracking data. Academic Press: New York.

Whitten P. 1982. Female reproductive strategies among vervets. Ph.D. dissertation, Harvard University.

Whitten T, Mustafa M, Henderson G. 1987. The ecology of Sulawesi. Yogyakarta: Gadja Mada University Press.

World Wildlife Fund. 1980. Cagar Alam Gunung Tangkoko Dua Saudara Nature Reserve Sulawesi Utara Management Plan 1981–1986. Bogor: Indonesia.

CHAPTER 11

The Natural History
of the Philippine Tarsier
(*Tarsius syrichta*)

Marian Dagosto, Daniel L. Gebo, and Cynthia N. Dolino

Tarsius syrichta was the first tarsier to become known to Western science, the first to be studied in captivity, and the first to be exhibited in Western zoos. It is well represented in museum collections. Nevertheless, the Philippine tarsier has not been well studied in the wild. There are only limited published observations concerning its ecology and behavior, and it has never been the subject of a long-term ecological study. With the aim of rectifying this situation, in May and June of 1997 and 1998 we carried out behavioral observations of radio-collared tarsiers on the island of Leyte in the southern Philippines. Although these data are preliminary, we report our observations on positional behavior, substrate use, group size, and home range, and make comparisons to the other species in the genus.

We give a brief account of the history of discovery of this species. The taxonomic status of *Tarsius syrichta* and its island populations is reassessed using a much larger sample than in previous works. Current morphological evidence cannot sustain the hypothesis of subspecies within the Philippine tarsier, although specimens from Dinagat, Surigao, Samar, and Leyte are generally larger than southern Mindanao tarsiers. Scenarios of deployment to the Philippines are discussed. Geological, biogeographical, and paleontological evidence indicate that *Tarsius syrichta* is a relatively recent (late Miocene to mid-Pleistocene) immigrant to the Philippines from Borneo.

History of Discovery

In the late seventeenth century, Georg Joseph Camel, a Jesuit missionary, described a Philippine tarsier to James Petiver, who published Camel's description and fanciful drawings in his *Gazophylaceum* (Hill, 1955). Unfortunately, the very monkeylike depiction in the *Gazophylaceum* led Linnaeus (1758) to the mistaken conclusion that Camel's animal was indeed a monkey, and this account became the basis for his *Simia syrichta*. This error went uncorrected until 1923, when Cabrera (1923) pointed it out. In the meantime, both Buffon (1765) and Nau (1791) described what are almost certainly Philippine tarsiers (Hill, 1955), although they did not recognize the primate affinities of their specimens. Meyer's work (1894–95) first gave a specific name, *Tarsius philippensis*, to the Philippine variety. Two other species, *Tarsius carbonarius* (Heude, 1898) and *Tarsius fraterculus* (Miller, 1911), were later identified from Mindanao and Bohol, respectively.

Early descriptions of the behavior of Philippine tarsiers were made by Camel, Cuming (1838), and Whitehead (in Thomas, 1896). Cook (1939) and Lewis (1939) made more extensive observations of captive animals over sixty years ago. *Tarsius syrichta* was the first species whose physiology was studied in captivity (Catchpole and Fulton, 1939) and the first species to be exhibited in Western zoos. A live pair was brought to the laboratory of John Fulton at Yale in 1938 and became the basis for the study of respiration and the estrus cycle (Catchpole and Fulton, 1939, 1943; Fulton, 1939, 1943; Clarke, 1943). In 1947, Wharton (1948, 1950) brought thirty one tarsiers to the United States that were exhibited at the National Zoo, the Bronx Zoo, and the Brookfield Zoo. The Zoological Society of London received three of Wharton's animals and they became the basis for a series of studies by Hill (Hill, 1951, 1953a,b,c,d; Hill et al., 1952). More Philippine tarsiers were imported for study or exhibit at other American and European institutions (Ulmer, 1960, 1963; Evans, 1967; Schreiber, 1968; Haring and Wright, 1989; Haring et al., 1985).

Many of these specimens ended up in museum collections; however, the best and largest series of Philippine tarsiers was collected in 1946–47 by Harry Hoogstraal as part of the Field Museum's Philippine expedition (Hoogstraal, 1947, 1951; Schmidt, 1947; Sanborn, 1952). These specimens, all from the Davao region of Mindanao, are housed in the Field Museum of Natural History in Chicago and the Philippine National Museum in Manila.

Taxonomy

The convoluted history of tarsier taxonomy and nomenclature is thoroughly reviewed by Hill (1953a, 1955) and Niemitz (1984a). There have been episodes of extreme splitting (the same taxa named several times; all varieties recognized at the specific level) and extreme lumping (all species of tarsiers included under the name *Tarsius spectrum*). Hill (1953a, 1955), Niemitz (1977, 1984a), and Corbett and Hill (1986) conclude that three species exist: the Philippine tarsier (*Tarsius syrichta*), the Bornean tarsier (*T. bancanus*), and the Sulawesi tarsier (*T. spectrum*). Since these publications, three other species have been recognized or returned to specific status: *T. pumilus* (Miller and Hollister, 1921; Musser and Dagosto, 1987; Niemitz, 1985), *T. dianae* (Niemitz et al., 1991), and *T. sangirensis* (Meyer, 1896–97; Feiler, 1990; Shekelle et al., 1997; Groves, 1998), all from Sulawesi. The recent recognition of multiple species of tarsiers on Sulawesi and in other small nocturnal primates such as galagos and *Microcebus* (Bearder et al., 1995; Groves, 2000; Rasoloarison et al., 2000) has spurred us to reanalyze the taxonomic status of the Philippine tarsier. First, we examine its position relative to those from other island groups (i.e., is the Philippine tarsier a distinct species?), and then within the Philippines (i.e., are there geographically limited, morphologically distinct taxa within the Philippine tarsier?). We examined ninety three Philippine tarsiers, thirty eight Bornean tarsiers, and sixty Sulawesi tarsiers in the collections of the American Museum of Natural History, the Cleveland Museum of Natural History, the Delaware Museum of Natural History, the Field Museum in Chicago, the Philippine National Museum in Manila, and the United States National Museum of Natural History in Washington, D.C. Cranial and dental measurements were taken as defined in Musser and Dagosto (1987). Of the total sample, sixty two Philippine tasiers, thirty five Sulawesi tarsiers, and fifteen Bornean tarsiers were from adults and complete enough to be included in a discriminant function analysis.

Is the Philippine Tarsier a Distinct Species?

The conclusion of recent revisers that there are at least three species (or species groups) of tarsiers (i.e., the Sulawesi tarsier, the Bornean tarsier, and the Philippine tarsier) has recently been supported by a multivariate analysis of cranial and dental morphology conducted by Groves (1998; this volume, chapter 8). His work shows that the three forms can be clearly discriminated from one another even though the former species were represented by only a few individuals. Our larger sample completely confirms Groves' assessment (Dagosto, 2001). The discriminant function (fig. 11.1a) totally separates the three species (*T. pumilus* and *T. dianae* are not included in the sample), and no individuals are misclassified (table 11.1).

Some of the discrimination may be due to size. Based on both skull length and body mass, *T. spectrum* is slightly smaller than the Philippine or Bornean tarsiers (tables 11.2 and 11.3; Dagosto, 2001). Therefore, data

Table 11.1. Jack-knifed Classification Matrix of Discriminant Function of Raw Data

	T. bancanus	*T. spectrum*	*T. syrichta*	% correct
T. bancanus	15	0	0	100
T. spectrum	0	36	0	100
T. syrichta	0	0	62	100

NOTES: See figure 11.1. All specimens are correctly identified.

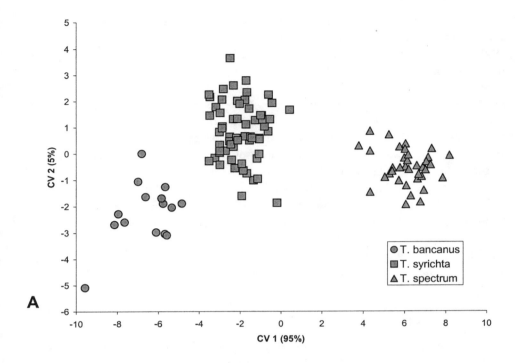

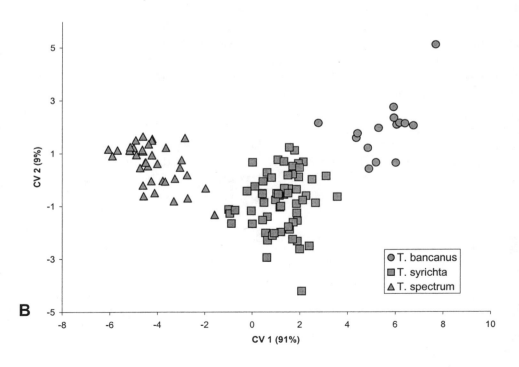

Table 11.2. Skull Length in Tarsier Species

	T. syrichta	T. bancanus	T. spectrum
Mean	38.93	38.38	37.22
SD	1.07	1.03	1.06
Range	36.60–41.93	36.40–40.55	35.50–39.50
N	67	22	47
Vs. T. bancanus	.044	—	.000
Vs. T. syrichta	—	.044	.000

NOTE: The last two rows give the p value of a t-test against the species named in the row heading.

were size "adjusted" by dividing by greatest skull length (GLS). After this transformation, the three groups are still almost completely separated (table 11.4; fig. 11.1); only one specimen of *T. spectrum* is misclassified as *T. syrichta*. In addition to size, the three species are distinguished by a cline in relative orbit size and tooth size from smallest (*T. spectrum*) to largest (*T. bancanus*) (Musser and Dagosto, 1987; Groves, 1998).

There are also differences among the species in aspects of postcranial anatomy. Sulawesi tarsiers have relatively short limb elements, *T. bancanus* has the relatively longest limb elements, and *T. syrichta* is intermediate (Niemitz, 1979a, 1984a; Musser and Dagosto, 1987; Dagosto et al., 2001; see "Positional Behavior" section below for further discussion). Based on analyses of both crania and postcrania, the Philippine tarsier is clearly morphologically distinct from both Sulawesi and Borneo tarsiers.

Are There Species or Subspecies within T. syrichta?

Within the Philippines, tarsiers are restricted to the Greater Mindanao faunal region (Heaney, 1985, 1986). Changes in sea level during the Pliocene and Pleistocene have separated the islands of this region several times, allowing opportunity for differentiation of populations. Three subspecies have been proposed: *T. syrichta syrichta*, from Leyte and Samar; *T. s. fraterculus* (Miller, 1911) from Bohol; and *T. s. carbonarius* (Heude, 1898) from Mindanao (see fig. 11.2). A specimen from Dinagat is larger than other Philippine tarsiers and may also be taxonomically distinct (Heaney and Rabor, 1982). On the other hand, the morphological grounds originally given for

Figure 11.1. (facing page) Discriminant function analysis (DFA) of log-transformed skull and dental dimensions as listed in Musser and Dagosto (1987). For *T. bancanus,* N = 15; for *T. syrichta,* N = 62; for *T. spectrum,* N = 36. (A) DFA of non-size corrected dimensions; (B) DFA of data size corrected using greatest length of skull. See also tables 11.1 and 11.4.

Table 11.3. Body Weights (in Grams) of Adult Tarsiers as Reported in the Literature

Species	Male mean	N	Range	Female mean	N	Range	Location	Reference
T. spectrum	110	1		113	1		Field	Niemitz et al., 1991
T. bancanus		6	120–132		11	94–114	Field	Gursky, 1997
		3	113–138		2	101–106	Captive	Roberts, 1994
	128	3		126.6	3		Captive	Wright et. al., 1986; Kappeler, 1990
				106.5	4		Captive	Stephan, 1984
	122							Spatz, 1984
	127.8	21	120–140	116.9	16	105–140	Field	Niemitz, 1979b, 1984a
		5	119–140		2	112–123	Field	Crompton and Andau, 1987
	112.5	2	100–125	117.5	2	110–125	Field	Davis, 1962[a]
T. syrichta	95	2		104	8		Davao	Wharton, 1950[b]
	125–145	1		100–122; 115	1		Davao Captive	Hill et al., 1952
				120–125	2		Davao, Captive	Ulmer, 1963
	120–135	1			1		Leyte, Field	Rickart et al., 1993
	123	1					Leyte, Field	Dagosto et al., 2001
	140.5	4	126–153				Surigao, Captive	Catchpole and Fulton, 1939; Clarke, 1943; Kennard and Willner, 1941[c]
	112–174	1		164–224	1		Mindanao and Bohol	Catchpole and Fulton, 1943
		11	85–183		13	83–182	Location unknown, captive	Evans, 1967
	122.6	5		110	3		Captive	Kappeler, 1990
	134.3	10		116.7	17			

NOTE: For each sex, the "mean" column gives the mean value for several animals, the singular value if N = 1, or the range of values for a single individual.

[a] Females were gravid.

[b] May have been young adults, weighed after 3–6 months in captivity.

[c] Listed as *Tarsius spectrum* in Kennard and Willner (1941) and Stephan (1984).

Table 11.4. Jack-knifed Classification Matrix of Discriminant Function with Data Size Corrected by the Geometric Mean (fig. 11.1)

	T. bancanus	T. spectrum	T. syrichta	% correct
T. bancanus	15	0	0	100
T. spectrum	0	35	1	97
T. syrichta	0	0	62	100

recognition of these subspecies are rather unconvincing, and both Hill (1953a, 1955) and Niemitz (1984a) doubt the existence of identifiable subspecies. With small samples, none could be established on morphometric grounds (Musser and Dagosto, 1987).

Our sample of Philippine tarsiers does not settle this issue since the vast majority of the specimens come from the Davao Gulf of Mindanao (see table 11.5). There are not enough specimens from other islands to perform a useful discriminant function analysis. The Davao collection, however, gives an excellent picture of a population of tarsiers against which specimens from other Philippine islands can be compared. To do this, a principal-components analysis on the covariance matrix of log-transformed variables was performed (Dagosto, 2001). This analysis indicates that the non-Davao tarsiers have higher PCA1 scores than the majority of the Davao Gulf specimens (fig. 11.3). The first principal component is often a size factor (Jolicoeur, 1963), meaning, perhaps, that non-Davao tarsiers are larger. Indeed, specimens from Leyte, Samar, Surigao (northeastern Mindanao), and Dinagat have skulls in the upper range of values for length (table 11.5). First axis score, however, correlates only poorly (r = .45) with geometric mean and even less well (r = .10) with GLS. In addition to (or perhaps instead of) size, high PCA1 scores indicate relatively large upper and lower first molar area.

To assess shape differences separately from size, we also performed the PCA on data size "adjusted" using the geometric mean (Jungers et al., 1995). This analysis (fig. 11.3) indicates that the Dinagat specimen is not shaped differently from Davao tarsiers, but the Bohol, Samar, and Leyte specimens score higher on PCA1 than the majority of Davao specimens. The variable that contributes most to the high score of these specimens on the first axis is the area of the first molar tooth.

Tarsiers from Leyte, Samar, Surigao, and especially Dinagat may on average be larger than Davao tarsiers. Tarsiers from Bohol are similar in size to Davao tarsiers (contra Miller, 1911), but may be shaped differently; like tarsiers from Samar and Leyte, they have relatively larger molars than Davao tarsiers. At present, the morphometric evidence is insufficient to support

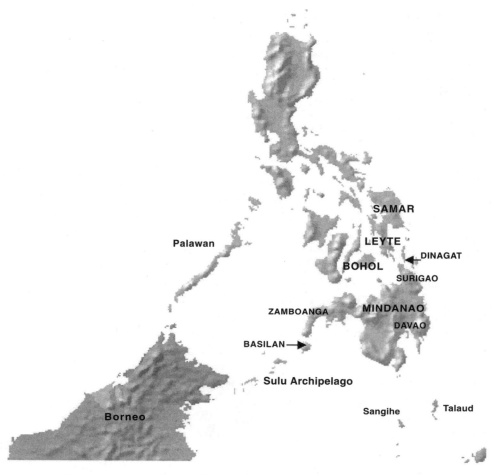

Figure 11.2. Map of the Philippine Islands showing areas inhabited by *Tarsius syrichta* (all capital letters) and other geographical features discussed in the text.

the recognition of morphologically distinct geographic subpopulations at any taxonomic level, but it does suggest that further investigation is warranted. It also implies that morphological variation is not simply island based. There may be a north-south cline in relative molar size and body size (with a secondary size decrease in the Bohol population).

Figure 11.3. (facing page) Principal components analysis of log-transformed skull and dental dimensions of *Tarsius syrichta* specimens. (A) Non-size-corrected data. (B) Data size corrected with the geometric mean. See figure 11.2 for location of samples.

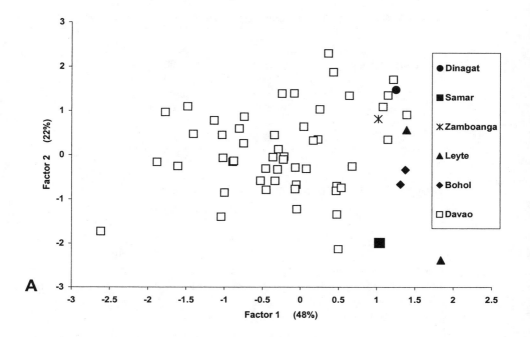

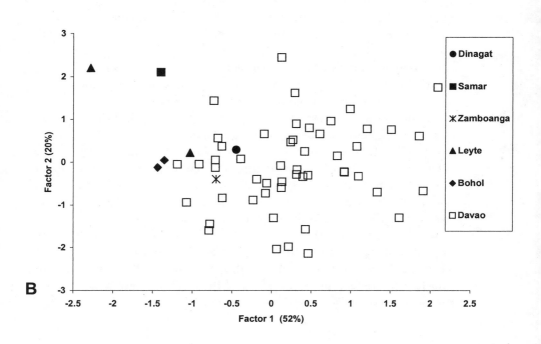

Table 11.5. Skull Length in *T. syrichta* from Different Geographic Locations

Area	Mean	Range	N
Davao, Mindanao	38.90	36.90–41.08	51
Bohol	38.22	38.00–38.44	2
Leyte	39.61	38.60–40.62	2
Samar	39.64	39.60–40.00	3
Surigao	39.80		1
Dinagat	42.91		1

Geographic Origin of Philippine Tarsiers

When and How Did Tarsiers Get to the Philippines?

Tarsiers are an ancient lineage. The oldest-known fossil tarsiers are from the Middle Eocene of China (Beard, 1998). By the middle Miocene, they are known from mainland Southeast Asia (Ginsburg and Mein, 1986). The fossil record of *Tarsius* indicates that this genus did not originate on any of the islands it currently inhabits but is of mainland Asian origin and reached them through dispersal or vicariance (Beard, 1998). Unfortunately, there is no fossil record of tarsiers from Sulawesi, Borneo, or the Philippines that can date a first occurrence. Geological evidence, however, helps constrain possibilities.

The geological history of the Philippines is extremely complex, but it is clear that the current configuration and location of land masses was not attained until quite recently (Hall, 1996, 1998). During the early Cenozoic the islands where tarsiers are currently found were located much farther east and south of mainland Asia than they are today, and there is no evidence of land bridges or other connections between these islands and other Asian islands or the mainland. There is virtually no possibility that tarsiers reached the Philippines by a vicariant event. Although the most western part of Mindanao (the Zamboanga peninsula) is geologically a part of the Sunda shelf onto which the main body of Mindanao was later (5 Ma) accreted (Hall, 1996), there is no evidence of emergent land until the late Miocene (Hamilton, 1979; Hall, 1998). It is also unlikely that there was emergent land before the Pliocene in Leyte and Samar (Heaney, 1986; Hall, 1998). Given the geologic history of the Philippine Islands and the amount and placement of emergent land during the Cenozoic, dispersal of tarsiers to the Philippines is very unlikely to have occurred before the late Miocene, and may have been much later. Detailed study of the Philippine rodent fauna (Heaney and Rickart, 1990; Musser and Heaney, 1992) implies several colonization events most likely occurring from 10 mya to 1 mya.

Where Did They Come From?

Previous authors have hypothesized both Borneo and Sulawesi as the likely place of origin of the Philippine tarsier. Like other islands of Southeast Asia, the Philippines have a rich endemic fauna; *Tarsius syrichta* is but one of many native species (Heaney and Regalado, 1998). The phylogenetic relationships of the majority of Philippine mammals, amphibians, reptiles, and freshwater fish lie strongly with Borneo (Brown and Alcala, 1970; Heaney, 1985, 1986). This is also true of tarsiers. The Philippine species is phenetically and cladistically more similar to the Bornean tarsier than to the Sulawesi forms (Groves, 1998; Musser and Dagosto, 1987; Niemitz, 1977). This strongly suggests that, like virtually all other Philippine mammals, *Tarsius syrichta* is an immigrant to the Philippines from Borneo. Davis (1962) and Musser and Dagosto (1987) have suggested that the current distribution and phylogenetic relationships of tarsiers results from successive waves of dispersal from the Sunda shelf with continuing evolution of the Sunda form. The earliest dispersal was of a primitive form to Sulawesi, a later dispersal sent a moderately evolved form to the Philippines, and the Sunda species continued to differentiate into the current Bornean form.

Although sea levels were lower during the late (12–16,000 years ago) and middle (160,000 years ago) Pleistocene, allowing narrower water barriers between Borneo and the Philippines, there is no evidence of a continuous land bridge during these times other than to Palawan (Heaney, 1985, 1991). The phylogenetic relations of freshwater fish suggest a Pliocene or early Pleistocene land bridge from Borneo to Mindanao, but the presence of this bridge has not been demonstrated geologically (Kornfield and Carpenter, 1984; Heaney, 1985). Likewise, the existence of land bridges between Borneo and the Philippines during the Pliocene or Miocene has not yet been established or refuted. It seems clear from analyses of both the rodent and shrew fauna that seawater barriers had to be crossed for colonization of the Philippines to occur (Heaney, 1986; Ruedi, 1996). Heaney (1986: 151) concludes that "it is conceivable, and indeed likely, that the Philippine fauna (except that of Palawan) is derived entirely from over-water immigrants."

It is most likely that tarsiers, like many other Philippine mammals, reached the Philippines through the Sulu rather than the Palawan archipelago (Groves, 1984). Although Palawan was directly connected to Borneo during the middle Pleistocene (Heaney, 1985, 1986) and shares many genera (96%) and species of mammal with it, none of the Bornean fauna of Palawan has reached other islands of the Philippines, and tarsiers do not occur on Palawan. Dispersal from Palawan to Greater Mindanao would likely involve transit through Mindoro or Panay, islands that do not harbor tarsiers today. No complete land bridge connected Borneo and Mindanao through the Sulu archipelago during the Pleistocene, but there were fewer and larger islands during periods of lower sea levels, including one that "extended from present day Tawi-Tawi to Jolo [which] filled most of the gap between Borneo and Mindanao" (Heaney, 1986: 143). The distance between this island and Greater Mindanao was probably less than 25 km.

Some authors (Niemitz, 1977; Crompton, 1989) have been critical of a derivation of Philippine tarsiers from Borneo, because (1) there are no tarsiers on the Sulu archipelago today, and (2) Sulawesi tarsiers are more primitive than the Bornean and Philippine forms, suggesting derivation from that island instead. However, tarsiers are found on Basilan, the island closest to Mindanao (Lawrence, 1939). The absence of tarsiers on the Sulu islands could be explained by a number of factors: rising sea levels resulting in reduced size of islands and probable changes in climatic conditions both of which could have resulted in local extinction (there are only seven species of mammals on the Sulu archipelago today). Heaney (1985, 1986) has documented the rapid extinction that takes place on small Philippine islands after isolation from a larger land mass.

Because Sulawesi tarsiers are morphologically more primitive than the Philippine and Bornean forms, Groves (1976), Niemitz (1977), and Crompton (1989) have also argued that a more likely scenario for tarsier dispersal involves the origin of tarsiers on Sulawesi, with later migration from this island group to Borneo and the Philippines. Niemitz (1977) proposed independent dispersal of Sulawesi tarsiers to Borneo and the Philippines; Groves (1976) hypothesized immigration from Sulawesi to Borneo, and then to the Philippines. These scenarios receive some support from Brandon-Jones (1998), who suggests the possibility of extirpation of tarsiers on Borneo during episodes of Pleistocene deforestation, and resupply from Sulawesi. This scheme, however, requires a rather late time of dispersal and thus very rapid morphological evolution of the Bornean tarsier, which is quite morphologically distinct from the Sulawesi species.

A route through the Sangihe and Talaud islands is mentioned as a possible Sulawesi-Philippines connection (Crompton, 1989). This possibility is based on the proximity of these islands to Mindanao (about 200 km) and on the presence of *Tarsius spectrum sangirensis* on the Sangihe islands. Meyer (1896–97) thought *T. s. sangirensis* had characteristics of both *T. spectrum* and *T. syrichta,* but this contention was refuted by Hill (1955), Niemitz (1977, 1984a) and Musser and Dagosto (1987), all of whom found this tarsier to be most similar to typical *T. spectrum. T. s. sangirensis* is larger than the average *T. spectrum* and has shorter lateral canines and incisors, a longer tooth row, a more sloping coronoid process, and finer fur (Feiler, 1990; Shekelle et al., 1997; Groves, 1998). Therefore, some authors recognize it as a species separate from *T. spectrum* (Shekelle et al. 1997; Groves, 1998), others as a sub-species of *T. spectrum* (Feiler, 1990; Musser and Dagosto, 1987). Regardless of its specific status, it does not seem to be more closely related to Philippine tarsiers than to Sulawesi tarsiers.

The Sangihe island route has all the same problems as other possibilities (Groves, 1976). The faunas of both Sangihe and Talaud are depauperate; there are no tarsiers on Talaud. There is no evidence of any past land bridge between these islands and the Philippines, although a more extensive island chain may have existed in the past (Late Miocene) (Moss and Wilson, 1998); it has been suggested as the dispersal route for the shared Sulawesi-Philippine clades of fanged frogs (Emerson et al., 2000). No other nonvolant Philippine mammal, with the possible exception of the rodent *Crunomys*, is uniquely shared between Sulawesi and the Philippines (Musser, 1982). *Crunomys*, however, does not occur on Sangihe or Talaud (Feiler, 1990).

Current geological, paleontological, biogeographic, and phylogenetic evidence favors the origin of tarsiers on mainland Asia with subsequent, almost certainly overwater, dispersal to Sulawesi and the Philippines. Philippine tarsiers are most probably immigrants from Borneo, through the Sulu archipelago, arriving sometime in the late Miocene to mid-Pleistocene.

Ecology and Behavior

Our study site is located at Mount Pangasugan, immediately east of the campus of the Visaysas State College of Agriculture on the island of Leyte. This site offers a variety of habitats from agricultural land, second growth, and disturbed forest at the base of the mountain (50–100 m), to undisturbed primary forest at higher elevations (300 m–1200 m) (Rickart et al., 1993). This work was done in the lower elevations (~100 m) in secondary forest. Using trained guides with hunting dogs, four male tarsiers were hand captured at their sleep sites. The tarsiers were weighed and fitted with radio-collars manufactured by Telonics. The collars weighed approximately 8 g, about 5–6% of the animal's body weight. Three tarsiers were recaptured and none lost or gained more than 2 g during the time of observation.

We conducted four to five dusk-to-dawn (17:30–05:00) and several (1–5) half-night (dusk to 22:00) follows of three of the collared tarsiers and collected some information on an uncollared individual. Survey time totaled 182 hours. During each of the all-night follows, we marked the location of the tarsier every fifteen minutes in order to determine home range and nightly minimum travel distance. When the animal was in view, bout sampling was employed to record the locomotor or postural mode, substrate type, height above the ground, and associated behavior (e.g., grooming, calling, feeding). Additional information can be found in Dagosto et al. (2001).

Group Size and Home Range

Field observers have reported that *T. syrichta* is often found in pairs (Cook, 1939; Rickart et al., 1993). In the captive setting, *T. syrichta* forms small groups (Hill et al., 1952; Haring et al., 1985), although Wharton (1950) reports that males do not tolerate other males well. The Leyte tarsiers, however, were solitary. The seven tarsiers we captured or otherwise observed were almost always alone at their sleep sites. Three of the four collared males were never observed within close proximity (< 3 m) to another tarsier. One male did share a sleep site with another (presumably female) tarsier, but only for three of eight nights of observation. At the sleep site, the two tarsiers were never closer than a meter and did not engage in grooming or other contact. We were able to follow both animals for about one hour after dusk, and for all three nights they went in different directions during the initial foraging period. We note here that our observations were made during the dry season only. Gursky (2000) has observed reduced intragroup encounters during the dry season in spectral tarsiers, and it is possible that the same phenomenon is at work in Leyte tarsiers. Nevertheless, the incidence of close physical encounters with other tarsiers during nightly travels or at sleep sites is extremely low in Leyte tarsiers.

Home range, as estimated by the "minimum home range method" (as defined in Bearder and Martin, 1979; most similar to the minimum convex polygon method), ranged from 0.6 ha to 2.0 ha. This is smaller than the 3–12 ha reported for *T. bancanus* (Crompton and Andau, 1987; Crompton, 1989), but similar to the 0.5–4 ha reported for *T. spectrum* and *T. dianae* (MacKinnon and MacKinnon, 1980; Niemitz, 1984b; Tremble et al., 1993; Gursky, 1995, 1997, 2000). In *T. bancanus* and *T. spectrum*, male home ranges are generally larger than those of females (Crompton and Andau, 1987; Crompton, 1989; Gursky, 1997, 2000). In *Tarsius spectrum*, home ranges reach their maximum in the dry season (Gursky, 2000). If the same trends hold for Leyte tarsiers, our values are likely to be maximum estimates.

Minimum nightly travel distance of the Leyte tarsiers ranged from 260 m to 556 m, which is less than the 1500–2000 m reported for *T. bancanus*, and the 300–900 m reported for *T. spectrum* (Gursky, 1997, 2000). The great difference between the Leyte tarsiers and the Bornean tarsiers may be partly due to the method of data collection, since we recorded location only every fifteen minutes, compared to every five minutes in Crompton's study. The difference between Leyte tarsiers and *T. spectrum*, however, cannot be attributed primarily to method, since Gursky's data is also based on fifteen-minute samples. In fact, since our data were collected during the dry season, and only on male tarsiers, a more appropriate comparison is the mean of approximately 825 m nightly path length for male spectral tarsiers in the dry season (Gursky, 2000).

It is impossible to determine precisely the social organization of *T. syrichta* with our data since our home-range calculations are limited to males and our study was conducted during a limited time frame. It seems very probable that this species does not have the small family groups typical of *T. spectrum* (Gursky, 1995), but instead a noyau system like *T. bancanus* (Crompton and Andau, 1987). It does, however, have small home ranges more similar to those of *T. spectrum* than to *T. bancanus*.

Positional Behavior

Compared to *T. bancanus*, *T. syrichta* has a relatively shorter hindlimb and hindlimb components and a relatively shorter hand (Niemitz, 1979a, 1984a; Musser and Dagosto, 1987). These elements are even shorter in *T. spectrum*. There is a significant difference between *T. bancanus* and *T. syrichta* in intermembral and humerofemoral indices (Dagosto et al., 2001). Thus, in terms of postcranial morphology, *T. syrichta* is intermediate in structure between *T. bancanus* and *T. spectrum*. Niemitz (1979a, 1984a) has proposed that these anatomical differences reflect differences in positional behavior. *T. bancanus* should be the most specialized, leaping the most frequently and using almost exclusively vertical supports. He expected that *T. syrichta* would use predominantly oblique supports, using leaping less frequently and quadrupedalism more frequently. MacKinnon and MacKinnon (1980) also stress the "less specialized" locomotor patterns and habitat use of *T. spectrum*. Preliminary data weakly support these conclusions: *T. spectrum, T. dianae*, and *T. syrichta* (captive data) leap slightly less than *T. bancanus* and use vertical supports less (table 11.6). It is, however, not possible to tell if these differences are significant, especially given the variation in frequency of behavior in a preliminary and longer-term study in *T. bancanus* (Crompton and Andau, 1986; Crompton, 1989). Data from *T. syrichta* in its natural habitat was collected to help test if these behavioral trends are indeed correlated with the anatomical trends (Dagosto et al., 2001).

A total of 1259 bouts of locomotor behavior was collected during our study (table 11.6). Leaping made up 58% of these bouts, climbing 26%, and quadrupedalism 10.8%. These values are generally similar to those of other tarsiers (MacKinnon and MacKinnon, 1980; Crompton and Andau, 1986; Crompton, 1989; Tremble et al., 1993). Assuming that behavioral data collection methods are similar, differences in frequency of locomotor behaviors does not seem to explain the anatomical differences among tarsier species.

Support use is estimated from 971 observations (table 11.6). Vertical supports predominate (64%), followed by obliques (20%) and horizontals

Table 11.6. Locomotor and Support Use Frequencies for Tarsiers

	T. spectrum	T. dianae	T. bancanus	T. syrichta (in captivity)	T. syrichta (wild)
Leap	63	58	66/54	61	58
Climb	14	22	27/35	28	26
Quad	?	11	3/3	?	10.8
Other/Unaccounted for	23	9	4/8	11	5.2
Vertical	~58	59	72/65	32	64
Oblique	~28	32	21/30	37	20
Horizontal	~15	10	1/5	30	10
Other/Unaccounted for	0	0	6/0	1	6

SOURCES: Data for *T. spectrum* are from MacKinnon and MacKinnon (1980); for *T. dianae* from Tremble et al. (1993); for *T. bancanus* from Crompton and Andau (1986) and Crompton (1989); for *T. syrichta* (captive) the locomotor data are from Gebo (1987) and the support data from Reason (1978); the *T. syrichta* wild data are from this study.

(10%). All tarsiers occasionally come to the ground to capture prey but rarely spend more than a few seconds there. The vast majority of supports used were very small (less than 5 cm). The preference of Leyte tarsiers for small vertical supports is similar to what has been described for *T. bancanus* in captivity (Roberts and Kohn, 1993; Roberts and Cunningham, 1986) and in the wild (Crompton and Andau, 1986). These data contrast with Reason's (1978) observations of *T. syrichta* in captivity, where he found them to use verticals much less (37.7%) and horizontals more (23.7%) than the Leyte tarsiers.

Sleep Sites and Ranging Behavior

Each tarsier used several (3–4) sleep sites located near the borders of its home range. Sleep sites are typically located in the dense undergrowth of ferns, saplings, and bamboo surrounding a large tree or trees (primarily *Artocarpus, Pterocarpus,* or *Ficus*). The undergrowth provides a cool, dark, moist, and sheltered sleeping area. The sleep sites of the Leyte tarsiers were generally low to the ground (all less than one meter high, except one site which was 3–4 m high in a tangle of palm fronds). Mindanao tarsiers are also reported to sleep close to the ground (Hoogstraal, 1951; Wharton, 1950). In contrast, Bornean tarsiers use sleep sites that are 3.5–5 m high (Niemitz, 1979b; Crompton and Andau, 1987), and those of spectral tarsiers may be even higher (Gursky, 1997).

Foraging and traveling generally took place very low to the ground. Sixty-five percent of observations were at 2 m or less, and 23% at less than 1 m (Dagosto et al., 2001). These numbers are somewhat biased because tarsiers are nearly impossible to observe when they climb higher than 5 m into a tree, which they often do to rest or when they are disturbed. This emphasis on the near-ground niche is also reported for other species. *Tarsius bancanus* spent 76% of time below 1.5 m (Crompton and Andau, 1986), and *T. dianae* 50% (Tremble et al., 1993). Niemitz (1984b) reports that *T. spectrum* was observed mostly between 1 and 3 meters. There seems to be a cline in tarsiers from the most specialized for the near-ground niche (*T. bancanus*) to the least (*T. spectrum* and *T. dianae*), with *T. syrichta* being intermediate.

Diet

In captivity, Philippine tarsiers have survived successfully on a diet of crickets and small lizards (Cook, 1939; Catchpole and Fulton, 1939; Haring and Wright, 1989). On Leyte, most of tarsier foraging activity took place near to the ground. The tarsiers clung to a small vertical support and scanned the ground and nearby foliage for prey. Although we observed numerous strikes, it was extremely difficult to determine the exact nature of the prey, because most items taken were small enough to be grabbed and completely eaten within a second or two. One exception was the capture of a rather large orthopteran. We note again that our observations are limited to the dry season. There is evidence for seasonal dietary shifts in spectral tarsiers (Gursky, 2000). During the dry season, spectral tarsiers forage more from the ground and eat less orthopterans and lepidopterans and more beetles, ants, and termites.

Conservation

Philippine tarsiers are currently classified as "data deficient" by the IUCN. The extensive degree of habitat destruction in the Philippines is cause for concern for the continued survival of its endemic fauna (Heaney, 1993; Wildlife Conservation Society of the Philippines, 1997). Given the difficulty of surveying small, nocturnal, arboreal animals, it is difficult to estimate population density or the remaining numbers of *T. syrichta*. A brief study conducted in Bohol calculated densities of 1–3 tarsiers per hectare from transect sampling (Lagapa, 1993). The home-range size of Leyte tarsiers yields a similar density estimate of 0.5–2 tarsiers per hectare. If tarsier densities are similar in primary forest—and 80,000 hectares of primary forest remain in Leyte (Margraf and Milan, 1996)—the contention of Heaney and Utzurrum (1991) that tarsiers may be fairly common is supported.

There are several factors, however, that might significantly influence such an estimate. First, our home-range estimates are very preliminary; we have no data on female home ranges, nor any knowledge of the degree of overlap of individual home ranges. Second, much of the remaining primary forest in the Philippines is at high elevations on steep slopes, and it is not known if tarsiers even inhabit these areas. Previous workers report that tarsiers are more common at low elevations in secondary forest than in primary forest (Fulton, 1939; Wharton, 1948; Hoogstraal, 1951; Heaney et al., 1989). Despite at least three vertical transect studies (on Mount Apo and Mount McKinley [Hoogstraal, 1951] and Mount Pangasugan [Rickart et al., 1993]), there are no collection records of Philippine tarsiers from altitudes greater than 800 m. Unfortunately, no good estimates of secondary forest area exist, and, in any case, all secondary forest is not suitable tarsier habitat. In our study area, large trees (trunks > 40 cm diameter) always anchor the sleeping and feeding areas and are used as rest areas during the night. Areas of dense undergrowth are the preferred foraging areas. We were unable to locate tarsiers in areas where most big trees and underbrush were cleared. Both Mindanao and Samar have large tracts of primary and secondary forest (Caldecott et al., 1997). Bohol, however, has virtually no remaining primary forest, and tarsiers may have become locally extinct in some areas (Evenhouse, pers. com.). Tarsiers inhabit many other small islands in the Philippines (e.g., Maripipi, Dinagat, Basilan). Surveys of these small and large islands need to be conducted in order to accurately determine the conservation status of *T. syrichta*.

Acknowledgments

We thank the editors of this volume for the opportunity to contribute. We gratefully acknowledge the cooperation and aid of personnel at ViSCA-GTZ, DENR Section 8, PAWB, and the PTFI. Special thanks go to Jesus B. Alvarez, Peter Balzer, Lawrence Heaney, and Paciencia Milan. The National Geographic Society, Primate Conservation, Inc., and Northwestern University provided funds for the study.

References

Beard KC. 1998. A new genus of Tarsiidae (Mammalia: Primates) from the middle Eocene of Shanxi Province, China, with notes on the historical biogeography of tarsiers. In Beard KC, Dawson M, editors, Dawn of the age of mammals in Asia, 260–277. Pittsburgh: Carnegie Museum of Natural History.

Bearder SK, Martin RD. 1979. The social organization of a nocturnal primate revealed by radiotracking. In MacDonald D, Amlaner C, editors, A handbook on biotelemetry and radio tracking, 633–648. Oxford: Pergamon Press.

Bearder SK, Honess PE, Ambrose L. 1995. Species diversity among galagos with special reference to mate recognition. In Alterman L, Doyle GA, Izard MK, editors, Creatures of the dark, 331–352. New York: Plenum Press.

Brandon-Jones D. 1998. Pre-glacial Bornean primate impoverishment and Wallace's line. In Hall R, Holloway JD, editors, Biogeography and geological evolution of SE Asia, 393–403. Leiden: Backhuys Publishers.

Brown WC, Alcala AC. 1970. Zoogeography of the herpetofauna of the Philippine islands, a fringing archipelago. Proc Calif Acad Sci 38: 105–130.

Buffon GLL. 1765. Histoire naturelle, génerale, et particulière, vol. 13. Paris: L'Imprimerie du Roi.

Cabrera A. 1923. On the identificiation of *Simia syrichta* Linnaeus. J Mammal 4: 89–91.

Caldecott JO, Piczon EC, Aliposa JS, de la Cruz M. 1997. Towards a biodiversity management for Samar Island. Paper presented at the 5th annual Philippine Vertebrate Biodiversity Symposium, April 22–25, 1997, Los Baños, Philippines.

Catchpole HR, Fulton JF. 1939. Tarsiers in captivity. Nature 144: 514.

Catchpole HR, Fulton JF. 1943. The oestrus cycle in *Tarsius:* observations on a captive pair. J Mammal 24: 90–93.

Clarke RW. 1943. The respiratory exchange of *Tarsius spectrum.* J Mammal 24: 94–96.

Cook N. 1939. Notes on captive *Tarsius carbonarius.* J Mammal 20: 173–177.

Corbett, Hill. 1986

Crompton RH. 1989. Mechanisms of speciation in *Galago* and *Tarsius.* J Hum Evol 4: 105–116.

Crompton RH, Andau PM. 1986. Locomotion and habitat utilization in free-ranging *Tarsius bancanus:* a preliminary report. Primates 27: 337–355.

Crompton RH, Andau PM. 1987. Ranging, activity, rhythms, and sociality in free-ranging *Tarsius bancanus:* a preliminary report. Int J Primatol 8: 43–71.

Cuming H. 1838. Letter to the Zoological Society of London. Proc Zool Soc London 1838: 67–68.

Dagosto M. 2001. Geographic variation of skull size and shape in *Tarsius syrichta.* Am J Phys Anthropol 114(S32): 57.

Dagosto M, Gebo DL, Dolino CN. 2001. Positional behavior and social organization of the Philippine tarsier (*Tarsius syrichta*). Primates 42: 101–111.

Davis DD. 1962. Mammals of the lowland rain-forest of North Borneo. Bull Nation Mus Singapore 31: 53–57.

Emerson SB, Inger RF, Iskandar D. 2000. Molecular systematics and biogeography of the fanged frogs of Southeast Asia. Mol Phylo and Evol 16: 131–142.

Evans CS. 1967. Maintenance of the Philippine tarsier in a research colony. Int Zoo Yrbk 7: 201–202.

Feiler A. 1990. Über die Säugetiere der Sangihe- und Talaud-Inseln der Beitrag A. B. Meyers für ihre Erforschung. Zool Abh Statt Mus für Tierkunde, Dresden 46: 75–94.

Fulton JF. 1939. A trip to Bohol in quest of *Tarsius.* Yale J Biol Med 11: 561–573.

Fulton JF. 1943. A trip to Bohol in quest of *Tarsius.* Am Sci 31: 151–167.

Ginsburg L, Mein P. 1986. *Tarsius thailandica* nov. sp., Tarsiidae (Primates, Mammalia) fossile d'Asie. Comptes Rendu 304: 1213–1215.

Groves CP. 1976. The origin of the mammalian fauna of Sulawesi (Celebes). Z Saugtierkunde 41: 201–216.

Groves CP. 1984. Mammal faunas and the paleogeography of the Indo-Australian region. Cour Forsch-Inst Senckenberg 69: 267–273.

Groves CP. 1998. Systematics of tarsiers and lorises. Primates 39: 13–27.

Groves CP. 2000. The genus *Cheirogaleus:* unrecognized biodiversity in dwarf lemurs. Int J Primatol 21: 943–962.

Gursky S. 1995. Group size and composition in the spectral tarsier, *Tarsius spectrum:* implications for social organization. Trop Biodiv 3: 57–62.

Gursky S. 1997. Modeling maternal time budgets: the impact of lactation and gestation on the behavior of the spectral tarsier, *Tarsius spectrum.* Ph.D. dissertation, State University of New York at Stony Brook.

Gursky S. 2000. Effect of seasonality on the behavior of an insectivorous primate, *Tarsius spectrum.* Int J Primatol 21: 477–496.

Hall R. 1996. Reconstructing Cenozoic SE Asia. In Hall R, Blundell D, editors, Tectonic evolution of Southeast Asia, 153–184. Geological Society Special Publication no. 106.

Hall R. 1998. The plate tectonics of Cenozoic SE Asia and the distribution of land and sea. In Hall R, Holloway JD, editors, Biogeography and geological evolution of SE Asia, 99–124. Leiden: Backhuys Publishers.

Hamilton W. 1979. Tectonics of the Indonesian region. Geol Survey Prof Papers 1078: 1–345.

Haring DM, Wright PC. 1989. Hand raising a Philippine tarsier, *Tarsius syrichta.* Zoo Biol 8: 265–274.

Haring DM, Wright PC, Simons EL. 1985. Social behavior of *Tarsius syrichta* and *Tarsius bancanus.* Am J Phys Anthro 66: 179.

Heaney LR. 1985. Zoogeographic evidence for middle and late Pleistocene land bridges to the Philippine Islands. Mod Quat Res SE Asia 9: 127–143.

Heaney LR. 1986. Biogeography of mammals in SE Asia: estimates of rates of colonization, extinction, and speciation. Biol J Linn Soc 28: 127–165.

Heaney LR. 1991. A synopsis of climatic and vegetational change in Southeast Asia. Climatic Change 19: 53–61.

Heaney LR. 1993. Biodiversity patterns and the conservation of mammals in the Philippines. Asia Life Sci 2: 261–274.

Heaney LR, Rabor DS. 1982. Mammals of Dinagat and Siargao Islands, Philippines. Occas Pap Mus Zool, University of Michigan 699: 1–30.

Heaney LR, Regalado JC. 1998. Vanishing treasures of the Philippine rainforest. Chicago: Field Museum Press.

Heaney LR, Rickart EA. 1990. Correlations of clades and clines: geographic, elevational, and phylogenetic distribution patterns among Philippine mammals. In Peters G, Hutterer R, editors, Vertebrates in the tropics, 321–332. Bonn: Museum Alexander Koenig.

Heaney LR, Utzurrum RCB. 1991. A review of the conservation status of Philippine land mammals. Assoc Syst Biol Philippines, Commun 3: 1–13.

Heaney LR, Heideman PD, Rickart EA, Utzurrum RB, Klompen JSH. 1989. Elevational zonation of mammals in the central Philippines. J Tropical Ecol 5: 259–280.

Heude PM. 1898. Mémoires concernant l'Histoire Naturelle de l'empire Chinois par des Peres de la compagnie de Jesus. Mem Hist Nat Emp Chin 4: 155–208.

Hill WCO. 1951. Epigastric gland of *Tarsius*. Nature 167: 994.

Hill WCO. 1953a. Note on the taxonomy of the genus *Tarsius*. Proc Zool Soc Lond 122: 13–16.

Hill WCO. 1953b. The blood-vascular system of *Tarsius*. Proc Zool Soc Lond 123: 655–694.

Hill WCO. 1953c. Caudal cutaneous specializations in *Tarsius*. Proc Zool Soc Lond 123: 17–26.

Hill WCO. 1953d. The female reproductive organs of *Tarsius*, with observations on the physiological changes therein. Proc Zool Soc Lond 123: 589–598.

Hill WCO. 1955. Primates: Comparative anatomy and taxonomy, vol. 2: Haplorhini, Tarsioidea. Edinburgh, UK: Edinburgh University Press.

Hill WCO, Porter A, Southwick MD. 1952. The natural history, endoparasites and pseudoparasites for the tarsiers (*Tarsius carbonarius*), recently living in the Society's menagerie. Proc Zool Soc Lond 1952: 79–117.

Hoogstraal H. 1947. The inside story of the tarsier. Chicago Natural History Museum Bulletin 18: 7–8, 4–5.

Hoogstraal H. 1951. Philippine Zoological Expedition 1946–1947: Narrative and Itinerary. Fieldiana: Zoology 33: 1–86.

Jolicoeur P. 1963. The multivariate generalization of the allometry equation. Biometrics 19: 497–499.

Jungers WL, Falsetti AB, Wall CE. 1995. Shape, relative size, and size-adjustments in morphometrics. Yrbk Phys Anthro 38: 137–161.

Kappeler PM. 1990. The evolution of sexual size dimorphism in prosimian primates. Am J Primatol 21: 201–214.

Kennard MA, Willner MD. 1941. Weights of brains and organs of 132 New and Old World monkeys. Endocrinology 28: 977–984.

Kornfield I, Carpenter KE. 1984. The cyprinids of Lake Lanao, Philippines: taxonomic validity, evolutionary rates, and speciation scenarios. In Echelle AA, Kornfield I, editors, Evolution of fish species flocks, 69–84. Orono: University of Maine Press.

Lagapa EPG. 1993. Population density, estimate, and habitat analysis of the Philippine tarsier (*Tarsius syrichta*, Linnaeus) in Bohol. Bachelor's thesis, University of the Philippines, Los Baños.

Lawrence B. 1939. Collections from the Philippine Islands: Mammals. Bull Mus Comp Zool Harvard 86: 28–73.

Lewis G. 1939. Notes on a pair of tarsiers from Mindanao. J Mammal 20: 57–61.

Linnaeus C. 1758. Systema Naturae per Regna Tria Naturae, Secundum Classes, Ordines, Genera, Species cum Characteribus, Differentiis, Synonymis, Locis. 10th ed. Halmiae: Laurentii Salvii.

MacKinnon J, MacKinnon K. 1980. The behavior of wild spectral tarsiers. Int J Primatol 1: 361–379.

Margraf J, Milan P. 1996. Ecology of Dipterocarp forests and its relevance for island rehabilitation in Leyte, Philippines. In Schulte A, Schöne D, editors, Dipterocarp forest systems: towards sustainable management, 124–154, New Jersey: World Scientific.

Meyer AB. 1894–95. Eine neue *Tarsius* art. Abh. u. ber. der K Zool u Anth RG-Ethn Mus Dresden 1: 1–12.

Meyer AB. 1896–97. Säugethiere vom Célebes-und Philippinen-Archipelago. Abh uber der K Zool u Anth RG-Ethn Mus Dresden 6: 1–36.

Miller GS. 1911. Descriptions of two new genera and sixteen new species of mammals from the Philippine islands. Proc USNM 38: 391–404.

Miller GS, Hollister N. 1921. Twenty new mammals collected by H. C. Raven in Celebes. Proc Biol Soc Wash 34: 67–76.

Moss SJ, Wilson MEJ. 1998. Biogeographic implications of the Tertiary palaeogeographic evolution of Sulawesi and Borneo. In Hall R, Holloway JD, editors, Biogeography and geological evolution of SE Asia, 133–155. Leiden: Backhuys Publishers.

Musser GG. 1982. Results of the Archbold expeditions. No. 110. *Crunomys* and the small-bodied shrew-rats native to the Philippine Islands and Sulawesi (Celebes). Bull Am Mus Nat Hist 174: 1–95.

Musser GG, Dagosto M. 1987. The identity of *Tarsius pumilus,* a pygmy species endemic to the montane mossy forests of Central Sulawesi. Am Mus Nov 2867: 1–53.

Musser GG, Heaney LR. 1992. A review of Philippine Muridae. Bull Am Mus Nat Hist 211: 1–138.

Nau B. S. 1791. Naturforscher (Halle) 25: 551–556.

Niemitz C. 1977. Zur Funktionsmorphologie und Biometrie der Gattung *Tarsius,* Storr, 1780. Cour Forsch-Inst Senckenberg 25: 1–161.

Niemitz C. 1979a. Relationships among anatomy, ecology, and behavior: a model developed in the genus *Tarsius,* with thoughts about phylogenetic mechanisms and adaptive interactions. In Morbeck M, Preuschoft H, Gomberg N, editors, Environment, behavior, and morphology: dynamic interactions, 119–138. New York: Gustav-Fischer-Verlag.

Niemitz C. 1979b. Results of a field study on the Western tarsier (*Tarsius bancanus borneanus* Horsfield) in Sarawak. Sarawak Museum Journal 27: 171–228.

Niemitz C. 1984a. Taxonomy and distribution of the genus *Tarsius* Storr 1780. In Niemitz C, editor, The biology of tarsiers, 1–16. Stuttgart: Gustav-Fischer-Verlag.

Niemitz C. 1984b. An investigation and review of the territorial behaviour and social organization of the genus *Tarsius.* In Niemitz C, editor, Biology of tarsiers, 117–128. New York: Gustav-Fischer-Verlag.

Niemitz C. 1985. Der Koboldmaki. Naturwiss Rund 38: 43–19.

Niemitz C, Nietsch A, Warter S, Rumpler Y. 1991. *Tarsius dianae:* A new primate species from Central Sulawesi (Indonesia). Folia Primatol 56: 105–116.

Rasoloarison RM, Goodman SM, Ganzhorn JU. 2000. Taxonomic revision of mouse lemurs (*Microcebus*) in the western portions of Madagascar. Int J Primatol 21: 963–1020.

Reason RC. 1978. Support use behavior in Mindanao tarsiers (*Tarsius syrichta carbonarius*). J Mammal 59: 205–206.

Rickart EA, Heaney LR, Heideman PD, Utzurrum RCB. 1993. The distribution and ecology of mammals on Leyte, Biliran, and Maripipi Islands, Philippines. Fieldiana: Zoology 72: 1–62.

Roberts M. 1994. Growth, development, and parental care in the Western tarsier (*Tarsius bancanus*) in captivity: evidence for a "slow" life history and non-monogamous mating system. Int J Primatol 15: 1–28.

Roberts M, Cunningham B. 1986. Space and substrate use in captive Western tarsiers, *Tarsius bancanus*. Int J Primatol 7: 113–130.

Roberts M, Kohn F. 1993. Habitat use, foraging behavior and activity patterns in reproducing Western tarsiers, *Tarsius bancanus* in captivity. Zoo Biol 12: 217–232.

Ruedi M. 1996. Phylogenetic evolution and biogeography of Southeast Asian shrews (genus *Crocidura*: Soricidae). Biol J Linn Soc 58: 197–219.

Sanborn CC. 1952. Philippine Zoological Expedition 1946–1947: Mammals. Fieldiana: Zoology 33: 87–158.

Schmidt KP. 1947. Pangolins, tarsiers, and flying lemurs of the Philippines. Chicago Nat Hist Mus Bull 18: 1–3.

Schreiber GR. 1968. A note on keeping and breeding the Philippine tarsier at Brookfield Zoo Chicago. Int Zoo Yrbk 8: 114–115.

Shekelle M, Leksono SM, Ichwan LLS, Masala Y. 1997. The natural history of the tarsiers of North and Central Sulawesi. Sulawesi Primate Newsletter 4.

Spatz WB. 1968. Die Bedeutung der Augen für die sagittale Gestaltung des Schädels von *Tarsius* (Prosimiae, Tarsiiformes). Folia Primatol 9: 22–40.

Stephan H. 1984. Morphology of the brain in *Tarsius*. In Niemitz C, editor, The biology of tarsiers, 319–344. Stuttgart: Gustav-Fischer-Verlag.

Thomas O. 1896. Trans Zool Soc Lond 14.

Tremble M, Muskita Y, Supriatna J. 1993. Field observations of *Tarsius dianae* at Lore Lindu National Park, Central Sulawesi, Indonesia. Trop Biodivers 1: 67–76.

Ulmer FA. 1960. A longevity record for the Mindanao tarsier. J Mammal 41: 512.

Ulmer FA. 1963. Observations on the tarsier in captivity. Deutsche Zool Garten 27: 106–121.

Wharton CH. 1948. Seeking Mindanao's strangest creatures. Natl Geogr 94: 389–399.

Wharton CH. 1950. The tarsier in captivity. J Mammal 31: 260–268.

Wildlife Conservation Society of the Philippines. 1997. Philippine red data book. Makati City, Philippines: Bookmark, Inc.

Wright PC, Izard MK, Simons EL. 1986. Reproductive cycles in *Tarsius bancanus*. Am J Primatol 11: 207–215.

CHAPTER 12

Can We·Predict Seasonal Behavior and Social Organization from Sexual Dimorphism and Testes Measurements?

Patricia C. Wright, Sharon T. Pochron, David H. Haring, and Elwyn L. Simons

In the literature, no agreement exists concerning the social organization or breeding system of tarsiers. Tarsiers were first described as organized into a "noyau system," like other nocturnal primates such as galagos with one male overlapping several territorial females (Niemitz, 1979; Bearder, 1987). Primatologists were surprised in the early 1980s, when, after following marked animals, MacKinnon and MacKinnon (1980) described tarsiers as gibbon-like models of monogamy. Others, using radio-collared or semicaptive animals, then observed tarsier groups as having one male with one or two females (Niemitz, 1984; Crompton and Andau, 1986). Recent papers suggest that tarsiers are polygynous, forming almost baboonlike multimale, multi-female groups (Gursky, 1995, 1997, in press; Bearder, 1999). Although *Tarsius bancanus* and *T. spectrum* have been studied in the wild, extensive behavioral field studies of *T. syrichta* have not been conducted. *T. bancanus* is described as monogamous, polygamous, or solitary (Niemitz, 1984; Cromption and Andau, 1986), while *T. spectrum* is described as both monogamous and polygynous (MacKinnon and MacKinnon, 1980; Gursky, in press). From group censuses of sleeping sites, *T. syrichta* organization is described as pairs (Cook, 1939; Rickart et al. 1993) or solitary and always alone at sleep sites (Dagosto et al., chapter 11, this volume).

   One objective of this study was to use morphological data collected in the wild and in captivity to describe the social system of two species of tarsiers: *Tarsius bancanus* and *T. syrichta*. First, we compared the body masses of both species. Sexual-selection theory predicts that differences in body mass between males and females are correlated, with strong male-male competition associated with highly polygynous mating systems (Darwin, 1871; Alexander et al., 1979). If sexual dimorphism exists in tarsier body size, we expect a polygynous social system; but if males and females are monomorphic, we expect a monogamous system.

Next, we examined testicles of both *Tarsius bancanus* and *T. syrichta* for evidence of sperm competition. Larger testes relative to body size have been associated with sperm competition in polygynous mating systems, whereas small testes relative to body mass are associated with monogamy (Harvey and Harcourt, 1984; Glander et al., 1992). If *T. syrichta* and *T. bancanus* have small testes relative to body size, we predict a monogamous system; while if they have large testes, we predict a polygynous system.

A second objective of this paper is to present for the first time the provenance of the individuals with the body mass measurements. Rarely have males and females been captured at the same site for a comparison of sexual dimorphism. In this paper, we compare wild-caught weights of both males and female *Tarsius bancanus* from Sabah, East Malaysia, and *T. syrichta* from Leyte, Philippines.

A third question posed in tarsier biology is: Are tarsiers seasonal breeders? Because testicle size changes with season in seasonal breeders (Pochron et al., 2002), this is an important question. By examining an extensive collection of preserved tarsier specimens, Hubrecht (1908) suggested that tarsiers give birth during nearly every month of the year and were not seasonal breeders. Ulmer (1963) suggested that tarsiers are seasonal breeders with births in May and December. To further gain insights into whether *Tarsius syrichta* or *T. bancanus* breed seasonally, we examined births and testicular volume taken monthly over six years in captivity at Duke University Primate Center.

Methods and Study Sites

Sandakan, East Malaysia

From September 13, 1983, until November 13, 1983, PCW visited Sepilok Reserve to capture twelve *Tarsius bancanus*. Ten Sepilok Park game rangers assisted in mist netting the tarsiers using ten mist nets for twenty eight netting nights. For each site, nets were erected at dusk and left open until dawn. Nets were checked for tarsiers every ninety minutes until dawn. These mist-netting sites were located in forested areas outside but surrounding the National Reserve of Sepilok.

Each tarsier was placed in a cotton drawstring bag and weighed with a Pesola spring balance (maximum reading: 500 g). Testicles were measured with calipers, and the reproductive condition of females was checked by palpation and condition of the nipples. Age estimation is based on dental eruption and wear. It should be noted that the fires that raged over Borneo in 1983 had burned secondary forests adjacent to many of the forests where these tarsiers were caught.

The Philippines

The Philippine tarsiers were captured by Mr. Baquilod during May on one of the Visaya islands. These twenty one *T. syrichta* were transported by PCW on board an airplane in the passenger section in two cat carriers. The carriers were partitioned into ten compartments each, one compartment for each tarsier, except one which held a mother and an infant. The tarsiers were fed insects and lizards and given eye droppers filled with water during the flight (Haring and Wright, 1989; Wright et al., 1989).

Duke University Primate Center, Durham, N.C., USA

The tarsiers were kept as male-female pairs. Each pair was housed in a room measuring 2 m long × 2 m wide × 2.75 m high, each of which was furnished with ten vertical bamboo poles, a small living fig tree, and vines harvested from the surrounding forest and replaced every six months. The tarsiers were housed under fluorescent lights with a nonfluctuating photoperiod (12-12 light/dark cycle). The tarsiers were fed only live animal prey including a diet of crickets and mealworms, supplemented monthly with anole lizards (*Anolis carolinensis*) and seasonally with large, wild-caught North Carolina orthopterans and an occasional laboratory-reared newborn mouse weighing less than 8 g. They drank fresh water from aluminum bowls. The room was misted three times a day to increase humidity and provide drinking water for the tarsiers and their prey (Wright et al., 1989).

Tarsius bancanus individuals were weighed and measured upon release into the DUPC enclosures on November 13, 1983, and periodically after that until their death. *T. syrichta* individuals brought back from the expedition to the Philippines were weighed and measured before their release into enclosures May 13, 1985, and periodically until they died. Two *T. syrichta* from Skansen Aquarium, Sweden, the first tarsiers to arrive at the DUPC, were weighed and measured before release into their cages at DUPC on May 13, 1983, (Amo Dulom) and February 26, 1984 (Saimon). The animals were kept on a twelve-hour timer schedule as wild populations of both are found within a few degrees of the equator (Haring and Wright, 1989; Wright et al., 1989).

Measurements

We measured body weights and testicular volume of both species of tarsiers at capture and in captivity over a six-year period. Both males and females of each species were caught at the same time in the same region in Sandakan or Philippines, avoiding the bias of geographic variation. See table 12.1.

Testes measurements and body weights of males were taken periodically

Table 12.1. Wild-Caught Weights of *Tarsius bancanus* in Sabah, East Malaysia, October 11–November 10, 1983

No.	Sex	Capture (gm)
52	F	126
53	M	116
54	F	124
55	M	125
56	F	122
57	M	128
57	F	140
58	M	136
59	F	125
60	M	134
61	F	132
91	M	126

SOURCE: Wright et al. 1987.
NOTE: The first column represents DUPC numbers of individuals.

over a six-year period. Three male *Tarsius bancanus* and two male *T. syrichta* were the subjects. The length and width of each testes were measured in millimeters with calipers. Volumes were calculated for comparisons.

Statistical Analysis

T. bancanus

Two wild-caught, adult males contribute data to the analyses. A third male (Kapis) was excluded from analyses because he contributed only one record. We have the following number of records (N = 2 males): Sempoladan (11) and Chani (11). An ANOVA indicates that body weights differ significantly between these two males (DF = 1, 20, F = 41.30, P < 0.0001) and that average testicle volume does not significantly vary (DF = 1, 20, F = 0.01, P = 0.92). Because we have an equal number of records from both males, the data can be combined; however, the effect of individual differences cannot be ruled out of any relationship.

T. syrichta

Five wild-caught, adult males contribute data to the analyses. One male (Bilar) became obese after capture, and we have excluded his records from the data set used here. We have the following number of records from each male (N = 5 males): Amo Dulom (11), Saimon (6), Bohol (4), Maasin (2), Davao (4). Multiple observations from the same tarsier are likely to be correlated. Ignoring this correlation by lumping all records together can potentially cause problems in statistical analysis if the record numbers are uneven (as here). We used an ANOVA to determine if data from the individual tarsiers could be safely combined or whether each individual should be treated separately. The results indicate that body-weight data from all males could be combined together (df = 4, 22, F = 2.08, P = 0.12). Like body-weight data, the average testicle volume for each male can also be combined (DF = 4, 26, F = 0.36, P = 0.84).

Results

Sexual Dimorphism in Body Mass

T. BANCANUS

T. bancanus exhibits no sexual dimorphism (F = 2.10, DF = 67, P = .15). The average male weight is 122.61 g (SE = 2.06, N = 23), while average female weight is 126.30 g (SE = 1.48, N = 45).

T. SYRICHTA

T. syrichta exhibits significant sexual dimorphism (F = 65.97, DF = 190, P < .0001), with females weighing 13.6% less than males. The average male weight is 131.78 g (SE = 2.05, N = 27), while average female weight is 113.86 g (SE = 0.85, N = 164).

Seasonal Effects: Females

T. BANCANUS

When captured in the wild during the months of October and November, one of the six female *Tarsius bancanus* was pregnant, and two gave birth within a month (50% were pregnant).

T. SYRICHTA

When captured in the wild in May, none of the six females was pregnant nor gave birth in the months after capture.

Seasonal Effects: Males

T. BANCANUS

A linear regression shows that the average testicle volume in *T. bancanus* changes with season. Month explains a significant percentage (20.2%) of the variation in average testicle volume (F = 5.050, DF = 21, P = 0.04). Body weight, however, does not fluctuate with season (F = 0.01, DF = 21, P = 0.91). Males remain at constant body weight regardless of season. A regression shows that testicle volume per body weight (Gonadosomatic Index) peaks in August through January and is lowest in February through

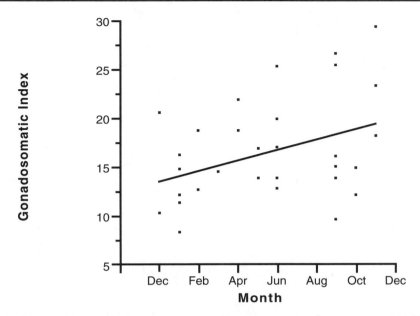

Figure 12.1. Seasonal changes in average testicle volume (mm^3) per body weight (g) in *Tarsius bancanus.* This population appears to breed August through January.

July, indicating that this species breeds seasonally. Month explains a significant percentage (19.1%) of the variation in Gonadosomatic Index (GI) (F = 4.72, DF = 21, P = 0.04). See figure 12. 1.

Based on these regressions, measurements taken in August through January were labeled "breeding season," and measurements from the other months were labeled "non-breeding season." The average testicle volumes of these two groups differed significantly (F = 5.69, DF = 21, P = 0.03).

T. SYRICHTA

A linear regression shows that average testicle volume in *T. syrichta* increases as January approaches. Month explains a significant percentage (16.6%) of the variation in average testicle volume (F = 5.76, DF = 30, P = 0.02). Body weight, however, does not fluctuate with season (F = 1.27, DF = 26, P = 0.27). Males remain at constant body weight regardless of season. GI may increase as January approaches. See figure 12.2. Month explains 12.2% of the variation in the GI, and the regression approaches significance (F = 3.49, DF = 26, P = 0.07). Males exhibit the highest GI measures from September through November, indicating that this species might breed seasonally. Based on these regressions, measurements taken in September through November were labeled "breeding season," and measurements from the other

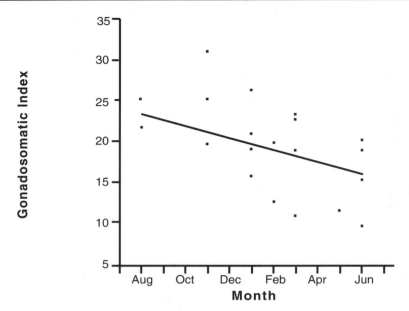

Figure 12.2. Seasonal changes in average testicle volume (mm³) per body weight (g) in *Tarsius syrichta*. This population appears to breed September through November.

months were labeled "nonbreeding season." The average testicle volumes of these two groups approached a significant difference (F = 3.63, DF = 35, P = 0.07).

Sperm Competition

Testicle Volume and Body Weight

T. BANCANUS

In the nonbreeding season, body weight explains 0.15% of the variance in average testicle volume (F = 0.02, DF = 11, P < 0.91). This relationship holds in the breeding season, where body weight explains 2.57% of the variance in average testicle volume (F = 0.21, DF = 9, P = 0.66). In *T. bancanus,* body weight cannot predict testicle volume in either season.

T. SYRICHTA

In the nonbreeding season, body weight explains 36.5% of the variance in average testicle volume (F = 10.36, DF = 19, P < .005). This relationship holds in the breeding season, where body weight explains 30.3% of the variance in average testicle volume (F = 3.91, DF = 10, P = 0.08). In the nonbreeding season, body weight predicts testicle volume significantly; the relationship approaches significance in the breeding season.

Relative Testes Size

Following Fietz (1999), we compared testes size and body mass of *T. syrichta* and *T. bancanus* with equivalent data from eighteen species of captive strepsirhine primates (*Cheirogaleus medius, Microcebus murinus, Mirza coquereli, Hapalemur griseus, Lemur catta, Eulemur coronatus, E. fulvus, E. macaco, E. mongoz, E. rubriventer, Varecia variegata, Galago moholi, Otolemur garnettii, Otolemur crassicaudatus, Perodicticus potto, Loris tardigradus, Nycticebus coucang,* and *N. pygmaeus;* Kappeler, 1997). As with the strepsirhine species, data from the two tarsier species were taken during the breeding season (Kappeler, 1997). A simple regression using Kappeler's data set yields an equation describing the relationship between body mass (g) and testicle volume (mm^3): testicle volume = 1.66 (body mass) + 467.97. The relationship is significant (F = 17.3, N = 18, P = 0.0007) and explains 51.9% of the variance. Fietz (1999) confirms this relationship.

For *T. syrichta* in the breeding season, the observed value for average testes volume (N = 13) was 2,626.35 mm^3. Using an average body weight of 136.32 g (N = 11), the regression predicts a testicle volume of 694.26 mm^3 for this species. Testis volume of *T. syrichta* exceeds the predicted value by 378.3%.

The breeding season values for *T. bancanus* are just as remarkable. For *T. bancanus* in the breeding season, the observed value for average testes volume (N = 13) was 2,592.21 mm^3. Using an average body weight of 121.33 g (N = 12), the regression predicts a testicle volume of 669.38 mm^3 for this species. Testis volume of *T. syrichta* exceeds the predicted value by 399.3%.

While these observed volumes exceed the predicted by remarkable percentages, the corresponding differences in testicle length and width are not so dramatic. For instance, *T. bancanus* in the breeding season demonstrates an average testes length of 10.6 mm and width of 7.9 mm. For *T. bancanus* to demonstrate the predicted volume, the testes length would need to decrease to 6.6 mm and the width to 4.9 mm. Using eighteen captive, seasonally breeding strepsirhine species as a model, both *T. syrichta* and *T. bancanus* have remarkably higher testicle volume than expected, indicating that sperm competition characterizes both tarsier species.

Discussion

As table 12.2 summarizes, these two tarsier species differ in sexual dimorphism, breeding-season length, and their relationship between season,

Table 12.2. Characteristics of *T. syrichta* and *T. bancanus* Based on Body Weight and Average Testicle Volume

	T. syrichta	*T. bancanus*
Sexual dimorphism	Yes	No
Seasonal breeder	Perhaps	Yes
Breeding-season length	3 months	6 months
Breeding season	Sep.–Nov.	Aug.–Jan.
Testes volume predict weight in breeding season?	Yes	No
Testes volume predict weight in nonbreeding season?	Yes	No
Sperm competition	Yes	Yes

average testicle volume, and body weight. Both are characterized by testicle volume that exceeds the expected. In these wild-caught, captive populations, *T. syrichta* is clearly sexually dimorphic in body weight, and *T. bancanus* is clearly not. The relationship between testicle volume, body weight, and season indicates that *T. syrichta* may breed seasonally and that *T. bancanus* clearly breeds seasonally in these populations. Assuming that both breed seasonally, *T. syrichta* mates in September, October, and November; *T. bancanus* mates in August through January. For *T. syrichta,* body weight predicts average testicle volume in both the breeding season and the nonbreeding season. For *T. bancanus,* body weight does not predict average testicle volume in either season.

If these species experience sperm competition, small males should increase their testes volume in the breeding season to equal that of large males (see Pochron and Wright, 2002). In *T. syrichta,* where body weight predicts average testes volume in both seasons, sperm competition cannot be confirmed by this method. However, we cannot rule it out, either. For instance, an evolutionary arms race could have large-bodied males increasing testicle volume in response to an increase in testicle volume in the small-bodied males. In this case, body weight would predict testicle volume in a species characterized by sperm competition. The fact that average testicle volume of this species in the breeding season exceeds the predicted by 378.3% provides a convincing argument for the strength of sperm competition in this species.

In *T. bancanus,* body weight fails to predict average testes volume in either season. This may be a true relationship or an artifact of small sample size. Our ability to detect seasonal breeding patterns in this small data set implies that the sample size is not necessarily responsible for the lack of relationship between body weight and testes volume. In this species, body weight may fail to predict average testicle volume, regardless of sample size. This indicates sperm competition. The existence of sperm competition in this species is bolstered by the finding that the observed average testicle volume exceeds the predicted by 399.3%.

In *T. bancanus*, average testes volume is independent of body size. Assuming that male fertility is positively correlated with testes volume via ejaculate size (Bercovitch and Nurnberg, 1996), males of small body size and males of large body size are equally likely to father offspring in this species, since testes volume does not increase with body weight. Future research that tests how well male body size (compared to other males) predicts access to estrous females and paternity may help elucidate male mating strategies. Larger males might not have a correspondingly large share of paternity.

The Effects of Captivity

For lemurs, captivity affects body weight (Fietz, 1999), testicle size (Fietz, 1999; Kappeler, 1997), and breeding system (Kappeler, 1997) in unpredictable ways. Therefore, using captive data to infer sperm competition in wild-caught tarsiers may be problematic. The results suggested by this study should be replicated in wild populations. It should be noted that females in captivity in both species cycled throughout the year, and that the light cycle of 12:12 was the same for both species in captivity (Wright et al., 1986a, b).

Predicting the Mating System

Intrasexual selection theory predicts that certain suites of characteristics accompany breeding systems, and this suite of characteristics may help us elucidate the breeding systems of *T. syrichta* and *T. bancanus*. Increased body size is seen and expected in polygynous species, while monogamous species are expected to demonstrate reduced sexual dimorphism. Intrasexual competition among polygynous male primates may lead to a single-male or multi-male breeding system. In the single-male groups, sperm competition should be very low since one male monopolizes all or nearly all females. The multi-male groups should experience intense sperm competition because more than one male has access to each estrous female. As a result, dimorphism should be highest in polygynous primates (both multi-male and single-male) and lowest in monogamous primates. Conversely, multi-male groups should exhibit sperm competition, and both single-male and monogamous primates should not (Harvey and Harcourt, 1984). As presented in table 12.3 sperm competition is expected (and often associated with) sexual dimorphism and multi-male breeding groups. Given this suite of characteristics, we might expect *T. syrichta* to demonstrate a polygynous breeding system and *T. bancanus* to demonstrate a less polygynous system.

Intrasexual competition between male mammals can occur as male-male contests, resulting in larger males with weapons such as antlers or canines (Darwin, 1871; Clutton-Brock et al., 1977). In comparisons of primate

Table 12.3. Patterns Predicted by Intrasexual Selection Theory

	Multi-male	Single-male	Monogamous	*T. syrichta*	*T. bancanus*
Sperm competition	Yes	No	No	Yes	Yes
Body dimorphism	Yes	Yes	No	Yes	No

NOTES: Intrasexual selection theory expects certain suites of traits. Multi-male groups (e.g., savannah baboons) are characterized by sperm competition and sexual dimorphism. Gorillas exemplify single-male traits, and gibbons exemplify the monogamous traits.

species, intermale competition is very strongly associated with canine dimorphism and/or body weight (Gaulin and Sailor, 1984; Kay et al., 1988; Shine, 1989; Plavcan and van Schaik, 1992). Alternatively, males that mate with multiple females can also compete after copulation by producing large amounts of sperm (Short, 1977; Harcourt et al., 1981). Sperm competition has been measured by comparing testicular volume to body mass, with larger testes being positively correlated with more females inseminated (Harvey and Harcourt, 1984). Promiscuous and polygynous anthropoid males show both types of male-male competition (Clutton-Brock et al., 1977; Harcourt et al., 1981). However, in his study of intrasexual selection and testis size in thirty five species of strepsirhine primates at Duke University Primate Center, Kappeler (1997) found an association between mating system and comparative testicular measurements. Prosimian species with promiscuous or polygynous mating systems had relatively larger testes than those with monogamous mating systems (Kappeler, 1997), confirming what had been documented in a comparative study of six sympatric species measured in the rain forest (Glander et al., 1992).

Previous field reports of the two tarsier species *Tarsius spectrum* and *T. bancanus* that have been followed by using radio-collared and marked animals conflict each other in descriptions of grouping patterns and social organization. There are several possibilities to explain these discrepancies. There are constraints in observing nocturnal primates in the wild, and field observations may give various results because of limited observations. Another possible explanation is that different species differ in their grouping structure. Our results using morphometrics gathered from a small sample of living individuals captured from the same geographic areas show that two species differ significantly in sexual dimorphism, a character used as a measure of social organization in other primates.

In these populations, *Tarsius syrichta* is clearly sexually dimorphic in body weight, and *T. bancanus* is clearly not. From these data, we predict that the Bornean tarsier (*T. bancanus*) may have a more monogamous social organization, while the Philippine tarsier (*T. syrichta*) would either have multi-male, multi-female groups, or the noyau social organization, with a dominant male visiting the separated territories of two or more females. Our morphometric analysis cannot distinguish these two patterns of social organizations.

The relationship between testicle volume, body weight, and season indicates that *Tarsius syrichta* may breed seasonally and that *T. bancanus* clearly breeds seasonally. Seasonal breeding has been associated with strong male-male competition in some New World monkeys and lemurs (Boinski, 1987; Wright, 1999) and can have many repercussions on social behavior (Paul, 1997; Gursky, 2000). These social effects should appear strongest in *T. syrichta.*

It has been suggested that prosimians and anthropoids differ in their response to sexual selection (Kappeler, 1991; Smith and Jungers, 1997). Do tarsiers conform to the anthropoid model of polygynous and promiscuous males being both larger in body mass and having larger testicular volume, or to the prosimian model of no sexual dimorphism in body or canine size, but testicular enlargement in more promiscuous mating systems? Our results give mixed signals. *Tarsius syrichta* fits a more anthropoid model, with both sexual dimorphism and testicular enlargement on the scale of a chimpanzee, while *T. bancanus* has a more prosimian pattern of large testes, but equal body size, as seen in prosimian primates.

Acknowledgments

We would like to acknowledge Patrick Andau, Chief Game Warden, for his support and assistance with logistics and authorizations in Sepilok and the other sites in Sabah, East Malaysia. Sepilok Reserve game rangers are thanked for assistance with mist netting. Jonas Wahlstrom, Director, Skansen Aquarium, is thanked for his assistance with Philippine tarsiers including the gift of Amo Dulom, the first tarsier at Duke, and Saimon. Skip Sivertson enthusiastically assisted with logistics and hospitality in the Philippines. We are grateful to Miles Roberts and Devra Kleiman for the *Tarsius bancanus* collaboration with the National Zoo. NSF grant BNS 81 20529 and NSF grant BNS83-10913 provided support for the field expeditions. M. K. Izard and A. Katz have contributed their expertise with tarsier management, care, and research. We thank Jukka Jernvall and two anonymous reviewers for excellent comments on the manuscript. This is DUPC publication #734.

References

Alexander RD, Hoogland JL, Howard RD, Noonan KM, Sherman PW. 1979. Sexual dimorphism and breeding system in pinnipeds, ungulates, primates and humans. In Chagnon N, Irons W, editors, Evolutionary biology and human social behavior, 402–435. Belmont, CA: Wadsworth.

Bearder SK. 1987. Lorises, bushbabies and tarsiers: diverse societies in solitary foragers. In Smuts BB, Cheney CL, Seyfarth RM, Wrangham RW, Struhsaker TT, editors, Primate societies, 11–24. Chicago: University of Chicago Press.

Bearder SK. 1999. Physical and social diversity among nocturnal primates: a new view based on long term research. Primates 40: 267–282.

Bercovitch FB, Nurnberg P. 1996. Socioendocrine and morphological correlates of paternity in rhesus macaques (*Macaca mulatta*). J Reprod Fertil 107: 59–68.

Boinski S. 1987. Mating patterns in squirrel monkeys (*Saimiri oerstedii*). Behav Ecol Sociobiol 21: 13–21.

Clutton-Brock TH, Harvey PH, Rudder B. 1977. Sexual dimorphism, socionomic sex ratio and body weight in primates. Nature 269: 191–195.

Cook N. 1939. Notes on captive *Tarsius carbonarius*. J Mammal 20: 173–177.

Crompton RH, Andau PM. 1986. Locomotion and habitat utilization in free-ranging *Tarsius bancanus:* a preliminary report. Primates 27: 337–355.

Darwin C. 1871. The descent of man and selection in relation to sex. London: Murray.

Fietz J. 1999. Mating system of *Microcebus murinus*. Am J Primatol 48: 127–133.

Gaulin SJC, Sailer LD. 1984. Sexual dimorphism in weight among the primates: the relative impact of allometry and sexual selection. Int J Primatol 5: 515–535.

Glander KE, Wright PC, Daniels PS, Merenlender AM. 1992. Morphometrics and testicle size of six rainforest lemur species from southeastern Madagascar. J Hum Evol 22: 1–17.

Gursky S. 1995. Group size and composition in the spectral tarsier, *Tarsius spectrum:* implications for social organization. Trop Biodiv 3: 57–62.

Gursky S. 1997. Modeling maternal time budgets: the impact of lactation and gestation on the behavior of the spectral tarsier, *Tarsius spectrum*. Ph.D. dissertation, State University of New York at Stony Brook.

Gursky S. 2000. The effect of seasonality on the behavior of an insectivorous primate, *Tarsius spectrum*. Int J Primatol 21: 477–496.

Gursky S. In press. Determinants of gregariousness in the spectral tarsier (Prosimian: *Tarsius spectrum*). J Zool.

Harcourt AH, Harvey PH, Larson SG, Short RV. 1981. Testes weight, body weight and breeding system in primates. Nature 293: 55–57.

Haring DM, Wright PC. 1989. Hand raising a Philippine tarsier, *Tarsius syrichta*. Zoo Biol 8: 265–274.

Harvey PH, Harcourt AH. 1984. Sperm competition, testes size, and breeding systems in primates. In Smith RL, editor, Sperm competition and the evolution of animal mating systems, 589–600. San Diego: Academic Press.

Hubrecht AAW. 1908. Early ontogenetic phenomena in mammals and their bearing on our interpretation of the phylogeny of the vertebrates. Quan J Microsc Sci 53: 1–181.

Kappeler P. 1991. Patterns of sexual dimorphism in body weight among prosimian primates. Folia Primatol 57: 132–146.

Kappeler P. 1997. Intrasexual selection and testis size in strepsirhine primates. Behav Ecol 8: 10–19.

Kay RF, Plavcan JM, Glander KE, Wright PC. 1988. Sexual selection and canine dimorphism in New World primates. Am J Phys Anthro 77: 385–397.

MacKinnon J, MacKinnon K. 1980. The behavior of wild spectral tarsiers. Int J Primatol 1: 361–379.

Niemitz C. 1979. Results of a field study on the Western tarsier (*Tarsius bancanus borneanus* Horsfield) in Sarawak. Sarawak Museum Journal 27: 171–228.

Niemitz C. 1984. An investigation and review of the territorial behaviour and social organization of the genus Tarsius. In Niemitz C, editor, 1–16. Biology of tarsiers, Stuttgart: Gustav-Fischer-Verlag.

Paul A. 1997. Breeding seasonality affects the association between dominance and reproductive success in non-human primate males. Folia Primatol 68: 344–349.

Plavcan JM, van Schaik CP. 1992. Intrasexual competition and canine dimorphism in anthropoid primates. Am J Phys Anthro 87: 461–477.

Pochron ST, Wright PC. 2002. Dynamics of testes size compensates for variation in male body size. Evol Ecol Res 4: 577–585.

Pochron ST, Wright PC, Scaentzler E, Ippolito M, Rakotoririna G, Ratsimbazafy R, Rakotosoa R. 2002. Effect of season and age on the gonadosomatic index of the Milne-Edwards sifakas (*Propithecus diadema edwardsi*) in Ranomafana National Park, Madagascar. Int J Primatol 23: 355–364.

Rickart EA, Heaney LR, Heideman PD, Utzurrum RCB. 1993. The distribution and ecology of mammals on Leyte, Biliran, and Maripipi Islands, Philippines. Fieldiana: Zoology 72: 1–62.

Shine R. 1989. Ecological causes for the evolution of sexual dimorphism: a review of the evidence. Q Rev Biol 64: 419–461.

Short R. 1977. Sexual selection and the decent of man. In Proceedings of the Canberra symposium on reproduction and evolution, Australian Academy of Sciences, 3–19.

Smith RJ, Jungers WL. 1997. Body mass in comparative primatology. J Hum Evol 32: 523–559.

Ulmer FA. 1963. Observations on the tarsier in captivity. Deutscher Zool Garten 27: 106–21.

Wright PC. 1999. Lemur traits and Madagascar ecology: Coping with an island environment. Yrbk Phys Anthro 42: 31–72.

Wright PC, Haring D, Izard MK, Simons EL. 1989. Psychological well-being of nocturnal primates in captivity. In Segal E, editor, Housing care and psychological well-being of captive and laboratory primates, 61–74. New York: Noyes Publications.

Wright PC, Haring D, Simons EL, Andau P. 1987. Tarsiers: a conservation perspective. Primate Conserv 8: 51–52.

Wright PC, Izard MK, Simons, EL. 1986a. Reproductive cycles in *Tarsius bancanus*. Am J Primatol 11: 207–215.

Wright PC, Toyama L, Simons EL. 1986b. Courtship and copulation in *Tarsius bancanus*. Folia Primatol 46: 142–148.

PART 3

FUTURE:
Conservation

CHAPTER 13

History of Captive Conservation of Tarsiers

Helena M. Fitch-Snyder

Historically, tarsiers have not thrived in captivity. The vast majority of imports did not survive long after their arrival, and captive-born offspring have suffered high mortality rates. This chapter is, in essence, a documentation of the extinction of captive tarsier populations in North America and Europe since the first known import in 1850. Although at least 146 *Tarsius syrichta, T. bancanus,* and *T. spectrum* tarsiers were imported into these regions, only one *T. bancanus* currently survives. Captive tarsier populations in Asia have also experienced very limited success.

This paper reports on the tarsier's import and reproductive history. Data were collected primarily from the International Species Inventory System (ISIS, 1998–2000), the North American Regional Studbook for Asian Prosimians (Fitch-Snyder, 1998), and published articles. These sources provided fairly complete information about captive tarsiers in North America, most of Europe, and some parts of Asia. However, in some cases, there are missing or conflicting reports about certain individuals. Institutional records and personal communication were used to help complete the data and resolve conflicting information.

Tarsiers present a real challenge to zoos and institutions that are interested in maintaining colonies for captive breeding. In the wild, habitat destruction is becoming a threat to tarsiers (MacKinnon, 1997; Collins et al., 1991; Hamer et al., 1997; Oliver and Heaney, 1996), yet maintaining healthy colonies in captivity to provide a second line of defense against extinction looks infeasible at this time. Part of the problem may be that tarsiers are the only primate that eats exclusively live animal prey in the wild, and there may be some mineral or nutritional component in wild orthopterans that we have not replicated in zoo diets. Life-history factors may also influence the failures in captivity. Female tarsiers give birth to a singleton that weighs a quarter of her body weight, weight gain of infants is rapid, and weaning occurs within two months (Haring et al., 1985; Haring and Wright, 1989). Diet and captive management conditions may play critical roles during their developmental process. This historical view of tarsiers in captivity, as disappointing as it is, may give us some insights that will be useful in future management plans of these species.

Tarsier Imports

Tarsius syrichta

The earliest husbandry and captive behavior information about tarsiers was reported by Captain Norman Cook, Retired, U.S. Army (1939). Cook kept fifteen tarsiers (*T. syrichta carbonarius*) as pets while he was living in the Philippines. He was successful in maintaining one of these tarsiers for sixteen months.

All the known T. *syrichta* imports from 1850 to 1986 are listed in table 13.1. These data show that 130 T. *syrichta* have been imported from the Philippines to Europe, North America, and Japan. Kohn et al. (1984) reported that the first documented tarsier import occurred in 1850. An animal identified as *Tarsius tarsius* (now known as T. *syrichta*) of unknown sex lived at Amsterdam Zoo in 1882.

The first tarsier imported into North America was a female T. *syrichta* that was received by Yale University's Peabody Museum in 1938 (Roberts et al. 1984). However, no additional information is available about this animal.

In 1947, a group of Philippine tarsiers were captured in Mindanao and imported by Charles Wharton (1948). At that time, Philippine tarsiers were quite numerous in the wild. Wharton (1950) reported that ninety four live tarsiers were easily captured in less than a week, and forty five were captured in a single day. Wharton's 1948 account of this event stated that thirty one living tarsiers arrived in Oakland, California. However, subsequent reports by Ulmer (1963), and Wright et al. (1987) account for only thirty. Wharton reported that part of the tarsier shipment was delivered to Chicago's Brookfield Zoo and New York's Bronx Zoo (now known as Wildlife Conservation Park) before the remaining animals were placed at the National Zoological Park in Washington, D.C. Records from these zoological institutions show that sixteen T. *syrichta* were received by the National Zoo, two went to the Bronx, and two were sent to Philadelphia. Crandall (1964) reported that some of these tarsiers also went to the Brookfield, St. Louis, and Detroit zoos. However, the only tarsier listed in these zoos' records was a male T. *spectrum* found in Brookfield's 1948 inventory. Since Wharton was listed as the source of this tarsier, this animal was likely to have been part of the T. *syrichta* group imported the previous year. This tarsier is therefore included with the T. *syrichta* listed in table 13.1.

The lack of zoo records for some of these tarsiers may have been because they died shortly after arrival or because the zoo did not own them. During these years, written records were not generally kept for animals loaned to an institution (Marvin Jones, pers. com.).

Most of the tarsiers from Wharton's first shipment did not survive more than a few years. Stud-book records show that only two infants were born,

Table 13.1. *Tarsius syrichta* Imports

Import Date	Origin	Numbers by gender			Receiving institution
		M	F	?	
17 October 1850	Philippines	0	0	1	Amsterdam[a]
1938	Philippines	0	1	0	Peabody Msm[b]
9 July 1947	Mindanao	1	0	1	Bronx[b]
28 July 1947	Mindanao	0	1	0	Philadelphia[b]
9 August 1947	Mindanao	1	2	13	Washington[b]
1947	Mindanao	1	0	0	Chicago B[b]
26 April 1952	Mindanao	0	0	2	Bronx[b]
18 October 1962	Mindanao	0	0	1	Osaka[c]
26 June 1963	Mindanao	1	1	2	Bronx[b]
8 May 1964	Mindanao	6	4	2	Frankfurt[d]
3 August 1965	Philippines*	1	1	0	Beaverton[b]
1 December 1965	Passay	1	0	0	Chicago B[b]
5 December 1965	Philippines*	5	5	0	Beaverton[b]
23 November 1966	Mindanao	0	1	0	Frankfurt[d]
20–22 Sept 1968	Cotabato	5	5	0	San Diego[b]
15 November 1969	Mindanao	1	5	0	Frankfurt[d]
1 March 1973	Philippines	0	0	8	Chicago LP[b]
August 1981	Philippines*	1	1	0	Skansen[e]
13 May 1983	Philippines[a]	1	1	0	Duke[b]
26 February 1984	Philippines[a]	1	1	0	Duke[b]
8 May 1985	Visaya Islands	4	3	0	Cincinnati[b]
13 May 1985	Visaya Islands	6	15	0	Duke[b]
17 November 1985	Visaya Islands	4	3	0	Chicago LP[b]
5 September 1986	Unknown	1	2	0	Cincinnati[b]
9 October 1986	Philippines*	1	3	0	Bristol[f]
21 November 1986	Philippines	0	3	0	San Diego[b]

*This animal was identified as *Tarsius spectrum* but was included in this table because it matched the import information published by Wharton (1948).

[a] Netherlands
[b] U.S.
[c] Japan
[d] Germany
[e] Sweden
[f] U.K.
[g] Manila
B = Brookfield Zoo
LP = Lincoln Park Zoo

but both died at birth (Fitch-Snyder, 1998). The maximum lifespan in this group of tarsiers was a female that was kept at Philadelphia Zoo for twelve years. Stud-book records show that Wharton arranged two more tarsier shipments to the Bronx Zoo in 1952 and 1963. These shipments contained six more tarsiers from Mindanao.

The Oregon Primate Center in Beaverton was the next organization to import Philippine tarsiers into North America. In 1965, they obtained two groups through Ravensden Zoological Company in Rushden, England. Philippine Animal Export-Import Company in Luzon originally exported these tarsiers from Manila. According to institutional records, the Oregon Primate Center had one tarsier birth the following year. However, the infant only survived three days.

C. S. Evans (1967) reported that the Oregon Primate Center received a total of thirteen *T. syrichta*. However, the primate center's records account for only twelve imports. One pair of *T. syrichta* was received in the first shipment in August 1965, and five more pairs arrived the following December. Chicago's Brookfield Zoo received a male Philippine tarsier only a few days before Oregon received its second shipment, which may account for the additional tarsier that Evans reported.

The European tarsier population (*T. syrichta*) was maintained primarily at the Frankfurt and Bristol zoos from the 1960s through the 1980s (Fitch-Snyder, 1998). According to Frankfurt Zoo's records, they received three shipments from Mindanao from 1964 through 1969 and imported a total of nineteen animals. They also had a pair that was transferred from Sweden, and had four births at their facility. One captive-born male lived for twenty two months.

Bristol had four individuals: one male and three females, imported in 1986. The longest-lived of these animals survived a little longer than two years.

Skansen Akvariet in Stockholm, Sweden, has kept several *T. syrichta* since the late 1960s, but import and birth records of those tarsiers are not available from that institution. Studbook records show that Skansen kept a pair of *T. syrichta* that was captured in June 1981 (Fitch-Snyder, 1998). This pair was at Manila Zoo before they were transferred to Skansen, and they were then sent to Frankfurt Zoo on June 3, 1982. The male died a few weeks later, but according to Frankfurt's records, the female lived there for an additional nine years. Haring and Wright (1989) reported that more than six *T. syrichta* infants were born at Skansen. Although none of them survived for long, one was hand-raised until eight weeks of age.

San Diego Zoo later received five pairs of *T. syrichta* through a dealer named Leonard Von Giese. They arrived from Cotabato, Philippines, from

September 20 to 22, 1968. Zoo records show one of these died in transit, and the last of the remaining tarsiers died three years later.

Frankfurt's records indicate that they purchased a female *T. syrichta* on November 23, 1966, and that it died on May 12, 1967. They also received one male and five females from Mindanao on November 15, 1969. Four of these lived for over six years.

According to Roberts (1985), Lincoln Park Zoo in Chicago received eight more Philippine tarsiers of unknown sex from a dealer named Villueneva on March 1, 1973. These tarsiers were not recorded in the zoo's records, possibly because the entire shipment was dead on arrival.

On May 13, 1983, Duke University Primate Center received a *T. syrichta* pair by way of Skansen Akvariet in Stockholm, Sweden (Wright, chapter 14, this volume). They also received another pair almost one year later, on February 26, 1984 (Fitch-Snyder, 1998). It is unclear where these tarsiers originated, as they were purchased through a dealer named Baquilod who had captured them on one of the Visaya islands, and they were at Skansen since 1981. One of these tarsiers, a male named Amo, lived at Duke until it was twelve years old. He was euthanized because he developed cataracts in both eyes (Patricia Wright, pers. com.).

In 1985, the largest group of *T. syrichta* was imported into North America since Wharton brought in the first group during the 1940s. Patricia Wright obtained and imported these tarsiers for breeding and research purposes after surveying the Philippine Island of Bohol in 1985 (Wright et al., 1987). The tarsiers were actually captured by Baquilod on the island of Samar or Leyte. Wright reported that twenty adult *T. syrichta* were taken to Duke University Primate Center on May 13, 1985. However, Duke's records and ISIS reports show that twenty-one tarsiers (one infant) were actually received.

According to their animal records, Cincinnati Zoo also brought in four males and three females (4.3) from Bohol as part of Wright's breeding and research effort. One wild-caught female from Duke's imports was later sent to Cincinnati, where she died after over fourteen years in captivity. Chicago Lincoln Park Zoo later received seven tarsiers (4.3) from Bohol in November of the same year.

The last imports into North America were made through Ravensden in 1986. Cincinnati purchased a male and two females, and San Diego acquired three females. One of San Diego's females died soon after arrival, and the remaining two were later transferred to Cincinnati, where they died within the next five years.

There are likely to be several institutions and private individuals in Asia who have maintained and bred this species. However, animal records are rarely kept there, and additional information is difficult to obtain. Richard

Tenaza (pers. com.) reported that University of the Philippines at Los Banos has maintained a breeding program for tarsiers.

Tarsius bancanus

In contrast to *T. syrichta,* there have been only twenty three known imports of *T. bancanus.* (table 13.2.) According to Marvin Jones (pers. com.), three of these were sent by Rotterdam Zoo to Rijks Museum at Leiden on April 3, 1926, presumably after they died. The tarsiers were caught in Borneo, but they were identified as *T. spectrum.* Since *T. spectrum* is restricted to Sulawesi (Haring and Wright, 1989), this specimen is included with the *T. bancanus* tarsiers listed in table 13.2. No information is available concerning how long these tarsiers lived at Rotterdam Zoo. There is a published account of a pair of tarsiers (presumably *T. bancanus*) that were housed at the Sarawak Museum throughout 1962 (Harrison, 1963). Both tarsiers died during that year, the female in her first trimester of pregnancy.

Roberts et al. (1984) reported that a male *T. bancanus* was kept at Max Planck Institute in Germany in 1966, but the death date for this animal was not recorded.

On only three occasions were *T. bancanus* imported into North America. Milwaukee received the first of these on July 17, 1974. This male originated in Sabah and was purchased from Charles P. Chase Co., a dealer who was based in Miami. Studbook records show that this tarsier was later transferred to International Animal Exchange in Ferndale on August 19, 1975; but there are no records concerning its survivorship.

Table 13.2. *Tarsius bancanus* Imports

| Import date | Origin | Numbers by gender | | | Receiving institution |
		M	F	?	
3 April 1926	Borneo[a]	0	0	1	Rotterdam[b]
~1966	Borneo	1	0	0	Max Planck[c]
17 July 1974	Sabah	1	0	0	Milwaukee[d]
12 November 1983	Sabah	3	3	0	Washington[d]
12 November 1983	Sabah	3	3	0	Duke[d]
10 January 1995	Sarawak	1	0	4	Singapore
8 June 1995	Sarawak	1	1	0	Singapore
5 February 1998	Sarawak	1	0	0	Singapore

[a] This animal was identified as *Tarsius spectrum* but was included in this table because its origin was reported as "Borneo."
[b] Netherlands
[c] Germany
[d] U.S.

In 1983, two years before Patricia Wright imported the Philippine tarsiers, she traveled to Sandakan, Sabah, in east Malaysia to capture twelve *T. bancanus* (Wright et al., 1987; Wright, chapter 14, this volume). Six pairs were caught between September 13 and November 13, and all six females were pregnant at the time of capture. Three of these pairs went to Duke, and the other three went to National Zoo. Roberts (1985) reported that one pair died within a few weeks of arrival at National Zoo. However, the other two pairs survived and produced offspring. One of these captive-born females is still living, and she is currently exhibited at Cleveland Zoo. The tarsiers that went to Duke lived from two to seven years, and five infants were born during this time (Wright, chapter 14, this volume). None of these infants survived longer than three days.

ISIS records (2001) report that Singapore Zoo presently has a male *T. bancanus* that was wild-caught from Sarawak and imported in June 1995. Singapore received an additional six during that year that either died or were transferred to an unreported location. According to ISIS records, a lone male was received in 1998, and it died the month after its arrival.

Taman Safari in Bogor Indonesia has also maintained *T. bancanus* at their facility (Sharmy Prastiti, pers. com.). However, numbers and histories of these tarsiers were not recorded.

Tarsius spectrum

Spectral tarsiers have only been documented in Rotterdam, London, Brookfield, and Singapore Zoos. However, they have also been reported at Taman Safari, Bogor, Indonesia (Myron Schekelle, pers. com.). The ten tarsiers shown in table 13.3 probably do not reflect the true history of this species in captivity, since complete written records are not available from Taman Safari and Singapore Zoo.

Table 13.3. *Tarsius spectrum* Imports

| Import date | Origin | Numbers by gender | | | Receiving institution |
		M	F	?	
1869	Sulawesi	0	0	1	Rotterdam[a]
1930–60 ?	Sulawesi	0	0	3	London[b]
12 March 1991	N. Sulawesi	1	0	0	Singapore
24 June 1991	Sulawesi	0	1	0	Singapore
27 June 1996	Sulawesi	1	1	0	Singapore
2 July 1998	Sulawesi	2	0	0	Singapore

[a]Netherlands
[b]U.K.

Table 13.4. Wild-Caught Tarsiers Surviving More than Five Years in Captivity

Institutions	Gender	Longevity
T. syrichta		
Cincinnati	M	5 years, 3 months
	M	5 years, 11 months
	M	6 years, 8 months
	F	10 years, 6 months
		11 years, 5 months
Chicago B	M	7 years, 7 months
	M	11 years, 10 months
	F	9 years, 3 months
Chicago LP, Duke	M	6 years
Duke	M	5 years, 1 month
	M	5 years, 1 month
	M	5 years, 10 months
	M	9 years, 8 months
	F	5 years, 10 months
	F	6 years, 7 months
	F	7 years, 7 months
	F	8 years, 4 months
Duke, Cincinnati	F	14 years, 1 month
Frankfurt	M	7 years, 2 months
	M	7 years, 4 months
	M	7 years, 11 months
	M	10 years, 7 months
	M	12 years, 5 months
	F	6 years, 4 months
	F	7 years
	F	8 years
Philadelphia	F	11 years, 11 months
Stockholm, Duke	M	11 years, 9 months
Stockholm, Frankfurt	M	10 years, 7 months
	F	10 years, 1 month
T. bancanus		
Washington, Cleveland	M	9 years, 7 months
Washington, Duke	M	5 years, 8 months
	M	7 years, 7 months
Duke	F	6 years, 3 months
Washington	F	8 years, 4 months
T. spectrum		
Singapore	F	> 5 years (still alive)

NOTE: B = Brookfield Zoo; LP = Lincoln Park Zoo.

Marvin Jones (pers. com.) reported that the earliest known *T. spectrum* was kept at Rotterdam Zoo in 1869, but no further information is available about this animal. The International Zoo Yearbook (Jarvis and Morris, 1960) lists three *Tarsius spectrum* that were at the London Zoo between 1930 and 1960, and one of these lived nearly three years. The only other *T. spectrum* reported in Europe was the possibly misidentified specimen that was noted in table 13.2.

In North America, one male *T. spectrum* is listed in Chicago Zoological Society's Brookfield Zoo's 1948 inventory. It is very likely that this tarsier was actually one of the *T. syrichta* imported by Wharton the previous year (table 13.1).

According to the 2001 ISIS Reports, Singapore Zoo lists three living *T. spectrum* at its facility. The female was received in 1991, and two confiscated males were obtained in July 1998. However, Richard Weigel (pers. com.) reported that Singapore Zoo had only one living pair of *T. spectrum* as of August 2001. The male was received in 1998, and the female was one of a pair that was acquired on June 27, 1996.

Other zoos and private individuals in Indonesia have worked with at least three different kinds of tarsiers, although exact numbers, species types, and animal histories are lacking. Tarsiers are also kept by a breeder in Indonesia named Danny Permadi (Sharon Gursky, pers. com.), and six adult *T. spectrum* have been reported at Bogor Agricultural Institute (Nietsch and Kopp, 1998).

Longevity of Wild-Caught Tarsiers

Despite the seemingly large numbers of tarsiers that did not survive long after they were imported, an encouraging number of these imports lived for many years. Table 13.4 lists all the wild-caught tarsiers that survived more than five years in captivity. There were twenty nine *T. syrichta* (seventeen males and twelve females) that lived for over five years. The oldest of these lived at the Philadelphia Zoo for nearly twelve years and at Duke University Primate Center for almost twelve years. Five *T. bancanus* are also listed as living over five years in captivity. The oldest of these survived for nine years and seven months. In both species, 22% of all wild-caught tarsiers with known capture and death dates survived captivity for over five years. Additionally, one female *T. spectrum* is still living at Singapore Zoo after over five years in captivity (Richard Weigel, pers. com.). Since actual capture and transfer dates were not always recorded, the information shown in table 13.4 is based on the first known transfer date after capture. Therefore, these tarsiers may have been considerably older than reflected on this table.

Table 13.5. Captive Births: *Tarsius syrichta*

Birth location	Birth date	Death date	Sex	Birth location	Birth date	Death date	Sex
Philadelphia	~1 June 1949	1 June 1949	M	Duke	28 May 1987	3 June 1987	F
Philadelphia	~1 June 1950	1 June 1950	M	Cincinnati	13 Sept 1987	18 Dec 1987	?
Chicago B	19 April 1966	1 May 1966	?	Duke	30 Oct 1987	30 Oct 1987	F
Beaverton	14 Sept 1966	17 Sept 1966	?	Duke	17 Nov 1987	17 Nov 1987	F
San Diego	18 May 1970	18 May 1970	M	Duke	2 Feb 1988	2 Feb 1988	F
San Diego	4 June 1971	5 June 1971	M	Cincinnati	19 March 1988	19 March 1988	?
San Diego	28 May 1972	28 May 1972	?	Duke	22 May 1988	22 May 1988	M
Frankfurt	29 July 1972	6 August 1972	?	Duke	22 May 1988	22 May 1988	F
Frankfurt	23 July 1973	26 July 1973	M	Cincinnati	14 July 1988	1 January 1999	M
San Diego	14 Sept 1973	29 Sept 1973	M	Duke	9 August 1988	9 August 1988	F
San Diego	9 April 1975	9 April 1975	M	Duke	15 August 1988	15 August 1988	?
Skansen	20 Feb 1983	??	F	Cincinnati	24 Nov 1988	11 Dec 1988	?
Frankfurt	3 March 1983	13 Jan 1985	M	Duke	28 Dec 1988	28 Dec 1988	F
Chicago LP	19 April 1986	25 July 1986	M	Cincinnati	7 June 1989	23 April 1991	M
Cincinnati	7 July 1986	12 July 1986	?	Cincinnati	7 Oct 1989	7 Oct 1989	?
Duke	29 Jan 1987	21 Dec 1991	F	Cincinnati	7 April 1991	7 April 1991	?
Duke	11 April 1987	11 April 1987	M	Cincinnati	29 June 1992	29 June 1992	?
Frankfurt	21 May 1987	21 May 1987	?	Cincinnati	18 March 1994	22 March 1994	?

NOTE: B = Brookfield Zoo, LP = Lincoln Park Zoo.

Table 13.6. Reproductive History of Successful Female Tarsiers

Institution	Dam's Studbook #	Birth dates
T. syrichta		
Beaverton	0625	9/66
Chicago LP	1071	4/76
Cincinnati[a]	1078	3/88, 11/88, 8/89
	1082	7/88, 6/89
Duke	1068	1/87, 10/87, 5/88, 12/88
	1070	5/87, 2/88, 8/88
	1084	11/87, 8/88
	1094	4/87, 5/88
Frankfurt	2110	3/83, 1/85
Philadelphia	1020	6/49, 5/50
San Diego[b]	1031	5/72, 9/73, 4/75
T. bancanus		
Duke[c]	1003	2/84
	1013	2/85, 10/88, 1/90
Washington	1005	12/83, 8/85, 4/86, 11/86, 6/87, 2/88, 6/89
	1007	3/84, 11/84, 9/85
	1028	3/91, 5/92

NOTE: LP = Lincoln Park Zoo.
[a] Cincinnati had five additional *syrichta* births from unknown dams.
[b] San Diego had two additional *syrichta* births from unknown dams.
[c] Duke had one additional *bancanus* birth from an unknown dam.

Captive Reproduction in North America and Europe

Tarsius syrichta

Although there have been thirty seven known *T. syrichta* births in North America and Europe (Fitch-Snyder 1998), twenty of these were stillbirths or died the same day (see table 13.5). Nearly 89% of captive-born *T. syrichta* did not live beyond the first year, and there were no recorded cases of successful second-generation reproduction in Philippine tarsiers.

Table 13.6. lists the reproductive history of successful female tarsiers as reported in their studbook records (Fitch-Snyder, 1998). In *T. syrichta*, eleven females gave birth between 1949 and 1989 in seven different facilities. One female *T. syrichta* at Duke Primate Center named Cotabato produced four infants within a two-year period. Her first infant was born on January 29, 1987, and it lived for nearly five years. Her second offspring was born nine months after the first, but it died the same day it was born. Cotabato's last two infants were born in May and December 1998, nearly twelve years after her first known reproduction. Her last infants also did not survive beyond their birth dates.

Table 13.7. Captive Births: *Tarsius bancanus*

Birth location	Birth date	Death date	Sex	Birth location	Birth date	Death date	Sex
Sabah (Duke)*	15 Oct 1983	15 Oct 1983	?	Washington	1 Nov 1986	17 Nov 1987	F
Washington	28 Dec 1983	16 Jan 1984	M	Washington	10 June 1987	13 June 1987	M
Duke	14 Feb 1984	17 Feb 1984	F	Washington	21 Feb 1988	21 Feb 1988	M
Washington	26 Mar 1984	13 April 1984	M	Washington	15 Sept 1988	ALIVE	F
Washington	2 Nov 1984	19 Feb 1990	M	Duke	16 Oct 1988	16 Oct 1988	?
Washington	11 Jan 1985	8 July 1988	M	Washington	13 June 1989	13 June 1989	M
Duke	21 Feb 1985	21 Feb 1985	M	Duke	29 January 1990	30 January 1990	M
Washington	15 August 1985	23 Oct 1985	M	Washington	26 March 1991	26 March 1991	F
Washington	8 Sept 1985	8 Sept 1985	M	Washington	2 May 1992	2 May 1992	M
Washington	21 April 1986	21 April 1986	M				

*This infant was born to wild-caught parents before they were exported.

Table 13.8. Captive-Born Tarsiers Surviving More than Three Years

Institutions	Gender	Longevity
T. syrichta		
Cincinnati	M	11 years, 6 months
Duke	F	3 years, 11 months
T. bancanus		
Washington	M	3 years, 6 months
Washington	M	5 years, 3 months
Washington, Cleveland	F	> 13 years (still living).

Studbook records show that Philadelphia Zoo was the first institution in which captive reproduction occurred in tarsiers (table 13.7). Two male infants were born in 1949 and 1950. Once again, neither of these lived beyond their first day of life. Ulmer (1963) reported that these two neonates both died of cerebral hemorrhage from head injuries. This apparently resulted from falling or hitting objects while clinging to their mother.

Brookfield Zoo had a male infant *T. syrichta* that was born in April 1966, and it died of undetermined causes at fifteen days (Schreiber, 1968). The mother died less than two weeks later of pneumonia.

The Oregon Primate Center also had an infant *T. syrichta* born in 1966 from one of their six tarsier pairs. This infant of unknown sex died three days later.

San Diego Zoo had five *T. syrichta* births from 1970 through 1975. Four died within twenty four hours, and one lived for fifteen days. One dam (1031) produced three of these infants, and the other two were born under grouped housing conditions, whereby the dams could not be identified (table 13.6).

Frankfurt Zoo had four *T. syrichta* births. Three survived only for a few days, but one lived nearly two years. Frankfurt's records show their later success was attributed to pregnancy watches, natural lighting, and nest boxes.

Studbook records show that one female *T. syrichta* was born at Skansen in Stockholm in 1983, but her death date was not reported. Lincoln Park Zoo had a male *T. syrichta* birth in 1986. This tarsier was conceived in the Philippines, and he lived for three months.

Cincinnati Zoo had a total of eleven *T. syrichta* births from 1986 to 1994. Five had unknown dams, and the remaining five were born to two different females. Six of their infants lived less than five days, though two males survived for over ten years. The longest-lived was over eleven years and ten months of age, the record for known captive-born tarsiers (see table 13.8).

Duke University Primate Center holds the record for captive tarsier births. Eleven *T. syrichta* were born at that facility. Three were stillbirths and six died in less than three days. However, Duke successfully hand-raised a tarsier infant (Haring and Wright, 1989), born on January 29, 1987. This female, named Mandarin, survived for nearly five years.

Tarsius bancanus

There have been nineteen known captive births of *T. bancanus*. Eight of these infants died the same day they were born. The first birth was in Sabah on the island of Borneo after its parents were captured for subsequent transport to Duke Primate Center. All the remaining *T. bancanus* born in North America were at National Zoo and Duke Primate Center.

National Zoo's Park (NZP) and its Conservation Research Center (CRC) had fourteen *T. bancanus* born between 1983 and 1992 (Fitch-Snyder, 1998). Of the fourteen infants born, six were stillborn or died on their day of birth. Of the first two births, one died from a fall and the other was killed by its father (Roberts, 1985). Five lived less than three months, and three reached adulthood (table 13.8). Roberts (1992) reported that the first conception can occur in female tarsiers at approximately two years of age, so the three-year-old tarsiers listed in table 13.8 were all mature enough to produce offspring.

One female born on September 15, 1988, at the National Zoological Park is currently living at Cleveland Zoo in Ohio. One tarsier was transferred to Duke at one year of age and died a year later (Roberts, 1992; DUPC Records). Two offspring were born to a captive-born *T. bancanus* female maintained at National Zoological Park (now living at Cleveland Zoo). However, neither survived past their birth dates. One *T. bancanus* female (1005) at National Zoo delivered seven infants in less than six years (see table 13.6).

Duke Primate Center's reports list five captive births, including one that was born and died in Sabah, east Malaysia, before being transported to their facility. Three of the four born at Duke did not survive beyond a few days, but one lived for nearly four years. One infant was killed by the male in the cage (probably not the father) (Patricia Wright, pers. com.). There were two known dams at the primate center, and one infant was born from an unknown dam (table 13.6).

Taman Safari in Indonesia has also maintained *T. bancanus*. According to Sharmy Prastiti (pers. com.), one infant was born at that facility and it died in 1998 (presumably the same year as its birth).

Tarsius spectrum

The only confirmed captive pairs of *T. spectrum* have been housed at Singapore Zoo, but there has been no known reproduction at this facility.

Taman Safari in Bogor, Indonesia, has also maintained this species, and it is possible that they have reproduced at that site (Myron Schekelle, pers.

com.). However, colony records are not available, and reproduction could not be confirmed through published information or communication with staff at Taman Safari.

General Causes of Death in Tarsiers

An examination of studbook data reveals some of the factors that have resulted in tarsier deaths (table 13.9). However, since only about one-third of all deaths have a reported cause, these results are very limited. For example, tarsiers that died in transit were not always reported to ISIS or the studbook keeper. Necropsies are not done in many cases, and even known causes of death are not always recorded.

The tarsier at the Philadelphia Zoo died at twelve years, not from old age but from pneumonia after a major storm that disrupted the electricity and chilled the zoo facilities (Patricia Wright, pers. com.).

Table 13.9. Known Causes of Death in Captive Tarsiers

Cause of Death	Numbers by Gender		
	M	F	?
T. syrichta			
Died in transit	2	0	1
Environmental or behavioral conditions	2	1	0
Euthanasia	1	3	0
Infection related	2	10	1
Injury from exhibit mate	1	0	1
Self-inflicted injuries	0	0	1
Stillbirth	3	2	1
Unknown			
Subtotal	11	16	5
T. bancanus			
Died in transit			
Environmental or behavioral conditions	3	0	0
Euthanasia	1	0	0
Infection related	2	2	0
Injury from exhibit mate	1	0	0
Premature/stillbirth	4	0	2
Unknown	0	1	0
Subtotal	11	3	2
Totals	22	19	7

NOTE: Cause of death was known for 48 captive tarsiers and unknown for 88 (136 deaths total). Due to the absence of details for the 88 cases of unknown cause, they are not included in this table.

Adult deaths in the large colony at Duke had many causes, including old age. One of the female *Tarsius bancanus* at Duke Primate Center died shortly after receiving Mebendazol, a treatment for parasites from the wild. Her necropsy revealed a severe infestation of parasitic cysts. The parasites that were killed by the medication may have formed a clot in her blood stream and caused the death. Other tarsiers died after exterminators sprayed their building and their insect food may have become contaminated. Another tarsier died in the winter when the cages were chilled from the breakdown of the primate center's heating system (Patricia Wright, pers. com.).

Husbandry Considerations

Information obtained from institutions with successful tarsier reproduction and longevity may provide a basis for future captive management efforts. Cincinnati Zoo attributes its success with *T. syrichta* to a warm enclosure with high humidity (see Ulmer, 1963), several vertical perching sites, and a selection of live food (Dulaney, 1985). The enclosure was kept between 80° and 90°F, and the relative humidity ranged from 80% to 90%. Humidity was maintained by splashes from a waterfall and a cypress bark mulch groundcover, which retains moisture without rotting. The tarsiers were fed gray-bird locust (*Schistocerca vaga*), crickets, mealworms, and hissing cockroaches (*Gomphadorina portentosa*). They also were occasionally offered newborn mice.

Haring and Wright (1989) reported that Duke Primate Center kept their tarsiers on a 12:12 light cycle with the white light provided by Vita-Lite florescent bulbs (Duro-Test Corporation, North Bergen, NJ). The tarsier rooms were maintained between 85° and 90°F, with relative humidity levels of 60%–80%. The primary diet of the *T. syrichta* and *T. bancanus* at Duke was crickets and anole lizards (*Anolis carolinensis*). Lizards were relished by *T. syrichta*, but *T. bancanus* would eat them only occasionally. The crickets were fed apples, crushed high-protein monkey chow, and a commercial cricket diet with a calcium supplement to make them more nutritious to the tarsiers. Whenever available, the tarsiers were also offered grasshoppers, dragonflies, cicadas, and mantids. All tarsiers at Duke refused any food that was not alive.

Although National Zoo offered a wide variety of food, their *T. bancanus* chose crickets as their main diet (Roberts and Kohn, 1993). The tarsiers foraged for crickets primarily in the arboreal locations in their enclosures. All successful facilities considered vertical substrates for clinging and leaping to be important environmental requirements for tarsiers.

Species differences are also important considerations in captive social management. Whereas *T. syrichta* have been successfully maintained in groups of up to five individuals including multiple-males, *T. bancanus* are much less tolerant of conspecifics (Haring et al., 1985). Roberts and Kohn (1993) recommend separating the sexes before parturition to prevent paternal aggression toward offspring.

Conclusions

Unlike the case with other small nocturnal primates (Glatson and Mazotta, 1998; Fitch-Snyder, 1998; Glatson, 2001), records show that the captive management of tarsiers has been largely unsuccessful. At the time of this writing, only one captive-born *T. bancanus* remains alive outside of Asia. Over 57% of wild-caught tarsiers survived less than two years. Although both *T. syrichta* and *T. bancanus* have reproduced, 81% of these offspring did not live beyond the first year. Tarsiers are the world's only primate that eats exclusively live food, primarily orthopteran insects (Gursky, chapter 10, this volume) and occasionally vertebrates such as lizards, snakes, and frogs (Niemitz, 1984). Despite extensive efforts to provide an adequate diet, especially at National Zoo, Duke, and Cincinnati, there has been consistently high infant mortality.

On an encouraging note, several tarsiers have lived from twelve to fourteen years in captivity. It is also promising that at least fifty six captive-born infants were produced. If captive tarsiers are to be successfully maintained in the future, basic husbandry requirements pertaining to diet and housing will need to be identified and provided (Wright et al., 1989). Factors such as the provision of additional calcium can make a critical difference in tarsier infant survival rate (Roberts and Kohn, 1993); or perhaps vitamin C (Pollock and Mullin, 1987). Furthermore, it will be essential to resolve the problems that have resulted in the high rate of infant death.

For the near future, captive conservation efforts should concentrate on developing successful tarsier husbandry in colonies within the habitat countries, near a natural food source. Only after successful offspring are produced and we have a second generation (F1) should colonies of captive tarsiers outside of Asia be considered. Colonies that reproduce in captivity are important as a second line of defense against extinction for the rarest species, including *T. syrichta* (Wright, chapter 14, this volume).

Acknowledgments

My sincere thanks are due to Barry Fass-Holmes and Joe Bussiere for reviewing the data and assisting with the tables and figures. Marvin Jones, Myron Shekelle, Richard Tenaza, Anita Rose Sebastian, Richard Weigl, Patricia Wright, David Haring, and Sharmy Prastiti kindly provided additional tarsier information that was not available in published sources. Darlene Rosemary, Patricia Wright, David Haring, and Marvin Jones reviewed the text and made helpful comments.

References

Collins NM, Sayer, JA, Whitmore TC. 1991. The conservation atlas of tropical forests: Asia and the Pacific. London: IUCN Macmillan.

Cook N. 1939. Notes on captive *Tarsius carbonarious*. J of Mammal 20: 173–178.

Crandall LS. 1964. The management of wild animals in captivity. Chicago: University of Chicago Press.

Dulaney M. 1985. Improved husbandry techniques for nocturnal animals. AAZPA Ann Proc 1985: 438–447.

Evans CS. 1967. Maintenance of the Philippine tarsier (*Tarsius syrichta*) in a research colony. Int Zoo Yrbk 7: 201–202.

Fitch-Snyder H. 1998. Asian prosimian North American regional studbook. San Diego: Zoological Society of San Diego.

Glatston AR. 2001. Relevance of studbook data to the successful captive management of grey mouse lemurs. Int J Primatol 22: 57–69.

Glatston AR, Mazzotta A. 1998. The European studbook for small nocturnal Madagascar prosimians. Rotterdam: Rotterdam Zoo.

Hamer KC, Hill JK, Lace LS, Langan AM. 1997. Ecological and biogeographical effects of forest disturbance on tropical butterflies of Sumba, Indonesia. J Biogeogr 24: 67–75.

Haring DM, Wright PC. 1989. Hand-raising a Philippine tarsier, *Tarsius syrichta*. Zoo Biology 8: 265–274.

Haring DM, Wright PC, Simons EL. 1985. Social behaviors of *Tarsius bancanus*. Am J Phys Anthro 66: 179.

Harrison B. 1963. Trying to breed *Tarsius*. Malay Nature J 17: 218–231.

ISIS. 1998–2001. SPARKS: Single Population Analysis and Record Keeping System. Apple Valley, NM.

Jarvis C, Morris D. 1960. International zoo yearbook. Zool Soc Lond 2: 288.

Kohn F, Roberts M, Keppel A, Maliniak E, Deal M. 1984. Management and husbandry of the western tarsier (*Tarsius bancanus*) at the National Zoological Park. Anim Keepers' Forum 11: 468–477.

MacKinnon K. 1997. The ecological foundations of biodiversity protection. In Kramer R, van Schaik C, Johnson J, editors, Last stand: protected areas and the defense of tropical biodiversity, 36–63. New York: Oxford University Press.

Niemitz C. 1984. Synecological and feeding behavior of Tarsius. In Niemitz C, editor, Biology of tarsiers, 59–75. Stuttgart and New York: Gustav-Fischer-Verlag.

Nietsch A, Kopp M-L. 1998. The role of vocalization in species differentiation of Sulawesi tarsiers. Folia Primatol 69 (suppl. 1): 371–378.

Oliver WLR, Heaney LR. 1996. Biodiversity and conservation in the Philippines. Int Zoo News 43: 329–337.

Pollock JI, Mullin RJ. 1987. Vitamin C biosynthesis in prosimians: evidence for the anthropoid affinity of Tarsius. Am J Phys Anthro 73: 65–70.

Roberts M. 1985. The management and husbandry of the western tarsier (*Tarsius bancanus*) at the National Zoological Park. AAZPA Ann Proc 1985: 466–475.

Roberts M. 1992. Growth, development, and parental care in the western tarsier (*Tarsius bancanus*) in captivity: evidence for a "slow" life-history and nonmonogamous mating system. Int J Primatol 15: 1–28.

Roberts M, Kohn F. 1993. Habitat use, foraging behavior, and activity patterns in reproducing western tarsiers, *Tarsius bancanus,* in captivity: a management synthesis. Zoo Biol 12: 217–232.

Roberts M, Kohn F, Keppel, A, Maliniak E, Deal, M. 1984. Management and husbandry of the western tarsier, *Tarsius bancanus,* at the National Zoological Park. AAZPA Ann Proc 1984: 588–600.

Schreiber GR. 1968. A note on keeping and breeding the Philippine tarsier, *Tarsius syrichta,* at Brookfield Zoo, Chicago. Int Zoo Yrbk 8: 114–115.

Ulmer FA Jr. 1963. Observations on the tarsier in captivity. Zool Gart 27: 106–121.

Wharton CH. 1948. Seeking Mindanao's strangest creatures. Natl Geogr 94: 388–408.

Wharton CH. 1950. The tarsier in captivity. J Mammal 31: 260–268.

Wright PC, Haring D, Izard MK, Simons EL. 1989. Psychological well-being of nocturnal primates in captivity. In Segal E, editor, Housing, care and psychological well-being of captive and laboratory primates, 61–74. New York: Noyes Publications.

Wright PC, Haring D, Simons EL, Andau P. 1987. Tarsiers: a conservation perspective. Primate Conserv 8: 51–54.

CHAPTER 14

Are Tarsiers Silently Leaping into Extinction?

Patricia C. Wright

The fossil record suggests that, long ago, tarsiers may have been widespread in Asia and Africa (Simons and Bown, 1985; Ginsburg and Mein, 1986; Beard, 1998; and Simons, this volume), but extant populations are restricted to a few islands in Southeast Asia. Are living tarsiers the remnants of a relict lineage in the final throes of extinction, or are they like didelphid possums, "living fossils" that are in no danger of disappearing in the near future? To answer this question, I review what we know about tarsier conservation status. Then, to better estimate future population viability, I examine life-history traits, geographic information, recent field data in both behavioral ecology and censuses of tarsiers, and known deforestation rates of tarsier habitat.

Current Conservation Status

There are currently seven recognized extant species of tarsiers, most of which are not classified in the IUCN Red List because they are "data deficient" (Hilton-Taylor, 2000) (table 14.1). Two tarsier species on the island of Sulawesi (*Tarsius dianae* and *T. spectrum*) are designated lower risk on the Red List, and a third Sulawesi species (*Tarsius pumilus*) has not been seen alive for over thirty years (Niemitz et al., 1991; Musser and Dagosto, 1987). The Asian Primate Conservation Action Plan (Eudey, 1992) mentions tarsiers only briefly. Reasons given for giving *Tarsius spectrum* and *Tarsius dianae* low conservation priority include nocturnal lifestyles, their ability to adjust to disturbed habitats, their presence in large, protected areas (*T. spectrum* in Tangkoko and *T. dianae* in Lora Lindi), and their high population density in studied areas (Gursky, 1998; Merker and Muhlenberg, 2000).

Why Is the IUCN Classification "Data Deficient"?

Because tarsiers are nocturnal and therefore difficult to census (see Dagosto, Gebo, and Dolino, chapter 11, this volume), their most basic behavior has been difficult to assess. Local difficulties have compounded the problem. For example, some surveys in areas where tarsiers may occur were restricted

Table 14.1. Tarsier Names, Red List Classification (Hilton-Taylor, 2000), and Distribution

Scientific name	Common name/s	Red List	Distribution
Tarsius bancanus ssp. borneanus (Horsefield, 1871)	Bornean tarsier, Western tarsier	DD	Sarawak, Sabah (Malaysia); Kalimantan (Indonesia); Darussalam (Brunei)
T. bancanus ssp. natunensis	Natuna Islands tarsier	DD	Unknown
T. bancanus ssp. saltator	Belitung Island tarsier	DD	Unknown
T. dianae (Niemitz, et al., 1991)	Dian's tarsier	LR/cd	Unknown
T. pelengensis (Sody, 1949)	Peleng Island tarsier	DD	Unknown
T. pumilus (Miller and Hollister, 1921)	Lesser spectral tarsier, Mountain tarsier, Pygmy tarsier	DD	Unknown
T. sangirensis	Sagihe Island tarsier	DD	Unknown
T. spectrum (Pallas, 1779)	Eastern tarsier, spectral tarsier, Sulawesi tarsier	LR/nt	Sulawesi
T. syrichta (Linnaeus, 1758)	Philippine tarsier	DD	Mindaneo, Visayas

NOTES: Red List categories are DD (data deficient), LR/nt (Lower Risk, near threshold), and LR/cd (Lower Risk, conservation dependent). *T. bancanus* may also exist in southern Sumatra (Rowe, 1996).

to daylight hours (Berenstain et al., 1986). Only very recently have field researchers focused on tarsier populations in the wild (Gursky, 1995, 1998, 2000, in press; Shekelle et al., 1997; Nietsch and Kopp, 1998; Nietsch, chapter 9, this volume). Recent improvements in technology have made nocturnal primate research a last frontier, and we have finally begun to better understand the social structures of night-active primates (Harcourt and Nash, 1986; Bearder, 1987, 1999; Sterling, 1993; Schmid, 1998; Muller, 1999). Consequently, within the last decade, primatology has exploded with descriptions of new taxa of nocturnal primates (Zimmerman et al., 1998; Bearder, 1999; Yoder et al., 2000).

Factors of Extinction Susceptibility: How Do Tarsiers Fit?

Some circumstances that influence susceptibility to extinction in primates include population density, body mass, age of first reproduction, interbirth interval, diet, home-range size, altitudinal range, maximum latitude, and geographic range (Harcourt and Schwartz, 2001). Furthermore, forces including human population density, human warfare, habitat destruction, and natural disasters influence extinction probabilities (Wright and Jernvall, 1999).

Factors that ameliorate extinction pressure in tarsiers include small body mass (80–150 g), early age of first reproduction (one or two years), a one-year interbirth interval, and small territories (MacKinnon and MacKinnon, 1980; Niemitz, 1984; Crompton and Andau, 1986; Wright et al., 1987; Roberts, 1994; Gursky 1995, 1998; Neri-Arboleda et al. 2002; chapter 10, this volume). These morphological and life-history parameters suggest that high population densities are still possible, supported by recent observations in the wild (MacKinnon and MacKinnon, 1980; Gursky, 1998; chapter 10, this volume; Dagosto, Gebo, and Dolino, chapter 11, this volume). Population density may be correlated with resource availability (Ganzhorn, 1995), and tarsiers opportunistically eat insects, an abundant resource in both disturbed and undisturbed habitats (Janzen, 1973; Lowman, 1982; Stork, 1987; Stork and Brendell, 1990). In general, nocturnal primates are usually less vulnerable to extinction than diurnal species (Jernvall and Wright, 1998). Night activity, as well as small body size, may be an advantage for tarsiers against human hunting. However, before we assume that tarsiers are not susceptible to extinction, several warning signs warrant consideration.

Life-History Factors

Although individual tarsier lifespans are unusually long for a small mammal—the record is fourteen years (Ulmer, 1960; Fitch-Snyder, chapter 13, this volume)—infant mortality rates, both in the wild and in captivity, are very high (Haring and Wright, 1989; Roberts, 1994; Gursky, 1997). High infant mortality is not rare in small, wild mammals, but tarsiers—like other primates—give birth to only a single offspring every year. They do not have large litters, and they gestate six months, compared to a few weeks in other similar-sized mammals (Izard et al., 1985; Wright et al., 1986a,b; Roberts, 1994). As wild tarsier populations diminish, these latter parameters may make it difficult for them to recover (Soule et al., 1979; Soule, 1980). And, since it appears that tarsiers are not likely to reproduce well in captivity (Haring and Wright, 1989; Roberts, 1994; Fitch-Snyder, chapter 13, this volume), few captive populations will be available for reintroduction as a second line of defense.

Specialized Diet

Primates consuming a specialized diet are more at risk than those able to choose a varied menu (Wright and Jernvall, 1999). Tarsiers are primarily insectivorous, specializing on large-bodied orthopterans, including walking sticks, cicadas, crickets, and moths (Niemitz, 1984; Crompton and Andau, 1986; Gursky, 2000). Although tarsiers opportunistically eat geckos, snakes, and frogs, these prey must be alive for tarsiers to consider them as food (Davis, 1962; Fogden, 1974; Niemitz, 1984; P. Wright, pers. obs.). Wild tarsiers have never been seen to eat vegetation, nectar, or fruits, and in captivity tarsiers refuse these items (Wright et al., 1989; Haring and Wright, 1989; Gursky, 2000). Although insects are typically viewed as an abundant dietary resource, the dispersion and availability of large-bodied insects could be patchy and limiting (Janzen, 1973; Atsalis, 1999)

Human invasion of tarsier habitats can cause habitat fragmentation, and deforestation can increase the exposure of tarsier habitat to clearings. This can increase sunlight in the habitats, which in turn may increase aridity. This change in ecology may affect the diversity and numbers of insect prey items available to tarsiers. For example, the diversity and density of large or-thopterans may decrease as forest habitat fragments decrease in size, as seen in some beetle species (Hanski, 1983; Klein, 1989; Estrada et al., 1998). Un-der this scenario, tarsiers seen in cultivated or more open areas may rep-resent a "sink population" (Pulliam, 1988) rather than a thriving genetic source for future generations. The fact that tarsiers have a low basal meta-bolic rate (McNab and Wright, 1987) may help them bridge a few days of prey depletion, but tarsiers cannot hibernate like small-bodied Mala-gasy primates (Wright and Martin, 1995; Schmid, 1998), and long-term cli-mate change (the droughts caused by El Niño, for example) that may nega-tively affect large-bodied insect abundance could be detrimental to tarsier populations.

Limited Geographic Range

Mammal and bird species with a small geographic range have been thor-oughly documented to be more vulnerable to extinctions than those with large ranges (Ceballos and Brown, 1995; Stotz et al., 1996; Wright and Jern-vall, 1999). Each tarsier species is limited to a small geographic range in Southeast Asia. In addition, species found on islands are more at risk than those on the mainland (Harcourt and Schwartz, 2001). The family Tarsiidae is now restricted to eleven islands. Furthermore, five out of the seven tarsier species are found on only one island. In a study of threatened birds in in-sular Southeast Asia, Brooks and colleagues (1997) found that single-island endemics are considerably more at risk than more widespread species. Five tarsier species are single-island endemics.

Rarity is often a characteristic of island species, including primate taxa (Happel et al., 1987). Only in two of nine major primate groups are most species rare (defined as having ranges smaller than 170,000 km^2). The fol-lowing are the only major groups restricted to islands: the lemurs of Mada-gascar (five genera, 40+ species) and the tarsiers of Southeast Asia (one genus, seven species) (Cowlishaw and Dunbar, 2000).

A further warning is indicated by Harcourt and Schwartz (2001). They found that the most tropically confined primate taxa are at higher extinction risk than more widely adapted taxa, and that maximum latitude correlates with ability to survive in altered habitats. Geographic range limits for the entire family of Tarsiidae are restricted to 10° N and 40° S from the equator.

High Human Population Density

The ever-increasing human population density correlates closely with the disappearance of primate species (Harcourt, 1996; Robinson et al., 1999). Hunting, slash-and-burn agriculture, and timber concessions increase when humans colonize an area (O'Brien and Kinnard, 1996; Peres, 1999; Tutin and White, 1999). Areas in Indonesia and the Philippines have rich volcanic soils and some of the highest human densities in the world (WRI Report, 2000; Heany and Regalado, 1998). The island of Borneo, with its poor soils and peat swamps, was less populated than the other islands that are inhabited by tarsier populations (WRI Report, 2000), but recent Indonesian transmigration policies have increased human settlement in Kalimantan (Indonesian Borneo).

Human Political Unrest

High human population density frequently brings war and human unrest, which create a conservation problem since protected areas are often targeted for exploitation during anarchy (Soule, 1991; Fimbel and Fimbel, 1997; Hart and Hart, 1997). Anarchy produces a rapid increase in all illegal activities in part because demand for forest products increases, and guards, unarmed and unsalaried, often lose their ability to protect wildlife preserves (van Schaik, 2002). Both Indonesia and the Philippines are engaged in civil war, which escalated after September 11, 2001, placing all species of tarsiers, except for *Tarsius bancanus,* at risk. Only East Malaysia and Brunei have remained calm; populations of *T. bancanus* exist there.

Primate species that are restricted to only one country are more at risk than species found in several countries, because any one political regime can devastate all forested areas in its realm (Wright and Jernvall, 1999). *Tarsius syrichta* is found only in the Philippines, and *T. pumilus, T. pelengensis, T. dianae, T. spectrum,* and *T. sangirensis* are found only in Indonesia. Only *T. bancanus* is found in multiple countries (East Malaysia, Brunei, and Indonesia).

Habitat Destruction

Over three quarters of the original habitat of tarsiers has been lost to deforestation (MacKinnon, 1986, 1997). In Indonesia, the annual rate of deforestation since 1985 has been estimated at up to 12,000 km² per year, and much of the remaining forest has been affected by logging and shifting agriculture (Collins et al., 1991; Hamer et al., 1997). On the island of Borneo,

67% of its 738,864 km^2 was forested in 1991 (Collins et al., 1991), and deforestation has increased exponentially in the past decade (WRI Report, 2001). In the Philippines, at least 94% of the total land area of approximately 300,000 km^2 was originally covered by tropical forests. However, by 1988, satellite imagery revealed that only 21% of the forest cover remained. According to World Bank figures, 90% of lowland forest in the Philippines has been lost in the last thirty years, with only 5% of the land area remaining in natural forest (Oliver and Heaney, 1996; Heaney and Regalado, 1998). The small island of Bohol has only remnants of forest; Leyte and Samar are also heavily deforested (See Dagosto, Gebo, and Dolino, this volume). Mindanao has lost most of its lowlands, and tarsiers may be restricted by altitude to live in forests below 800 meters. Unlike Indonesia, Brunei, and East Malaysia, which have designated many protected areas, the Philippines have serious inadequacies in the existing protected areas network (MacKinnon, 1986; Oliver and Heaney, 1996; Heaney and Regalado, 1998).

With such recent and rapid deforestation, predicting the long-term effect on tarsier populations is impossible. An "extinction debt" could lead to cascading tarsier population crashes in the future (Tilman et al., 1994). In addition, "edge effects"—the changes in temperature, moisture, and increased light levels at the periphery of forests—may be more destructive to tarsier populations than we have observed so far (Woodroffe and Ginsberg, 1998).

Natural Disasters—Wildfires

In general, islands are more prone to natural disasters such as monsoons and El Niño events than are continental environments (Wright, 1999). In 1983, and again in the 1990s, wildfires—probably facilitated by forest thinning through human timbering and the drying effects of droughts following El Niño—have destroyed large rain forest areas in Kalimantan, Sulawesi, and Sumatra (Berenstain et al., 1986; Leighton and Wirawan, 1986; Kinnaird and O'Brien, 1998). Small mammals like tarsiers cannot escape such monumental fires, and entire populations are destroyed (Wright et al., 1987). These primates with single births, long gestation lengths, and high infant mortality have difficulties repopulating devastated areas, and local extinction from these random catastrophes are a real risk (Lande, 1993; Lawrance, 1994).

Patchiness of Populations?

Tarsiers are difficult to census and many who conduct long-term field studies have not observed them at their study site (Peter Rodman, pers. com.; Carey Yeager, pers. com.). One explanation for this phenomenon is that researchers have not been out in the forest at night. However, other nocturnal primates such as *Aotus,* Malagasy lemurs, galagos, or lorises are seen occasionally during daylight by diurnal researchers. A second explanation is that tarsiers are not found uniformly in all forested areas, having instead a more patchy distribution than so far described. Recent evidence has been accumulating which documents that the rain forest is much more heterogeneous than previously assumed (Tuomisto et al., 1995; Hubbell et al., 1999). Perhaps the patchy distribution of tarsier populations within a rain forest is dependent upon undocumented heterogeneity in plants and their insect predators (MacKinnon et al., 1996), but further research including both ecology and tarsier surveys is needed to understand the relationship between ecology and local tarsier distributions.

Future Conservation Action

The review of the variety of factors that make tarsiers susceptible to extinction shows that there are reasons to be concerned for all seven species. The amount of rain forest left in the combined habitats in Brunei, East Malaysia, and Kalimantan may enable *Tarsius bancanus* to have viable populations (MacKinnon, 1997). Rates of deforestation on the smaller islands of Sulawesi, Mindanao, Leyte, and Samar are high, and we need to know the impact of this ongoing development on populations of tarsiers before we can accurately assess probable extinction rates (Heaney and Regalado, 1998).

The first step in tarsier conservation is to change their "data deficient" status. Within the next few years, we should prioritize a survey for all tarsiers, focusing especially on *Tarsius pumilus, T. sangirensis, T. pelengensis,* and *T. syrichta.* Presently, we lack estimates for *T. syrichta* populations on small islands such as Basilan, Dinagat, Maripipi, and surveys on these islands have been encouraged by Dagosto, Gebo, and Dolino (this volume).

Another step suggested by Heaney (1993) is to assist the Philippine tarsier by establishing protected areas, increasing public awareness of biodiversity values, and initiating more conservation and development projects in the Philippines. The Philippines have lagged behind other biodiversity hotspots such as Madagascar and Brazil in initiating active conservation programs (Heaney, 1993; Myers et al., 2000; Wright and Andriamihaja, 2002, in press).

A plan to assist Indonesian tarsier species during wartime might be modeled after other war-torn countries. The staff salaries and infrastructure of Indonesian protected areas—which are now under siege due to political unrest—might be considered for special funding, as has been accomplished by UNESCO and UNF in Congo (UNESCO Report, 2000).

The consistent failure of tarsiers to raise offspring in captivity prevents the use of captive-born individuals for reintroduction, a conservation effort used successfully in other primate species (Seal et al., 1990; Britt et al., 2000).

As discussed by Fitch-Snyder (chapter 13, this volume), no reproductively viable second generation in captivity has ever been achieved, and currently no tarsier breeding colonies exist anywhere out of the habitat countries.

Spectral tarsiers have been a popular tourist attraction in Tangkoko, northern Sulawesi, and generated local good will toward the Tangkoko Nature Reserve. But tourism is sensitive to political upheaval and cannot be relied on as a conservation tool (Kinnaird and O'Brien, 1996; Davenport et al., 2002).

In a comparison of tarsier densities in primary forest, selectively logged forest, and small plantations, *Tarsius dianae* populations were three times less abundant per hectare in the selectively logged area (Merker and Muhlenberg, 2000). The reduced density of locomotor supports and sleeping sites, as well as the noise of timber exploiters, could account for this. In the agro-forestry and plantation areas it was suggested that the high abundance of insects might have kept the tarsier populations as high as found in the primary forest (Merker and Muhlenberg, 2000). This result on the effect of land use on tarsier population densities may be useful for effective tarsier-conservation management plans.

Tarsiers are an ancient line of primates. In all other primate families and in the other geographic areas where primates occur, no comparable niche exists like the one that tarsiers inhabit. The uniqueness of their environment has also selected for morphological and behavioral uniqueness. Tarsiers, for instance, are the only primates that eat only live animal prey, that have eyeballs as big as their brain, and that give birth to offspring a quarter of the weight of the mother. Their vertical clinging and leaping locomotion is highly specialized; with tibia and fibula fused, they can turn their head 180 degrees, and their brain is simple and smooth. Losing these unusual creatures would be a tragedy. Are these seven species on eleven islands leaping rapidly to extinction? An urgent need exists for immediate fieldwork to answer that question and to counter this predicament.

Acknowledgments

I am most grateful to Sharon Pochron, Marian Dagosto, David Haring, Helena Fitch-Snyder, Jukka Jernvall, and Friderun Ankel-Simons for their comments on this manuscript.

References

Atsalis S. 1999. Diet of the brown mouse lemur (*Microcebus rufus*) in Ranomafana National Park, Madagascar. Int J Primatol 20: 193–230.

Beard, KC. 1998. A new genus of Tarsiidae (Mammalia: Primates) from the middle

Eocene of Shanxi Province, China, with notes on the historical biogeography of tarsiers. Bull Carnegie Mus Nat Hist 34: 260–277.

Bearder SK. 1987. Lorises, bushbabies and tarsiers: diverse societies in solitary foragers. In Smuts BB, Cheney DL, Seyfarth RM, Wrangham RW, Struhsaker TT, editors, Primate societies, 11–24. Chicago: University of Chicago Press.

Bearder SK. 1999. Physical and social diversity among nocturnal primates: a new view based on long-term research. Primates 40: 267–282.

Berenstain L, Mitani JC, Tenaza RR. 1986. Effects of El Niño on habitat and primates in east Kalimantan. Primate Conserv 7: 54–55.

Britt A, Welch C, Katz A. 2000. Ruffed lemur re-stocking and conservation program update. Lemur News 5: 36–38.

Brooks TM, Pimm SL, Collar NJ. 1997. Deforestation predicts the number of threatened birds in insular Southeast Asia. Conserv Biol 11: 382–394.

Ceballos G, Brown JH. 1995. Global patterns of mammalian diversity, endemism and endangerment. Conserv Biol 9: 559–568.

Collins NM, Sayer JA, Whitmore TC. 1991. The conservation atlas of tropical forests: Asia and the Pacific. London: IUCN Macmillan.

Cowlishaw G, Dunbar R. 2000. Primate conservation biology. Chicago: University of Chicago Press.

Crompton, RH, Andau PM. 1986. Locomotion and habitat utilization in free-ranging *Tarsius bancanus:* a preliminary report. Primates 27: 337–355.

Davenport L, Brockelman WY, Wright PC, Ruf K, Rubio del Valle FB. 2002. Ecotourism tools for parks. In Terborgh J, van Schaik C, Davenport L, Rao M, editors, Making parks work: strategies for preserving tropical nature, 279–306. Covelo, CA: Island Press.

Davis DD. 1962. Mammals of the lowland rain forest of north Borneo. Bulletin of the National Museum of Singapore. 31: 5–129.

Estrada A, Coates-Estrada R, Anzures Dadda A, Cammarano P. 1998. Dung and carrion beetles in tropical rain forest fragments and agricultural habitats at Los Tuxlas, Mexico. J Trop Ecol 14: 577–593.

Eudey A. 1992. Asian primate action plan. Gland, Switzerland: IUCN.

Fimbel C, Fimbel R. 1997. Rwanda: the role of local participation. Conserv Biol 11: 309–310.

Fogden MPL. 1974. A preliminary field study of the western tarsier, *Tarsius bancanus* Horsefield. In Martin RD, Doyle GA, Walker AC, editors, Prosimian biology, 151–165. London: Duckworth.

Ganzhorn JU. 1995. Low level forest disturbance effects on primary production, leaf chemistry and lemur populations. Ecology 76: 2084–2096.

Ginsburg L, Mein P. 1986. *Tarsius thailandica* nov. sp. Tarsiidae (Primates, Mammalia) fossile d'Asie. C R Acad Sci Paris Ser II 304, no. 19: 1213–1215.

Gursky S. 1995. Group size and composition in the spectral tarsier, *Tarsius spectrum:* implications for social organization. Trop Biodivers 3: 57–62.

Gursky S. 1997. Modeling maternal time budgets: the impact of lactation and gestation on the behavior of the spectral tarsier, *Tarsius spectrum.* Ph.D. dissertation, State University of New York at Stony Brook.

Gursky S. 1998. Conservation status of the spectral tarsier, *Tarsius spectrum:* population density and home range size. Folia Primatol 69: 191–203.

Gursky S. 2000. The effect of seasonality on the behavior of an insectivorous primate, *Tarsius spectrum.* Int J Primatol 21: 477–496.

Gursky S. In press. Determinants of gregariousness in the spectral tarsier (Prosimian: *Tarsius spectrum*). J Zool Soc Lond.

Hamer KC, Hill JK, Lace LS, Langan, AM. 1997. Ecological and biogeographical effects of forest disturbance on tropical butterflies of Sumba, Indonesia. J Biogeog. 24: 67–75.

Hanski I. 1983. Distributional ecology and abundance of dung and carrion-feeding beetles (Scarabaedae) in tropical rain forest in Sarawak, Borneo. Acta Zool Fennica 167: 1–45.

Happel RE, Noss JF, Marsh C. 1987. Distribution, abundance, and endangerment of primates. In Marsh CW, Mittermeier RA, editors, Primate conservation in the tropical rain forest, 83–107. New York: Alan R. Liss.

Harcourt AH. 1996. Is the gorilla a threatened species? How should we judge? Biol Conserv 75: 165–176.

Harcourt AH, Schwartz MW. 2001. Primate evolution: a biology of Holocene extinction and survival on the Southeast Asian Sunda Shelf Islands. Am J Phys Anthro 114: 4–17.

Harcourt CS, Nash LT. 1986. Social organization of galagos in Kenyan coastal forests: I: *Galago zanzibaricus.* Am J Primatol 10: 339–349.

Haring DM, Wright PC. 1989. Hand raising a Philippine tarsier, *Tarsius syrichta.* Zoo Biol 8: 265–274.

Hart T, Hart J. 1997. Zaire: new models for an emerging state. Conserv Biol 11: 308–309.

Heaney LR. 1993. Biodiversity patterns and the conservation of mammals in the Philippines. Asia Life Sci 2: 261–274.

Heaney LR, Regalado JC. 1998. Vanishing treasures of the Philippine rainforest. Chicago: Field Museum Press.

Hilton-Taylor C. 2000. 2000 IUCN red list of threatened species. Gland, Switzerland: IUCN.

Hubbell SP, Foster RB, O'Brien ST, Harms KE, Condit R, Wechsler B, Wright SJ, de Lao SL. 1999. Light-gap disturbances, recruitment limitation, and tree diversity in a neotropical forest. Science 283: 554–557.

Izard MK, Wright PC, Simons EL. 1985. Gestation length in *Tarsius bancanus.* Am J Primatol 9: 327–331.

Janzen DH. 1973. Sweep samples of tropical foliage insects: effects of seasons, vegetation types, elevation, time of day, and insularity. Ecology 54: 687–708.

Jernvall J, Wright PC. 1998. Diversity components of impending primate extinctions. Proc Natl Acad Sci USA 95: 11279–11283.

Kinnaird MF, O'Brien TG. 1996. Ecotourism in the Tangkoko DuaSudara Nature Reserve: opening Pandora's box? Oryx 30: 65–73.

Kinnaird MF, O'Brien TG. 1998. Ecological effects of wildfire on lowland rainforest in Sumatra. Conserv Biol 12: 954–956.

Klein BC. 1989. Effects of forest fragmentation on dung and carrion beetle communities in central Amazonia. Ecology 6: 1715–1725.

Lande R. 1993. Risks of population extinction from demographic and environmental stochasticity and random catastrophes. Am Nat 141: 911–927.

Lawrance WF. 1994. Rainforest fragmentation and the structure of small mammal communities in tropical Queensland. Biol Conserv 69: 23–32.

Leighton M, Wirawan N. 1986. Catastrophic drought and fire in Borneo tropical rainforest associated with the 1982–1983 El Niño Southern Oscillation event. In Prance GT, editor, Tropical rain forests and the world atmosphere, 75–102. Boulder, CO: AAAS, Westbury Press.

Lowman, MD. 1982. Seasonal variation in insect abundance among three Australian rain forests, with particular reference to phytophagous types. Australian J Ecol 7: 353–363.

MacKinnon J, MacKinnon K. 1980. The behavior of wild spectral tarsiers. Int J Primatol 1: 361–379.

MacKinnon J, MacKinnon K. 1986. Review of the protected areas system in the Indo Malayan Realm. Cambridge, UK: IUCN/UNEP.

MacKinnon K. 1986. The conservation status of nonhuman primates in Indonesia. In Benirschke K, editor, Primates: the road to self-sustaining populations, 99–126. New York: Springer-Verlag.

MacKinnon K. 1997. The ecological foundations of biodiversity protection. In Kramer R, van Schaik C, Johnson J , editors, Last stand: protected areas and the defense of tropical biodiversity, 36–63. New York: Oxford University Press.

MacKinnon K, Hatta G, Halim H, Mangalir A. 1996. The ecology of Kalimantan. The ecology of Indonesia series, vol. 3. Hong Kong: Periplus Editions.

McNab BK, Wright PC. 1987. Temperature regulation and oxygen consumption in the Philippine tarsier *Tarsius syrichta*. Physiol Zool 60(5): 596–600.

Merker S, Muhlenberg M. 2000. Traditional land use and tarsiers—human influences on population densities of *Tarsius dianae*. Folia Primatol 71: 426–428.

Muller AE. 1999. Social organization of the fat-tailed dwarf lemur (*Cheirogaleus medius*) in northwestern Madagascar. In Rasamimanana H, Rakotosamimanana B, Ganzhorn J, Goodman S, editors, New directions in lemur studies, 57–68. New York: Plenum.

Musser GG, Dagosto M. 1987. The identity of *Tarsius pumilus*, a pygmy species endemic to the montane mossy forests of Central Sulawesi. Am Mus Nov 2867: 1–53.

Myers N, Mittermeier RA, Mittermeier CG, da Fonseca GA, Kent J. 2000. Biodiversity hotspots for conservation priorities. Nature 403: 853–858.

Niemitz C. 1984. An investigation and review of the territorial behaviour and social organization of the genus *Tarsius*. In Niemitz C, editor, Biology of tarsiers, 117–128. New York: Gustav-Fischer-Verlag.

Niemitz C, Nietsch A, Warter S, Rumpler Y. 1991. *Tarsius dianae:* a new primate species from Central Sulawesi (Indonesia). Folia Primatol 56: 105–116.

Nietsch A, Kopp M. 1998. Role of vocalizations in species differentiation of Sulawesi tarsiers. Folia Primatol 69: 371–378.

O'Brien TG, Kinnaird MF. 1996. Changing populations of birds and mammals in North Sulawesi. Oryx 30: 150–156.

Oliver WLR, Heaney LR. 1996. Biodiversity and conservation in the Philippines. Int Zoo News 43: 329–337.

Peres C. 1999. The effects of subsistence hunting and forest types on the structure of Amazonian primate communities. In Fleagle JG, Janson C, Reed KE, editors, Primate communities, 268–283. London: Cambridge University Press.

Pulliam HR. 1988. Sources, sinks and population regulation. Am Nat 132: 652–661.

Roberts M. 1994. Growth, development, and parental care in the western tarsier (*Tarsius bancanus*) in captivity: evidence for a slow life history and nonmonogamous mating system. Int J Primatol 15: 1–28.

Robinson JG, Redford KH, Bennett EL. 1999. Wildlife harvest in logged tropical forests. Science 284: 595–596.

Schmid J. 1998. Daily torpor in mouse lemurs, *Microcebus* spp.: metabolic rate and body temperature. Folia Primatol 69: 394–404.

Seal US, Ballou JD, Padua CV. 1990. *Leontopithecus* population viability analysis workshop report. Captive breeding specialist group, Species Survival Commission. Belo Horizonte, Brazil: IUCN.

Shekelle M, Mukti S, Ichwan L, Masala Y. 1997. The natural history of the tarsiers of North and Central Sulawesi. Sulawesi Primate Project Newsletter.

Simons EL, Bown TM. 1985. *Afrotarsius chatrathi*, first tarsiiform primate (Tarsiidae?) from Africa. Nature 313: 475–477.

Soule MD. 1980. Thresholds for survival: maintaining fitness and evolutionary potential. In Soule ME, Wilcox BA, editors, Conservation biology: an evolutionary-ecological perspective, 151–169. Sunderland, MA: Sinauer.

Soule MD. 1991. Conservation: tactics for a constant crisis. Science 253: 744–750.

Soule ME, Wilcox BA, Holtby C. 1979. Benign neglect: a model of faunal collapse in the game reserves of East Africa. Biol Conserv 15: 259–271.

Sterling EJ. 1993. Patterns of range use and social organization in aye-ayes (*Daubentonia madagascariensis*) on Nosy Mangabe. In Kappeler PM, Ganzhorn JU, editors, Lemur social systems and their ecological basis, 1–10. New York: Plenum Press.

Stork NE. 1987. Guild structure of arthropods from Bornean rain forest trees. Ecol Entomol 12: 69–80.

Stork NE, Brenddell MJD. 1990. Variation in the insect fauna of Sulawesi trees with season, altitude and forest type. In Knight WJ, Halloway JD, editors. Insects and the rainforest of South East Asia (Wallacea), 173–190. London: Royal Entomological Society of London.

Stotz DF, Fitzpatrick JW, Parker TA III, Moskovits DK. 1996. Neotropical birds: ecology and conservation. Chicago: University of Chicago Press.

Tilman D, May RM, Lehman CL, Nowak MA. 1994. Habitat destruction and the extinction debt. Nature 371: 65–66.

Tuomisto H, Ruokolainen K, Kalliola R, et al. 1995. Dissecting Amazonian biodiversity. Science 269: 63–66.

Tutin C, White L. 1999. The recent evolutionary past of primate communities:

likely environmental impacts during the past three millennia. In Fleagle JG, Janson C, Reed KE, editors, Primate communities, 220–236. London: Cambridge University Press.

Ulmer F. 1960. A longevity record for the Mindanao tarsier. J Mammal 41: 512.

UNESCO Report. 2000. Paris.

van Schaik C. 2002. Anarchy and parks: dealing with political instability. In Terborgh J, van Schaik C, Davenport L, Rao M, editors, Making parks work: strategies for preserving tropical nature, 352–363. Covelo, CA: Island Press.

Woodroffe R, Ginsberg JR. 1998. Edge effects and the extinction of populations inside protected areas. Science 280: 2126–2128.

Wright PC. 1999. Lemur traits and Madagascar ecology: coping with an island environment. Yrbk of Phys Anthropol 42: 31–72.

Wright PC, Andriamihaja BA. 2002. Making conservation work in Ranomafana National Park, Madagascar. In Terborgh J, van Schaik C, Rao M, Davenport L, editors, Making parks work: strategies for preserving tropical nature, 112–132. Covelo, CA: Island Press.

Wright PC, Andriamihaja BA. In press. The conservation value of long-term research: a case study from Parc National de Ranomafana. In Goodman S, Benstead J, editors, The natural history of Madagascar. Chicago: University of Chicago Press.

Wright PC, Jernvall J. 1999. The future of primate communities: a reflection of the present? In Fleagle JG, Janson C, Reed KE, editors, Primate communities, 295–309. Cambridge, UK: Cambridge University Press.

Wright PC, Martin LB. 1995. Predation, pollination and torpor in two nocturnal primates: *Cheirogaleus major* and *Microcebus rufus* in the rain forest of Madagascar. In Alterman L, Doyle G, Izard MK, editors, Creatures of the dark, 45–60. New York: Plenum Press.

Wright PC, Haring D, Izard MK, Simons, EL. 1989. Psychological well-being of nocturnal primates in captivity. In Segal E, editor, Housing, care and psychological well-being of captive and laboratory primates, 61–74. Park Ridge, NJ: Noyes Publications.

Wright PC, Haring D, Simons EL, Andau, P. 1987. Tarsiers: a conservation perspective. Primate Conservation 8: 51–54.

Wright PC, Izard MK, and Simons EL. 1986a. Reproductive cycles in *Tarsius bancanus*. Am J Primatol 11: 207–215.

Wright PC, Toyama L, and Simons EL. 1986b. Courtship and copulation in *Tarsius bancanus*. Folia Primatol 46: 142–148.

WRI Report 2000. Washington, DC: World Resources Institute Press.

Yoder AD, Rasoloariaon RM, Goodman SM, Ganzhorn JU, Irwin JA, Atsalis S. 2000. A new species of mouse lemur. Proc Natl Acad Sci USA 97: 11325–11330.

Zimmerman E, Cepok S, Rakotarison N, Zietemann V, Radespiel U. 1998. Sympatric mouse lemurs in Northwest Madagascar: a new rufous mouse lemur species (*Microcebus ravelobensis*). Folia Primatol 69: 106–114.

The Editors and Contributors

The Editors

Patricia C. Wright received her B.A. from Hood College in 1966, and the Ph.D from City University of New York in 1985. Wright is a Professor of Anthropology and Adjunct Professor of Ecology and Evolution at the State University of New York, Stony Brook and Director of the Institute for the Conservation of Tropical Environments, Stony Brook and Madagascar. She was awarded a MacArthur Fellowship in 1989, an honorary doctorate in science from Hood College in 1991, a Chevalier's Medal of Honor from the Malagasy (Madagascar) government in 1995, and the Earthwatch Investigator of the Year award in 1998. She has been featured in several films, including *Me and Isaac Newton,* directed by Michael Apted. Patricia Wright has conducted field research in Peru, Paraguay, East Malaysia, the Philippines, and Madagascar. Her special research interests include field and captive studies of primate behavioral ecology, including nocturnal lifestyles, monogamy, parental care, female dominance, long-term demography, predation pressures, and reproduction. In 1986 Dr. Wright and colleagues discovered a new species of primate in Madagascar, the golden bamboo lemur. Her interest in conservation biology led her to assist the Malagasy government to develop Ranomafana National Park, of which she is presently International Coordinator of Research. During her eight years at Duke University, she and Dr. Simons established a tarsier colony at Duke Primate Center, and she has studied the behavior and reproduction of two species of tarsiers.

Elwyn L. Simons received his B.S. degree from Rice University in 1952, the Ph.D. from Princeton University in 1956, and D.Sc. from University College, Oxford, in 1959. Dr. Simons is a James B. Duke Professor of Biological Anthropology and Anatomy and Zoology at Duke University, and a member of the National Academy of Science and the American Philosophical Society. Dr. Simons received the Alexander von Humboldt Award from the German Federal Republic in 1975–76, honorary citizenship of Fayum Province, Egypt, in 1985, Knight of the National Order of Madagascar in 1998, and the American Association of Physical Anthropology's Charles R. Darwin Award for Lifetime Achievement in 2001. In 1994 two of Elwyn Simons's many prominent former students honored him as the creator of a new science: paleoprimatology. He has worked since the 1960s in Egypt, successfully searching for fossils and building the foundation of the study of anthropoid origins. In the 1980s he also masterminded the rise of the Duke University Primate Center into a world-renowned institution, in which he works with living and subfossil prosimians from Madagascar. Together with Dr. Patricia Wright he established a viable captive tarsier colony that ultimately gave rise to this book on tarsiers— their present, their past, and their future.

Sharon Gursky received her B.A. from Hartwick College in 1989, an M.S. from the University of New Mexico in 1991, and the Ph.D. in anthropology from the State University of New York at Stony Brook in 1997. She has been studying wild spectral tarsiers in Indonesia since 1990. Her dissertation research focused on the parental care patterns of spectral tarsiers, exploring questions concerning the infant parking strategy of the spectral tarsiers. Her current work focuses on the ecological and social factors leading to gregarious behavior in spectral tarsiers. Dr. Gursky is presently an Assistant Professor at Texas A&M University.

The Contributors

Robert L. Anemone is an associate professor at Western Michigan University, Kalamazoo. He has research interests in vertebrate paleontology and functional anatomy of living and fossil primates. He has been leading field teams to the Great Divide Basin in Wyoming to study Paleocene and Eocene primate fossils since 1994.

Friderun A. Ankel-Simons is a research associate at the Duke University Fossil Primate Facility at Duke University, Durham, North Carolina. She is well known for her textbook on primate anatomy.

John Czelusniak is a research professor of anatomy and cell biology at Wayne State University in Detroit. He is interested in molecular phylogenies of primates and databases for primate genetics.

Marian Dagosto is a professor at Northwestern University Medical School, Department of Cell and Molecular Biology in Chicago. Her areas of interest include primate evolution, anatomy, systematics, and positional behavior of fossil and living prosimians. She has conducted fieldwork on the Philippine tarsier.

Cynthia N. Dolino is a student of biology at Silliman University, Negros, Philippines. In addition to her studies of tarsiers, she has been active in the survey of Philippine herpe to fauna.

Helena M. Fitch-Snyder is a behavioral research associate at the Zoological Society of San Diego, Center for the Reproduction of Endangered Species (CRES). Her area of interest is communication, maternal behavior, and reproduction in Asian prosimians, lion-tailed macaques, and golden monkeys. She is in charge of the IUCN studbook for tarsiers.

Daniel L. Gebo is a professor in the Department of Anthropology at Northern Illinois University in Dekalb, Illinois. His specialty is primate anatomy and locomotion in both fossil and living primates. He is well known for his textbook on primate locomotion.

Morris Goodman is a professor of molecular genetics, anatomy, and cell biology at Wayne State University in Detroit. His specialty is molecular evolution, with emphasis on globin genes and the X-linked HPRT gene. Dr. Goodman is a member of the National Academy of Science and received the Darwin Award for Lifetime Achievement in Physical Anthropology in 2002.

Colin P. Groves is a professor of biological anthropology at the Australian National University School of Archaeology and Anthropology, Canberra. His textbook on primate taxonomy has become a classic. He has published extensively on primate taxonomy, evolution, conservation, and ethics.

David H. Haring is the Registrar at the Duke University Primate Center, Durham, North Carolina. He is well known for his photographs of prosimian primates and of wildlife in Madagascar.

Nina G. Jablonski is the Irvine Chair and Curator of Anthropology at the California Academy of Sciences, San Francisco. She has published extensively on living and fossil Asian primates, and has edited a book on Colobine monkeys.

Carla M. Meireles is a research associate in Brazil working on the molecular genetics of primates.

Brett A. Nachman is a research associate at Colorado University, Boulder, Colorado. His interest is in primate evolution and the biomechanics of locomotion.

Alexandra Nietsch is a research associate at the Free University, Berlin, Germany. Her special interest is taxonomy, acoustics, communication, and conservation of primates, especially tarsiers.

Scott L. Page is a research associate at the Molecular Genetics Laboratory at Wayne State University in Detroit. He is currently working on the molecular genetics of primates.

Sharon T. Pochron is a research associate in the Department of Anatomy, State University of New York at Stony Brook. Her interests are primate behavior and ecology and she has conducted field research on both baboons and prosimian primates.

Jeffrey H. Schwartz is a professor of anthropology at the University of Pittsburgh. He has published extensively on primates and primate evolution, including orangutans.

Cornelia Simons is a science coordinator at Shodor Education Foundation, Inc., in Durham, North Carolina. Her research interests include the morphology and anatomy of primates.

Derek E. Wildman is a research associate at the Molecular Genetics Laboratory at Wayne State University in Detroit, Michigan. He is currently working on the molecular genetics of primates.

Anne D. Yoder is an associate professor of biology at Yale University in New Haven, Connecticut. She is particularly interested in molecular systematics and primate phylogenies, especially of prosimians.

Index

Note: Page numbers in italics indicate figures.